# VALVE SELECTION AND SPECIFICATION GUIDE

# Valve Selection and Specification Guide

*Ronald C. Merrick*

VNR Van Nostrand Reinhold
New York

To my wife, DIAN TEINERT, who has published also, and to my daughter KATHARINE, both of whom have mostly seen only the back of my head for the last two years.



Library of Congress Catalog Number 90-12613
ISBN 0-442-31870-7



Printed in the United States of America

Van Nostrand Reinhold
115 Fifth Avenue
New York, New York 10003

Chapman and Hall
2-6 Boundary Row
London, SE1 8HN, England

Thomas Nelson Australia
102 Dodds Street
South Melbourne 3205, Victoria, Australia

Nelson Canada
1120 Birchmount Road
Scarborough, Ontario M1K 5G4, Canada

16 15 14 13 12 11 10 9 8 7 6 5 4 3 2

**Library of Congress Cataloging-in-Publication Data**

Merrick, R. C., 1951–
Valve selection and specification guide / R. C. Merrick.
p. cm.
Includes bibliographical references (p. ) and index.
ISBN 0-442-31870-7
1. Valves—Handbooks, manuals, etc. I. Title.
TS277.M47 1990
621.8'4—dc20

90-12613
CIP

# CONTENTS

# FOREWORD

Today, people who specify or select valves spend over two-thirds of their time researching literature for information on valve sizing, availability, materials, and standards. This is nonproductive time. Unfortunately, most companies do not have the luxury of a team of experts with the necessary experience and education in all of the different fields that apply to valves. The next best alternative is to understand what valves are and all the things they can do.

By definition, valves are devices that stop, start, mix, or change the direction and/or magnitude of the fluid flow, pressure, or its temperature. As a specifier or selector you will have to determine whether the valve is going to be used for flow control, throttling, or for on-off service. Then you will have to determine the cycle life or frequency of their operation.

You will discover that valves are classified into three categories: on-off valves, control or regulator valves, and fixed valves such as orifice plate, nozzle, duckbill, rupture disk, blind valve, etc. These valves represent approximately thirty different design configurations. It has been said that if cost and delivery were no problem, any one of the seven basic valve styles could do the job of any other one. But cost and delivery are very important factors in the real world. So you have to be able to distinguish among these seven styles: ball, butterfly, gate, globe, pinch/diaphragm, plug, and poppet valves.

You will also find that valves are usually arranged into two groups:

general purpose and special purpose. General-purpose valves are off-the-shelf items. Special-purpose valves are those made of special materials for special processes operating under extreme conditions of pressure, temperature, or flow. Valves are also classified according to function, size, fitting end, flow and pressure drop, and whether they are high or low recovery types. Still further classifications are by material, type of fluid carried, pressure and temperature rating, and actuating member.

Since most valves are capable of being operated manually, automatically, or remotely, you will have over fifty different basic actuators to select from and these are manufactured by at least 200 different companies. The actuators will also have a host of accessories with thousands of variations.

As if that wasn't enough, you are also confronted with over 2000 valve manufacturing companies that supply valves for over 400 valve classifications for various services such as oil, petrochemical, nuclear, cryogenics, vacuum, etc. More than fifty organizations produce over 1500 different standards and this does not even include local, city, state, and federal codes.

It should be obvious that valves require lots of attention.

The quality of any system is predicated on good equipment and hinges directly on the specification and selection of system components. The design, selection, and specification of valves can challenge the best educated and most experienced person. You will discover that you are in one of the few disciplines that truly requires you to become a "Jack of All Trades." It is a field that challenges every discipline of engineering and technology.

The process of valve selection has become more complex because so many valves are controlled by computers, making it crucial for these valves to be accurate and efficient. In addition, valve specifiers and selectors are now also responsible for meeting OSHA standards. Other agencies, such as EPA, are beginning to enforce emissions standards. Faulty valves can cause leaks that result in explosions and fires, or emit toxic liquids and gases into the atmosphere, endangering people and the environment.

Because valves can represent anywhere from 9 to 12 percent of total capital spent on a new oil or petrochemical plant, it will be extremely important for the valve specifier and/or selector to remember that the average valve will more than likely be repaired or replaced within 3 years. One of your most crucial tests will be your ability to specify only those valves that can be purchased with the quality that allows for safe operation over a projected period of time. The best valve selection will require you to select valves that can be serviced easily under all conditions, especially in an emergency. It is important that the valve and associated equipment and parts be readily available now and in the future. Consideration should also be given to future upgrade in system performance or to possible process change or modification. And finally, your most difficult task will be to convince management that the low initial cost of a cheap valve that does not meet minimum standards is not a long-term saving. The main reason for a plant's major expenses after a valve is installed is the replacement of valves of inferior quality. You will be required to review the history of a valve company's safety record when specifying or selecting a valve.

In our ever-changing world, valve technology is being challenged as never before. Although economics and availability still play a major role, concerns about product liability are changing the way people think about specifications and selection. Major catastrophes like Three-Mile Island, Bhopal, and the big oil spills point up the need for everyone involved in the selection of fluid-handling devices, such as valves, to understand the ramifications of their choices. It is extremely important that specification writers dictate the valve quality necessary based on information supplied by manufacturers, standards, and other data and that they prescribe the inspection and maintenance to be performed regularly.

*Valve Selection and Specification Guide* not only will tell you how valves and valve terminology evolved, it will also help you save time in writing specifications and will enable you to select that best valve, in a timely manner, for your needs, economically. It will help you toward a better understanding of the importance of valves in a system and how they affect our everyday lives.

Jerry L. Lyons
*President and Chief Executive Officer*
*Yankee Ingenuity, Inc.*

# PREFACE

This book is designed as a guide for engineers and others responsible for valves. It covers both manual and automated valves, beginning with defining valve requirements, including selection of valve types, through engineering, specification, procurement, installation, and operation. It includes a description of just about every commercially available valve type and actuator type, their respective advantages and disadvantages, and the history behind them.

There are many different industries that use valves, and the perspectives of the various fluid processing and refining industries, the power industry, the utility distribution industry and the general industrial market are all included.

The Introduction gives a set of background information, history, and other discussions to prepare the reader for what the book covers. Also included is an introduction to basic types of flow control and basic broad categories of valving. This part of the Introduction is intended to show why various types of flow control are used and how they differ from each other.

The first step toward obtaining valves is the act of specification. In Chapter 1, various sources for specifications are described, including national codes and standards, technical society specifications and guidelines, industry practices, and company policies. In addition, the selection of appropriate rating and material for individual valves is covered along with performance requirements. There is a list of common fluid services with a short guide to some of the valving and specification solutions that are most commonly used.

Chapter 2 is an extensive description of valve designs ranging from the commonplace to the exotic, with emphasis on the most useful types.

Topics covered include the way each valve works, its mechanical characteristics, and how those characteristics affect reliability, shutoff, and flow control, and the suitability of each type for various process uses. The advantages and disadvantages of each type and some of the reasons behind the choice of each type are also discussed.

Operators are the subject of Chapter 3. The chapter is primarily aimed at what the industry refers to as control valves, but all methods of operating or actuating valves are discussed, including all types of manual operators and all the accessories required on valves. The mechanics of a valve's response to a control signal are described.

Chapter 4 is a guide to the procurement process, how it works, what happens "behind the scenes," and how to effectively get what you want.

Maintenance of valves, diagnosing failures, and preventing problems are the topics of Chapter 5. There is also a guide to valve maintenance and a summary of the requirements for refurbishing valves.

Chapter 6 contains a brief description of sizing methods and a discussion of the resources available to size valves.

The appendix contains a list of codes and standards that the reader will find most helpful. The glossary is one of the most extensive valve glossaries available anywhere, listing hundreds of terms and definitions related to valving and materials.

This book took some time in coming to fruition, since the original discussions on its form and content occurred while I was on assignment with Fluor Arabia Ltd. in Yanbu al-Sinaiyah, Saudi Arabia. When we were exchanging proposals and outlines for the book, I felt sure that there would be time for me to devote to the book with few other distractions. That turned out not to be the case, since the writing coincided with one of the peaks in the valve and piping specification business. Somehow, after concluding this project, my sanity is reasonably intact, although there might be some of my colleagues and family who would debate the point.

While this book represents, at least in part, the experiences I have accumulated performing the tasks mentioned in some of the chapters and describing the more memorable valves I have encountered, there were times during the writing when I realized that what I knew was vastly overwhelmed by the number of things I did not know in any sort of detail suitable for publication. In other words, I am grateful for the opportunity to have written this book just from the standpoint of learning so much more about the subject as I was writing it. Of course, after publication of this book, I plan to continue learning more about the subject of valves, and I hope to hear from you, the readers, about anything you feel merits inclusion in a future work and anything you feel that I have misstated here.

*Ronald C. Merrick*

# ACKNOWLEDGMENTS

Many people have shared their knowledge of valves with me over the years. Many other people have learned about valves with me, by working with me on valve engineering, procurement, and repair. Among these, especially, would be the people of Fluor Daniel Inc., Houston Engineering Center, and its predecessor, Fluor Engineers and Constructors, Inc., Houston Division, which was its name when I came on board fifteen years ago. Of these people, the most influential have been the late C. E. (Frosty) Frost, who was my mentor, Jim Siedhoff, Jack Hutchison, Larry Jackson, and Carolyn Denning. There have been hundreds of others, including people from offices in Irvine, Greenville, Chicago, Calgary, Haarlem, and al-Khobar. In addition, there have been many clients, including Aramco, Texaco, Amoco, Mobil, Hoechst Celanese, and many others, code committee members, manufacturers, operators and mechanics who have shared knowledge with me.

The following people reviewed portions of the manuscript and made many valuable suggestions: Dr. Jerry Lyons, P.E.; Bernie Stoffel; Craig White, P.E.; Carolyn Denning; John Sharp; and H. C. Herrington. The Fluor Daniel Houston librarian, Doris Brooks, was also of great help. The manuscript was written in Volkswriter 3, and printed on Donna Fertitta's and Jan Roberson's computers.

I am also grateful to the many manufacturers, valve shops and distributors who furnished artwork, photographs, and illustrations for the book, who are listed below.

## Illustration Credits

Stan Miller, Adams Valves, Inc.; Mike Ford and Cindy Welch, American-Darling Valve, division of American Cast Iron Pipe Co.; Connie Lucas, Anderson, Greenwood & Co., and Ron Wrenn, The Rutherford Co.; Sandy Simpson, Apollo / Conbraco Industries, Inc., Jimmy Dunn and Knox Munnerlyn, Knox & Associates; Teresa Bush, Atomac, Durco GmbH; Lois Schepis, Automatic Switch Co. (ASCO); Jim Sheehan, Bray Valve and Controls, and Stan Simons, Baro Controls; Milton Winebrenner and Rusty Klasen, Flow Control Division, Cooper Industries, Inc.; Susan Janettas, Crane Co., and Tom Lockwood, Lockwood Co.; Cindy Sartain and Ron Kennedy, Daniel Valve Co.; J. G. Puls, DeZurik, a unit of General Signal, and Steve Flynn, Rainey Engineering; Mike Gully, Dunn's Valve Testers, Inc.; Myles Sweeney and Dick Medvid, Duriron Company, Inc.; Lynn Elliott and Steve Swaim, EIM Controls, Inc.; Jim Kiefer, Elkhart Brass Mfg. Co.; Peter Van Blarcom, Fetterolf Corporation, and Joe Henderson, Ekholm & Henderson; Bob Sogge, Fisher Controls Intl., Jim Azzinaro, Posi-Seal International (Fisher Controls Intl.), and Mike Taylor (Quarter-Turn Products) and Steve Shaver, Puffer-Sweiven; Puffer-Sweiven; David Edmonds and Vivian Fulks, Flowseal, a unit of Mark Controls Corp., and Doug Zimmerman, Valquip; Rob Joghems, Fluor Daniel Inc.; Jimmy Vowell, General Valve Company; Sharon Jackson, GH Bettis, and Dan Brock, Automated Valve Systems, Inc.; Kevin Speed, Hex Valve Co.; Rick Robertson, Hoerbiger Corporation of America, Inc.; Howard Kook, Kemutec, Inc., and Tom Brandt, Tom Brandt & Associates; Dick Conley, Kerotest Manufacturing Corp.; John Gray, Limitorque Corp.; John Palmerton and Georgene Harris, Marlin Valve, Inc.; Mike Ridey and Joe O'Sullivan, Marpac, Inc.; Sally Parker and Chuck Blanks, Masoneilan / Dresser Valve and Controls Div.; Delores Zimmerman and Wayne Grabowski, McCanna, Inc.; Leonard Wheeler, Media Valve Co., Inc.; Sarah Davis, Mogas Industries, Inc., and Rick Tompkins, Performance Valve; Tom Culbreth, Mueller Steam Specialty Co., and Patty Teague, Marshall J. Brown Co.; Jim Shea and Joe Davis, Neles-Jamesbury, Inc.; Betty Atkins and Kelly Kelsheimer, Newmans Inc.; Ray Williams and David Chasis, Nibco Inc.; Jack London and Dennis Bowling, Orbit Valve; Gene Tully, Don Cumming and Mike Bradbury, Pacific Valves, a unit of Mark Controls Corp.; Chris Raftis, Red Valve Co.; William Hayes, Crane Resistoflex Co.; Ron Brown, Roto Hammer Co., and Glen Otto, McJunkin; Britt Terrell, Schütte-Meyer, Inc.; Steven Kump and Martin Wilmore, Shafer Valve Co.; Robert Johnson and Ed Sadoski, Smith Valve Corp.; Jimmy Dutton, Stockham Valves & Fittings; Michael Phillips, SVF Inc.; Joey Shepard, Dan Levengood, David Simco and Paul Wright, Swagelok Co.; Sandra Pavlik and Phil Herrmann, Tri-Clover, Inc.; Phil Skousen and Frank Baronne, Valtek, Inc.; William Sandler and Carl Novak, Valve Manufacturers Association of America; Marion MacDonald and Greg Ganz, Velan Valve Corporation; Howard King, Vetco Gray Inc.; Ben Ruiz, Hank Yates, and Joe Delaney, Henry Vogt Machine Co.; Don Zito, Watts Regulator Co., and Dave Beckham, The Beckham Co.; Frank Sinclair, Westlock Controls Corp.; Rebecca Forstrom, Worcester Controls Corp., and John Bosco, Tech Controls Corp.; Diane Skillman and J. J. Court, Xomox Corporation; Doris Sheridan and John Voorhees, Yarway Corp., a subsidiary of Keystone International, Inc.; Greg Bowers, ITT Engineered Valves; Bill Medwig, Ed Majewski and Rich Anderson, Edward Valves Inc.; Stan Trapp, Waterman Industries, Bill Edwards and Vern Jones, Nordstrom Valves Inc.

# INTRODUCTION

This book is about valves—not only what they are and how they are made, but how to use them, how to choose them, how to buy them, how to keep them operating, and what to do when they do not work. The information here should make it easier for you to apply your knowledge to the subject of valves, and make better decisions about what valve to use and where to use it.

The book covers all types of manual and automated valves, along with the operators and actuators used with them, in each of three basic categories—shutoff valves, modulating valves, and nonreturn valves. The operating principles, the advantages and disadvantages, and the availability of each type are all discussed.

Some basic assumptions have been made about you, the reader. One is that you need to know something about valves, most likely as a user of valves in some piping system. Although there will be some general discussion about piping, especially as it pertains to valves and the way they relate to the piping systems they are installed in, the assumption is that you have at least a little knowledge of piping systems or can acquire it from another source. This is because there is not room here to talk about valving and also to discuss the requirements for pipe routing, line sizing, supports, weight, clearances, hydraulic considerations, heat transfer, stress analysis, and everything else that is involved with a complete piping system. This is also because the field of piping is broad enough that in most engineering organizations large enough to need detailed work in all of these fields, the functions are split up among several or many people.

Another assumption is that you are dealing with valves and piping in an industrial situation of some sort, whether it is a process plant, a chemical or hydrocarbon plant, a general industrial plant that has some piping systems, a power generation utility, a distribution utility, or any industrial installation that uses fluids of some sort and needs to move them from one spot to another. You may be the original specifier, the process designer, the person responsible for or participating in the detail engineering, the buyer, the inspector, or the operating or maintenance person who has valves under your jurisdiction.

What is not assumed is that you already know everything contained herein. Although it is likely that many of the readers of this book are engineers or are functioning in an engineering environment of some sort, it is not necessary to have an extensive knowledge of engineering theory to understand the topics here. Much of what is contained in this book is plain mechanics.

A final assumption is that you are not responsible for *designing* or *building* valves on the basis of the information contained here, but rather will be responsible for *selecting, specifying, purchasing, using,* or *maintaining* valves built by valve manufacturers. The skills required to make and sell valves run heavily toward machine design, casting technology, manufacturing engineering, reliability theory, and other skills not required for being able to specify and use valves. Those topics are covered much better in other sources.

Of course it is not possible to cover everything there is to know about valving either. For one thing, there is not likely to be one individual who can know all there is to know about such a diverse subject that spans such a variety of uses and industries. But the general discussions presented here should be sufficient to get you thinking about what it is that you need in your particular location. You, after all, are the person who decides what valve you want and what you can get for your particular application. This book should help you be a more effective valve user and specifier.

A brief explanation of the conventions used in this book: The book was written for the person familiar with piping terminology, with common terms and definitions that exist in the North American valve market and engineering practice. A few terms used in the English-speaking world outside North America are also referred to. The terminology commonly used in the English language to describe sizes and ratings has evolved over more than a century from multiple origins in North America, Great Britain, and Europe into an inconsistent hodgepodge of meanings. In the last 20 years, standards organizations in North America and Europe have each designated "official" terminology for pipe sizes and pressure classes, so that there are now two sets of standard nomenclature, one for the United States and some other locales that have a heritage of the inch-pound or imperial system, where the nomenclature is based on American National Standards Institute (ANSI) terminology, and the other one for the fully metric or International System of Units (SI) system, where the nomenclature is based on International Standards Organization (ISO) terminology. There are also many places in the world where both sets of terminology are in use.

In the portions of the world where ISO nomenclature is predominant, the terminology is clear. Dimensions are expressed in millimeters (usually), pressures are expressed in either bar or kilopascal (kPa), pipe sizes are called DN (for "nominal diameter"), and ratings are called PN (for "nominal pressure"). In the United States and various places where engineering installations were first derived from United States practice, the situation is less clear. Dimensions and pressures have never changed from the nineteenth century practices of feet/inches and pounds per square inch (psi). Yet pipe sizes that are colloquially referred to as "inches," meaning the nominal size, and that were formerly referred to as IPS (for "iron pipe size"), are now referred to by American standards organizations as NPS (for "nominal pipe size"), while pressure ratings that are colloquially referred to as "pounds," meaning psi, are designated "class" in the standards. Since most people familiar with American engineering technology use the ordinary terms first, they will be used first here also, followed by the others, even though "NPS" is practically never used in normal usage. Pure dimensions will be shown as "4½ in. (115 mm)," with each set of units rounded to reasonable values, not necessarily equivalent between the imperial and metric values. Actual pressures will be shown as "100 psi (7 bar)," with the same rounding practice. Nominal pipe sizes will be called "6-in. pipe size (DN 150)," without use of the NPS units. Pressure ratings will be called "class 150 (PN 16)," unless no close ISO rating exists.

## A Brief History of Valving

The first fluid to be conveyed by humans was water, and the first conduits constructed by human beings were for the conveyance of water to places where it was not available naturally. For thousands of centuries, humans had been able to live only where water fell or where the rivers flooded every year to provide the water needed for growing crops, and only as far away from the water source as they could carry vessels filled with water for daily use. The Mesopotamian civilizations were perhaps the first to dig canals to bring water to where it was convenient for people to live, and to bring water to the large areas of crops that were required to sustain cities. The Romans built many miles of aqueducts across large areas of Europe, bringing great quantities of water to sustain their civilization. These human-made watercourses contained the earliest valves, which consisted of pieces of wood or stone to divert or obstruct flow.

Also in this era, other flow-handling devices such as pumps (the Archimedes screw), power generation equipment (waterwheels), and the first piping were developed. The Romans made extensive use of piping, at first made of lead since lead was the most malleable metal at their disposal. The Latin name for lead, *plumbum,* gave rise to our word "plumbing." The first valves to be used in these piping systems were cocks, or plug valves, which were used into the Middle Ages and beyond, chiefly as bungs on barrels of consumable liquids.

Slightly more sophisticated piping systems were developed over the course of centuries, including cast-iron piping that could be laid for miles. The motive power was still mostly gravity. The development of coal mining in the British Isles and elsewhere gave rise to the need to remove water that collected in the mines. The Newcomen engine and other early steam engines provided the first source of power and pressure to pump water substantial distances, and the first use of pressurized and elevated-temperature fluids (steam) outside of laboratories.

For a long time, all of this machinery was hand built, one-of-a-kind construction. Each piece and each of what we would now think of as separate subassemblies was built on the spot by the same craftsmen. The development of mass production techniques, the growth of infrastructure such as municipal water and gas projects, the growth of industrial production of pressure equipment such as steam locomotives and fire engines, and the beginnings of the oil and chemical industries all led to the creation of the valve industry. In the United States, several valve manufacturing companies that were founded in the 1840s and 1850s are still active today.

The development of reliable valves and pipe led to the growth of many other industries that needed such material, from power stations to food canneries. Other valve requirements, such as designs capable of withstanding the rigors of shipping and of wartime hazards, led to improved designs and improved materials also. The gate, globe, and check valves had been around for many years, and were gradually supplemented by other types as more reliable sealing techniques and more accurate machining became common. The plug valve, the ball valve, and the butterfly valve all evolved over a period of many years. Later some other types, like the diaphragm valve, were invented by a single person. As designs evolved, some of these types eventually came to resemble each other or to overlap in their applications and useful ranges. Many of the older designs have been downgraded in their applications or have disappeared from the market almost entirely as new products and new design improvements have come along. The development of improved nonmetallic materials for sealing and chemical resistance has moved this process along rapidly.

In addition to the design and material improvements that have been made to the valves themselves in the last century and a half, there have been great advances in the control and automation of valves and complete industrial systems. Control systems have evolved, first to permit the controlling of individual flows from local instruments through pneumatic signal loops, to the use of remote actuated valves controlled from a central location, to the electronic revolution that has permitted total control of a complete process plant or hundreds of miles of pipeline from a single room. Electronic instrumentation and computer controls have permitted the near-total automation of process plants and factories and self-diagnosis of malfunctions.

In the years following World War II, the United States dominated the world market for manufactured goods, especially engineered products. Today, the lower cost of production in many countries has resulted in the shifting of much industrial production to other locations around the world. Foundry work has been especially affected. Much of the older technology and commodity production that has remained in the United States has moved to states where the cost of manufacturing is lower than in the older manufacturing centers where much of the valve industry, like much else of American industry, had been concentrated for many years.The United States, however, remains a strong contender in the engineering and design of industrial products, and has a very large share of the market of more advanced manufacturing and design.

World economics and exchange rates do change, though, and there has even been some return from overseas of industrial production that had departed the United States. There is a sizable market for the export of high-technology valve products, even to some of the countries that seemingly are taking the biggest share of the commodity markets. There is also a recognition, especially with engineered products, that the producers of the most successful finished goods are the ones who conduct their engineering in the United States and design their products around American codes and standards.

The valve industry in the United States is still a major contributor to domestic production and to the economy. The size of the valve market in the United States approaches $3 billion in yearly sales, with approximately 50,000 people employed in the manufacture and sale of valves. Additionally, many of the leading valve companies in the world are American-owned multinationals that make and sell valves around the world for international consumption.

Today there are valves capable of operating from near-vacuum to 100,000 psi (7000 bar) and beyond, and from the temperature of liquid helium to beyond 1500°F (800°C). There are valves more than 1000 feet (300 m) underwater, and valves in space. There are human-made valves implanted in thousands of human bodies. While this book cannot begin to cover the outermost reaches of valve technology, it is worthwhile to contemplate the extent and the possibilities of the valve universe.

## The Use of Valves

Where are all these valves used? At least some are being used in commercial or industrial settings. Most people who use valves, and for that matter most people who make or sell valves, see only a small portion of the total valve spectrum. There is a trade association for the valve industry, the Valve Manufacturers Association (VMA), which tracks the valve market in the United States. Their statistics show the sales of valves by market, in terms of dollars (Fig. I-1). The average value of valves sold

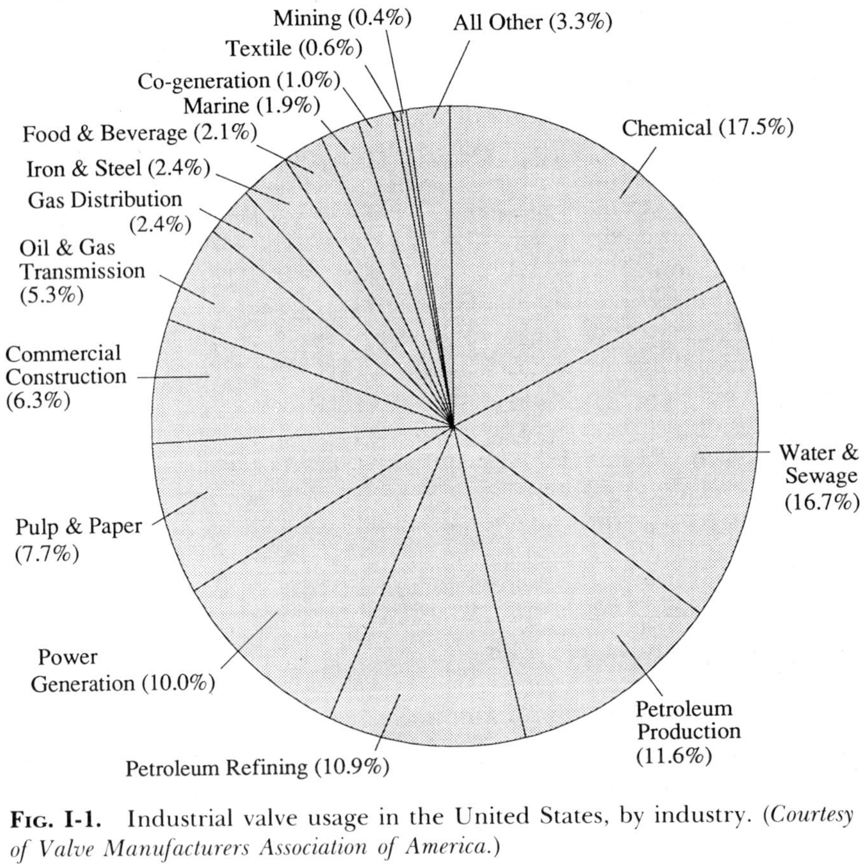


**Fig. I-1.** Industrial valve usage in the United States, by industry. (*Courtesy of Valve Manufacturers Association of America.*)

in each market varies somewhat. The proportion of the value of valves to the value of other material sold in these markets probably varies also. Some of these markets are quite closely related, and some of the categories might overlap such as petroleum production and oil and gas transmission, or petroleum refining and chemical.

Each of these categories consists of a number of smaller and sometimes almost unrelated industries. For instance, the oil and gas transmission category actually consists of two businesses with different economics, service conditions, and valve selections. Gas distribution, although having some of the same requirements as the transmission companies, is different and has different valve requirements since the territories it traverses are different—city streets and front lawns rather than mountains and deserts. The chemical market consists of petrochemicals, which are flammable but not very corrosive and handled in large volumes, and the specialty and fine chemicals market, where services may be very corrosive and contamination sensitive but not always flammable. Water valves and sewage valves also have requirements different from each other, as do fossil power valves and nuclear power valves. Several industries have unique requirements, such as food and beverage production.

Companies sometimes are engaged in several of these categories, but where this is true, almost invariably there are separate and unrelated business units. A large multinational oil company, for example, will have an exploration and production unit that buys valves for gas and/or oil production, a refining and marketing unit that makes and sells basic hydrocarbon products like gasoline, a pipe-line unit that transports production to the refineries and finished and semifinished hydrocarbons between refineries and to terminals, a marine transportation unit that buys valves suitable for shipboard use, perhaps a chemical unit that makes basic chemicals and plastics, and often a cogeneration unit that makes and sells power and steam.

Some of the valves available on the commercial market are suitable for several of these segments, but most, at least most of the big-ticket ones, are designed according to each industry's unique codes and customs and are not really suitable or acceptable for many of the others. Manufacturer's product lines reflect this fact, and so most manufacturers likewise participate in only a few or perhaps only one of these markets. The chemical and refining markets, which have broadly the same governing codes and overlapping design practices, have probably the largest selection of manufacturers, while the oil and gas production, nuclear power, and water and sewage markets probably have the smallest choice of manufacturers.

Each of the major types of valving (Fig. I-2) can be used in almost all of these markets, although the participation of a given valve type in some markets is very small. The ups and downs of sales mostly reflect

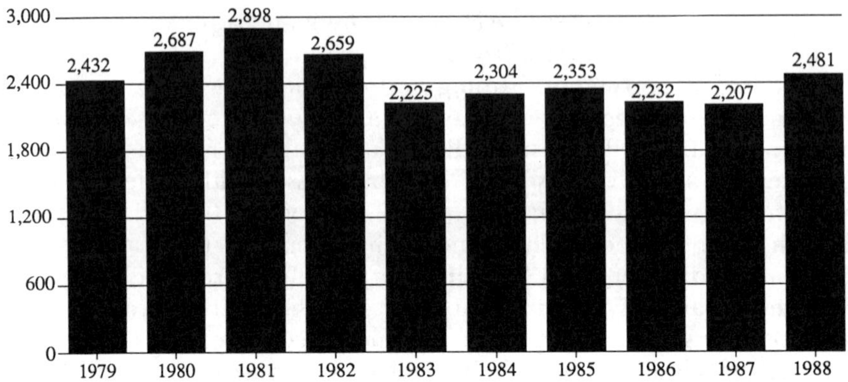


## Selected Valve Types, 1979-1988

| Year | Gate, Globe and Check Valves | IBBM Gate Hydrants and AWWA Butterfly Valves | Ball Valves | Industrial Butterfly Valves | Plug Valves | Automated Valves |
|---|---|---|---|---|---|---|
| 1979 | 577.5 | 205.6 | 370.1 | 242.0 | 206.2 | 546.2 |
| 1980 | 655.3 | 197.5 | 464.4 | 245.6 | 269.4 | 556.8 |
| 1981 | 651.3 | 209.8 | 521.4 | 252.3 | 302.0 | 610.3 |
| 1982 | 530.2 | 193.8 | 451.1 | 223.2 | 246.3 | 590.8 |
| 1983 | 374.6 | 233.9 | 322.1 | 200.9 | 194.2 | 525.6 |
| 1984 | 346.1 | 290.8 | 384.0 | 206.0 | 223.2 | 506.1 |
| 1985 | 343.7 | 294.2 | 406.2 | 208.1 | 229.0 | 528.0 |
| 1986 | 340.8 | 322.4 | 355.5 | 185.3 | 198.2 | 510.9 |
| 1987 | 354.7 | 315.9 | 344.0 | 176.7 | 199.7 | 480.3 |
| 1988 | 400.6 | 309.7 | 408.6 | 227.2 | 217.2 | 561.2 |

**Fig. I-2.** Sales of individual valve types. (*Courtesy of Valve Manufacturers Association of America.*)

the condition of the industries that buy these valves from year to year, rather than being evidence of any major trends in valve consumption.

## The Function of Valves

There are many basic types of valves in common use. There are a bewildering variety of styles, designs, configurations, advantages, and improvements offered in the valve market. But what do these valves actually do for you? Why would you want one type and not another? And why do you want a valve in the first place?

There are three basic functions that valves perform. When we speak of a valve, we mean a mechanical device that performs some action which has an effect on a flow stream. These three functions are:

- Allowing or shutting off flow (on-off or block)
- Changing the magnitude of flow (throttling or control)
- Allowing flow in one direction only (nonreturn or check)

The three functions may overlap somewhat, and do in the majority of valve designs. It is quite common for a throttling valve to be capable of shutting off flow, that is, to function also as a block valve. On the other hand, many block valves are not very good at throttling or controlling flows. The nonreturn valves typically are not intended to be operated by an outside force, or to provide a block or throttling capability, but in a few cases this is possible.

Some valve designs can easily be characterized as possessing one of these abilities, and not the others. Other types are thought of predominantly in terms of one of these functions, yet are perfectly capable of performing a different function. Still others are very difficult to characterize as only one, and are recognized as being equally capable of performing more than one function.

Each of these functions can be made use of in a number of ways. Observing a process flow diagram makes this apparent. Some of these ways are:

- *Block valves.* Controlling the presence or absence of fluids, preventing or allowing the mixing of flow streams, keeping fluids away from areas where people are working, shutting off flows in an emergency

- *Throttling valves.* Regulating the volume of a flow stream, regulating the pressure of a flow stream, regulating the temperature of a flow stream

- *Nonreturn valves.* Preventing return flow in a line, preventing flow in one line from entering another, preventing overpressuring of low-pressure components, preventing damage to equipment

Each of these abilities can be made use of in a number of ways. The ability of block valves to allow and shut off flows, for instance, makes them useful in maintenance work, since piping generally has to be drained before work can be done on it, and there has to be a means of assuring that fluids do not enter any enclosed spaces that humans are in. (For safety's sake, just one valve is not usually considered sufficient to isolate work areas. Usually, an additional precaution like a blank or a removable spool is required. But the valves must be closed first, to allow these additional devices to be made use of.) The mixing of batches can often be accomplished by opening the block valve on each of several inlet lines for a set period of time, then closing it. Flow streams can be mixed in a rough proportion just by having the lines sized to the right proportions, then allowing the product to flow through them with only an on-off valve on each line. Safety shutdown systems use block valves, actuated from a remote location or automatically, to shut down the flow of fluids around a plant or in a pipeline if a given condition occurs, to keep the fluids from escaping or from continuing to feed a reaction or a fire. An additional type of block valve, known as a relief valve or safety valve, *opens* automatically when the pressure below it exceeds a predetermined value.

Throttling valves, whether manually adjusted or equipped with an actuator that continually adjusts the position of the valve on the basis of an input signal from some sensing device, are used to control some property of the flow. Essentially, each of these fluid control actions is performed by controlling the mass flow rate through the valve. The differences lie in the way that the fluid responds, which is determined by the fluid characteristics, the valve configuration, or the piping and storage volume configuration. The level in a tank can be controlled by a level sensor determining if the level is below a certain point, and if so, sending a signal to open the inlet valve to allow more liquid flow in or to close the outlet valve to allow less liquid to flow out. Pressures can be controlled in a similar manner, or automatically in a self-regulating device, by sensing the pressure level in the region downstream of the valve. If the pressure level is too low, the valve opens, admitting more fluid mass, usually vapor, to the region and increasing its pressure, or vice versa, if the pressure level is too high, by opening the outlet valve somewhat and allowing some fluid mass to flow out, thereby decreasing the pressure. Temperature regulation is a little more difficult, and is directly accomplished only with vapors or vaporizing fluids that lose heat when their pressure is lowered, by discharging through a valve into a region of lower pressure. Temperature regulation can be accomplished indirectly by throttling valves that control the mass flow of different fluids with different temperatures into the same region.

Nonreturn valves control the fluid characteristics in a volume or section of piping in a more simplistic way, but still as a highly useful component. Most rotating machinery is intended to rotate in only one direction, to drive the fluid through it, whether it is a pump (liquids) or a compressor (gases). The flow of fluids backward through the machinery, which naturally would have the tendency to turn the rotating parts backward, is at best undesirable and usually damaging. Check valves are used to prevent this action. Check valves are used to prevent the flow of higher pressure fluids into regions of the piping that are designed for lower pressures, and can be used anytime that it is desirable to allow a fluid flow to move in only one direction.

The individual capabilities of each valve type will be discussed in Chapter 2, and the capabilities of each actuator type in Chapter 3, in terms of these three basic functions. Each valve type will be classified as one of these functions, but since in fact many valve designs can be more than one of the three, the alternate uses and the relative utility of each will also be mentioned.

# 1

# Valve Design and Specification

## Industry Codes and Standards

### *History of Standards*

Most valves used in industry are subject to the same codes, standards, and regulations that govern any other component. The various documents that exist have been developed over the years, some by industry organizations such as the American Petroleum Institute (API) or the Technical Association of the Pulp and Paper Industry (TAPPI); some by engineering societies such as the American Society of Mechanical Engineers (ASME), the Instrument Society of America (ISA), and the Manufacturers Standardization Society of the Valve and Fitting Industry (MSS); and some by governmental bodies such as the Department of Defense. Many of these documents have the force of law. In almost all cases, they are also invoked in contract documents. Sometimes a user may choose not to require certain standards, but manufacturer's production standards often are designed around these standards anyway. There are many standards-making organizations in the United States and in the world, some that cooperate with each other and others that stand alone. Each broad category of industry has its own selection of standards and standards-making organizations, with some industries much more concerned with codes and standards than others.

DEFINITIONS A *regulation* is a directive of the government, either a law passed by legislative action or a military or civilian directive that has the force of law, which covers some aspect of the requirements for either an entire engineered system, such as a ship, or a specific manufactured product, like a valve. A *code* is a set of rules for general guidance, issued by an engineering society but often adopted by governmental bodies and required by law, which covers the engineering and design of a given type of plant or engineered system, such as a refinery or a boiler. A *standard*, which is issued by an engineering society, a trade organization, or an individual company, sets down minimum requirements for the design and manufacture of one particular object, such as a valve. A *specification* is the set of documents, which may or may not include references to all of the larger documents (codes, standards, and/or regulations) above, plus specific descriptions or data sheets, that are assembled to describe the total requirements for a particular item or group of items.

Many groups use the terms "standard" and "specification" interchangeably, or define these two terms by transposing the definitions given above. But the usage here is the most common and probably the most descriptive.

DEVELOPMENT A brief history of the development of standards and the organizations that make them will help explain what they do and how they affect our professional and personal lives. For the first 100 years or so of the industrial revolution, there was no concept of standardization. For hundreds of years, machinery had been built by the same people who designed it, in "laboratories" where they labored. These people were often unschooled, naturally gifted mechanics, or artisans, as they were then known. An "engineer" was one who was involved with engines of some sort, an "engine" being an apparatus that accomplished some task, whether it was throwing stones over an enemy's walls or lifting water from a mine or hauling wagons along a railroad. Gradually, the educated elite who had always been schooled in physics and mathematics began to take an interest in practical mechanical applications, and later in electrical applications. Yet, for many years, in fact until early in this century, the primary method of acquiring engineering knowledge was by apprenticeship, and the primary method of developing new inventions was by experimentation.

The natural consequences of this state of affairs were, first, a trial-and-error method of developing products for people to use, and, second, an inability to interchange goods and services freely due to the lack of standardization. The use of trial and error—the way good mechanics made pieces to fit on their machines as they saw fit—never could have advanced the cause of standardization, and the lack of standardization meant that mechanics had no resources beyond what they could think up themselves. There was often no reason to see beyond trial and error because everything was handmade and many things were one of a kind.

The beginnings of the end of this situation, which had existed from medieval times when engines and machinery were actively employed to wage war and to construct buildings, lay in the scientific and engineering societies that sprang up to exchange information about what individual craftspeople had accomplished. These societies arose in the mid to late eighteenth century in many countries, modeled after the earlier royal societies for philosophy and science. Inventors like Benjamin Franklin played a prominent role in these societies. The impetus that led to the development of these societies also spawned the first companies, or groups of people organized to accomplish a task of benefit to all, like companies of soldiers.

In the late eighteenth and early nineteenth centuries, these groups developed naturally in the cities, which were the only locales with enough people in one place for a significant exchange of information, either by direct conversation or by the circulation of printed material. Also located in the cities was the humanpower required for the first "manufactories" where manufactured goods (such as cloth) could be made by the first manufacturing machinery. The details of machinery and chemical processes were recognized early on as being valuable, and exporting technology was looked on as a serious crime in England, China, and other places.

The factories, where large numbers of people made large quantities of goods, led to an accumulation of capital and wealth that could be allocated to the creation of more machinery and more wealth. Many of the principals in the first engineering and scientific societies were people who had acquired some wealth in just this manner. They wanted to trade information in order to acquire the ability to make and sell more things, and to get their products out to more places. Their attempts to optimize other people's designs naturally led to some "convergence," or consensus as to the best way to do a certain thing. Then as now, the government was the biggest single customer for some mechanical devices such as guns. Many of these factories, while collecting many artisans under one roof to make the same product at the same time, still consisted of many people working in parallel, making many or all of the parts themselves and fitting them together themselves. The development of the idea of mass-produced interchangeable parts was the second step toward standardization.

Transportation both directly and indirectly encouraged standardization and made it possible for larger numbers of people to communicate. For instance, all the locomotives and cars belonging to a railroad company had to run on the same track and be capable of coupling to one another, along with any equipment belonging to any other company with which they were interchanged. It soon became possible for large numbers of people from one city to hold meetings with large numbers of people from other cities. However, the very introduction of technology meant that civilization demanded that the technology be improved. The steam boiler was the first serious example.

The steam boiler was the first source of power not arising from the muscles of humans or animals. It took 150 years, from the early 1700s to the middle of the second half of the 1800s, for the steam boiler to evolve from a largely decorative object that also did work to the predominant source of industrial and motive power in the world. Yet the steam boiler was also one of the leading hazards of the industrialized world. The first explosion of a steamboat boiler took place just a few years after the first successful steamboat runs. The problem of steamboat boiler explosions in the 1820s led to a study by the Franklin Institute, which had been formed in Philadelphia in 1824 as a technical and engineering society. Their systematic testing of boilers and boiler explosions led to the first scientific knowledge of how boiler structures reacted under pressure. Their recommendations were enacted into law in the Steamboat Act of 1838, which was the first government regulation on the safety of industrial installations in the United States.

Such practices and regulations were slow to take hold, though. More than 10,000 boiler explosions a year occurred in the United States in the 1870s, killing and injuring many people. It was the explosion of a fire engine in New York City that focused the attention of the technological leaders of the time on ways to improve safety and reduce the danger to human life. The formation of the American Society of Mechanical Engineers (ASME) in 1880 was prompted as much by a desire to improve the safety and reliability of the steam boiler as by anything else, although the wish to improve the efficiency of manufacturing through standardization was certainly another goal.

The ASME definitely succeeded in its efforts. By 1900 the number of boiler explosions in the United States had been reduced to a small fraction of the number in the 1870s, as a result of the development of the ASME boiler code and the imposition of its requirements by law in many jurisdictions. Also by 1900 or shortly thereafter, many of the organizations that are active in the standards-making field to this day, or their direct ancestors, had been formed.

In general the structure of the standards organizations of the world today is one of cooperation between groups of quite different structure. The International Standards Organization (ISO) is the world body to which most national standards coordinating organizations belong. These national standards organizations are often government bodies, but in the United States a volunteer organization is recognized as the official coordinating body. In the United States, the American National Standards Institute (ANSI) is the body that serves this function. ANSI once was a standards-writing body as well, but has turned the actual writing of standards over to engineering and trade societies in order to be purely a coordinating and disseminating body. In other countries the relationship varies. In the United Kingdom the British Standards Institution is a standards-writing body as well as a coordinating body, and also serves as a testing organization. Many less well-developed countries such as Saudi Arabia have standards institutions, but often permit foreign (such as American) standards to be used in their entirety. Most European national standards organizations may fade away, or take a somewhat different role, after the 1992 European Community unified internal market plan. There is now the European Committee for Standardization (CEN, its initials in most European languages) which is coordinating on a Community-wide basis without United States participation.

The organizations that actually write codes and standards in the United States include engineering societies such as ASME, the Institute of Electrical and Electronics Engineers (IEEE), the Instrument Society of America (ISA), and the Society of Automotive Engineers (SAE). All or almost all of the standards that these organizations publish are also adopted by ANSI. Since the adoption of standards written by a society that is recognized by ANSI is just about automatic, the ANSI adoption principally means no more than providing a different number for the standard. Even the practice of multiple numbering is fading as ANSI becomes more and more a coordinating body for the work of the sponsoring societies.

Other standards-making organizations include industry organizations such as the American Petroleum Institute (API), the Technical Association of the Pulp and Paper Industry (TAPPI), and the American Water Works Association (AWWA). In addition to organizations made up of product users, there are others such as the Manufacturers Standardization Society of the Valve and Fitting Industry (MSS), made up of manufacturers who have formalized many of what began as common practices among several manufacturers. Originally these practices were intended to promote interchangability of one manufacturer's product for another. Many of these organizations also function as trade and lobbying organizations, but the standards they publish sometimes carry a lot of influence in industries that have very little to do with the original organization. Some industries find their standards-making expression purely through engineering societies, while others rely more on their trade organization.

At the next level are standards written by a corporation for itself and its internal use. Some large companies have so much power in the market as customers and have so much influence in engineering circles that the standards they originally wrote for themselves have been adopted as de facto standards for an industry; and in many cases these standards have been, after review by standards-making organizations, transformed into new national standards. An example is Exxon, whose standards for testing of valves for fire-safe behavior have been used or acknowledged by many other companies in the hydrocarbon and related industries.

The standards-making process is an involved one. It normally depends on the committee structure of the standards organization. Most organizations, which generally have many functions other than standards, have a permanent set of committees that have created various standards in the past and are now charged with keeping them up to date and considering additions and new areas of coverage. There is generally a small paid staff to perform clerical and publishing functions, but the engineering expertise is almost all volunteer. Sometimes the companies are the actual members, and when a committee person retires from the company the company replaces that person with another designee. More usually, the individuals are members and the companies employing the committee members provide funding for time and travel. Some committees meet on a scheduled basis regardless of business, some never meet but conduct all their business by mail, and some meet frequently prior to publication deadlines for standards but infrequently in off years.

The codes and standards produced by these organizations are of great importance to design and specification engineers. Almost all construction today is covered by some type of code, and a lot of manufactured products are covered by some published standard. In some industries such as food processing, a government agency is actually inspecting the construction to see that it is in compliance with the government regulations (in this case, the Food and Drug Administration [FDA] or the United States Department of Agriculture [USDA]). In most private industry, state law or local ordinance may require conformance to some codes such as the boiler code. More often, design and construction contracts specify that a given code is to be used. Many contracts also require the use of any industry standards that apply to a given item being specified or installed on the project. Most often company policy is the actual "trigger" that invokes the use of particular codes and standards for end users, in conjunction with insurance contracts that will not cover conditions in which a particular code is not followed. There is another set of organizations involved at this point, primarily concerned with industrial safety and loss prevention, which include the Factory Mutual Association (FM) and Underwriters Laboratories (UL), which do not actually write standards but do write performance codes and test or inspect products as to their ability to perform reliably.

What do the codes and standards mean to us, as people involved in specifying or maintaining a plant or process unit? First, they save us a lot of the basic engineering work that our predecessors had to do. The codes already assume for us the safety factors to be used in calculations. Very basic engineering decisions about what is safe and unsafe are already established in the codes. For instance, certain practices and material choices that are known to be unsafe are mentioned and either prohibited or advised against in the codes. The choice of basic materials is aided by the codes, which list materials known to be suitable (and usually widely available). The allowable stresses to be used in calculating the required material thicknesses are also given. In essence, the codes distill the knowledge of a century of good engineering practice. However, all codes will state that their rules are not to be used as a substitute for good engineering judgment. The practices and materials that they prohibit are the ones that, basically, are guaranteed to get you in trouble. There is still a lot of leeway in most sections of most codes to permit you to do the cheapest job that you can and still be fairly safe, if that is what you have to do, or to do the best project that money can buy. In the latter instance the code will give you a few hints, but it is up to you to figure out how to maximize your design.

Even with today's sophisticated techniques for engineering analysis, many codes and standards contain data that go back a number of years. Advances in technology that permit the user to do more with less are written into the codes as they occur, but the manner of updating codes has changed somewhat in recent years. Many codes and standards used to accumulate revision data for 8 to 10 years before being updated, and many standards that address material or technology that is mature, or in declining use, may still be updated only every 10 years or so. But many codes are now on a regular revision and update cycle of every 3 or 4 years, so that small advances get incorporated and distributed more often. Areas of research in, for instance, materials science that leads to improved steel-making capabilities and higher allowables, are generally not adopted until, first, the technology is widely available, to avoid promising something that manufacturers cannot deliver, and, second, the advance is dramatic enough to change the rules significantly. Some codes have pegged a 10 percent change in recognized allowable stresses as the point where the difference is significant enough to be published. This keeps changes in thousands of installed databases around the world to a minimum.

These codes and standards apply to valves as they do to any other component of an industrial installation. They impose limits on the pressures and temperatures for which a valve can be used; they impose some limits on the services for which certain materials of construction can be used; they define the standard materials, sizes, and ratings that are available; they impose rules for the calculation of wall thicknesses and other component sizes; sometimes they impose minimum acceptable sizes and thicknesses; they define the requirements for testing and acceptable limits on defects and performance; they define the modes of communication between signal sending and receiving devices. All of these functions and many more mean the specifier of valves is not working in a figurative vacuum. For valve specifiers who work in normal circumstances, engineering valves that will be used in a typical industrial situation, the standards represent a set of building blocks from which to choose the ones that are needed. Of course, not every case is like this. Valves for outer space or ultra-high or ultra-low pressure/temperature or other special situations that are outside the limits of published standards have to be designed from the ground up. Valves that are used in unregulated applications can be built any way. But not very many valve engineers often find themselves in either of these extreme situations.

The effect of these standards on the products you buy or make is, of course, significant. Compare an automobile of the late 1960s to one of the late 1980s, for example. Granted that most of the additional requirements for automobiles are created by government regulation rather than by industry action, but the changes are there nevertheless. Valves are not as dramatically affected as automobiles, but standards changes do cause changes in valve production over time. It is also true that, of the plethora of standards that could apply to a given product, the same item may and often does conform to many of them. This is even true of the international market to some extent, in that many of the ISO standards are identical or nearly identical to ANSI standards. Yet real differences do exist.

The names and abbreviations for standards-making organizations and their codes and standards are listed elsewhere in this book, along with a list of many of the most common American standards that have some bearing on valve design or installation.

## *Materials*

The field of materials science is the source of knowledge about material properties and their behaviors. Of course, the materials that were first used for engineered products gave rise to the first data about material properties, so this is more or less a "which came first, the chicken or the egg?" history. Of all the standards that apply to valves, the material standards are the most basic and the ones that evolved the most naturally.

Many of the materials that are in common use today, including cast brass and bronze, cast gray iron, and rolled plate, go back in one form or another for centuries. Many techniques of metalworking also go back for centuries. It is really only the scope that has changed, except for a few dramatic inventions such as the steel-making process (1850s) that replaced iron plate and the development of arc welding (1910s to 1930s) that replaced riveting and large integral castings. Other new developments have created or exploited materials that have permitted the emergence of technologies that were undreamed of before, such as the commercialization of aluminum (1880s), stainless steels (1910s), titanium (1950s), plastics (1920s to the present), and composites (1960s to the present).

During the nineteenth century, there was a gradual improvement in foundry and mill practice, resulting in better products over the years. A lot of this development was through systematic research, eventually developing into formal metallurgical and materials knowledge. In terms of commercial availability, for about 100 years there were "merchant" grades, meaning products made for general industrial use or run-of-the-mill products, and special grades for special purposes, such as "flange quality" and "firebox quality" steel. In fact, these latter two terms continued in use in specifications until the 1960s. Firebox quality plate was the higher grade, used in boilers for applications where the plate would be directly exposed to flame plus pressure, such as firebox walls, and flange quality plate or boiler plate was the lower grade suitable for boiler shells and other applications where it would experience pressure plus temperature, but not flame impingement. It was still a better grade of plate than that used for pressure-retaining applications without elevated temperatures, which was a better grade of plate than that used for structural work.

The designations that arose from commercial practice were formalized over the years by the American Society for Testing and Materials (ASTM), beginning in 1898. The first specifications that were published by ASTM were for these types of steel, iron, and brass/bronze products that had evolved. In fact, the first specification, ASTM-A1, on rolled railroad rail, first published in 1901, is still in effect today, as are the specifications for structural grade steel (ASTM-A36), and the specifications that provide general requirements for plate products (ASTM-A6 and A20). The original specification (ASTM-A212) for flange and firebox quality plate has been discontinued, though, along with many other early specifications, because advances in both steel-making technology and inspection practices have made them obsolete. Sometimes new standards are written for new, higher sets of requirements while leaving the old ones more or less intact because there is still a sizable market for the old product in markets that have less demanding requirements. In the worldwide steel market, however, uses for the old product eventually dry up and the standards are withdrawn. An example is wrought-iron plate and pipe, which is probably not produced anywhere in the world today. Today, also, even the terms "flange quality" and "firebox quality" have essentially disappeared from use in most industries.

You can see for yourself how material specifications have evolved if you have access to a library of specifications. Most libraries maintain at least one copy of all previous editions. This allows for the possibility of researching the exact requirements that were in effect at the time that a given object was manufactured. If you have a copy of the current set of

ASTM specifications, the footnote at the bottom of the first page of each specification shows the date that it was first issued, along with other information such as the committee that has jurisdiction over that specification. At the top of the first page, the other organizations that have adopted the ASTM specification in some way, other than just referencing it, are listed.

The ASTM specifications are discussed first because ASTM is the standards organization with the most far-reaching influence on materials. Originally, the scope of ASTM was to be, as its name implied, just material specifications and specifications for standard test methods and acceptance criteria. Over the years, its scope has broadened to include complete product specifications for items including plastic pipe and fittings, plus thousands of items in nonindustrial use such as textiles, medical items, consumer-related testing procedures including toy safety, and even some very personal products.

Each of the broad categories of standards that ASTM covers is under the jurisdiction of a committee. The material specifications of most interest to piping and valve specification engineers just happen to be listed first in the publication order, in volume 01.01, and are under the jurisdiction of committee A-1 on Steel, Stainless Steel, and Related Alloys. On average, two or three standards are revised every month in this volume, and maybe a hundred a month in the entire scope of ASTM concern. Revisions in steel standards, generally, are minor, most being concerned with fine tuning test and inspection criteria or covering new manufacturing technologies. Standards on plastics and other evolving industries are changing quite a bit more than the standards on steel and steel alloys.

Other standards organizations also discuss materials. First would be ASME, which covers in part II of the Boiler and Pressure Vessel Code many of the same materials and many of the same specifications as ASTM, but only the ones that pertain to the code or are referenced in the code itself. The ASME material specifications are designated by SA- or SB- with otherwise the same numbering system as the ASTM specifications. Occasionally the two are different in their requirements or in the number of grades of material that are covered. Where the two are the same, the ASME specification will so state. At one time, for general industrial use where no code applied, many ASTM specifications also had a different number assigned by ANSI. This practice has been discontinued. A number of different standards organizations with special interests, most of which are in the industrial sector and do not have very much to do with valves, also have assigned their own numbers to ASTM specifications and reference them as if they were their own specifications.

Many standards for manufactured items will reference the ASTM specifications for many or all the components. In the case of material built to a specific code, generally the code requires that ASTM specifications be used where a suitable material exists. Sometimes only specific ASTM standards are permitted, by means of an appendix that lists the acceptable standards. In some codes, the user has the privilege of using material not mentioned specifically by the code if certain rules are followed. The intent of these rules is to permit new materials (referred to as "unlisted" materials) to be used in code construction if there is some reason that nothing already listed in the code is suitable. Other codes require the approval of the code committees (by means of applying for a "code case"), which, in effect, gives the materials properties committee of the code an opportunity to consider making this new material a part of the code in future editions.

Generally, the codes are concerned with the principal components of a manufactured item. In the case of valves, this normally means the pressure-retaining parts and closure parts, including the body, bonnet, wedge, plug, ball or disc, stem, and bolting. Other minor parts, although perhaps covered by ASTM specifications, are often referred to in parts lists as "commercial," meaning they were not bought or specified to an ASTM standard. In nuclear work, under ASME section III or the Navy nuclear program, where certification of virtually everything is required by law, all these little parts will be identified to an ASME specification or military (MIL) specification, but not in valves intended for ordinary refining, chemical, fossil power, or common industrial use. In ordinary situations where radiation, very high pressures or temperatures, extreme reliability requirements, or other highly unusual circumstances are not involved, this is perfectly acceptable, since failure of any of these small parts is rarely critical or injurious, many of these parts are consumed by the thousands at fairly low unit costs, and many of them are too small to mark adequately anyway.

For generations, the commonly accepted names for materials were generic, such as the firebox quality plate referred to above. Although many ASTM specifications had been in place since the early years of the twentieth century, for a long time many valve manufacturers preferred not to make too much public mention of them, preferring instead to use the trade names they had developed. This was understandable, since sometimes these manufacturers had invented the particular alloy or process in question anyway, and their use of it often predated the establishment of the national specifications (ASTM or otherwise). In addition, the standards organizations themselves had over the years traded out coverage. Welding rods and filler metal compositions were first addressed by ASME, for instance. These standards became ASTM standards, also assigned a number by ANSI, but during the 1960s these standards were transferred to the American Welding Society (AWS). Today ASME retains only the requirements for welding procedures, the use of welding rods and the requirements relating to deposited welds, such as preheating and postweld heat treating, and AWS is in charge of the specifications for the welding rod and wire. AWS is the sponsoring organization for a couple of codes, covering structural steel and sheet metal welding.

To this day, many manufacturers prefer to refer to their alloy materials by their own trade names, although the practice is decreasing. Many manufacturers will, instead or in addition, use the trade names assigned by the steel manufacturer that developed the alloy, or the plastic or rubber compound used in a particular seal. This practice is not so confusing, since at least the general public knows these products by the same name. Many of these products are referred to so often by their trade names that people may not know that they are still trademarks. Some examples would be Stellite, used for hardfacing, which is a trade name of Stoody Deloro Stellite Inc., formerly owned by Haynes Stellite Corporation, formerly owned by Union Carbide Corporation, and Teflon, which is a trademark of E. I. du Pont de Nemours Co. Inc. You can ponder this the next time you use a Xerox machine (trademark of Xerox Corporation) or a PC (trademark of IBM Corporation).

The use of these trademarks made it very difficult to compare products of different companies, and made it very difficult to know if you were actually getting what you specified. Perhaps the move among the general public for more consumer information over the last 20 years had something to do with this. At any rate, most manufacturers are now much more likely to refer to their materials by generic names, such as 316 stainless steel, or by specific grades, such as ASTM-A351, grade CF8M (which is the same as 316 stainless steel).

Interestingly enough, both of the last examples are from other sets of specifications. The designation "316" is actually AISI 316, from one of several sets of steel grade numbering systems developed by the American Iron and Steel Institute and intended to apply to raw steel products, and the "CF8M" is from the Alloy Castings Institute (ACI) numbering system for alloy steel casting compositions. In each case, ASTM eventually used the numbering systems as the basis for their own designations. These same sets of nomenclature appear in several groups of ASTM specifications, including ones for industrial non-pressure-retaining components, and others for the production of raw castings and forgings whose purchasers are manufacturers who will make them into some other product form, such as valve parts. Any of these types of nomenclature is acceptable to some degree, but the best choice is the specification and grade that is directly applicable to the product you wish to use.

How can you tell what to call the specification and grade you have chosen, or what others will call it? For fair trade reasons, the standards organizations will not use trademarks or proprietary designations in their publications. The most common way is to look up each specification for the different product forms you need (forgings for valve bodies and flanges, castings for valve bodies and specialty components, wrought product for fittings, pipe, bar, and plate for large welded components, and so on), determine the actual chemical and physical requirements and other requirements, then cross-check until you find a match. Not all alloys have exact equivalents in each product form, though. Some never will, because some alloys are almost uncastable or not suitable for forging or some other reason. If you have the data from the original material developer (the founder or mill), they should indicate, besides their own reports of chemical and physical properties, what specification and grade applies. For that matter, anyone who rolls or pours the alloy in question should show that information in their catalog or other literature.

Once you have this data, how do you know that you have found the equivalent grades? One fairly good guide, which was developed for just this reason, is the Unified Numbering System (UNS) developed by the American Society for Metals (ASM). This numbering system took the hodgepodge of existing alloy numbering systems, some good and some not so good, and combined them into a six-character designator consisting of a letter that describes a broad material category, followed by five digits that usually are the old designations followed by enough digits to distinguish one from the other. For instance, the wrought 316 stainless steel products and the CF8M cast stainless steel products may now all be referred to as UNS S31600. Other common designation systems exist, although not as widely based as for metals. There are several thermoplastic material designation systems, a classification system for sheet gasket materials, in ASTM-F104, and several others.

How do you find the specification itself? One way is to simply start with the ASTM index (ASTM volume 00.01, formerly part 48), and look for alloys similar to what you need. The index is grouped by subject, although its indexing system is frankly obscure. Or, if you know you are looking for particular end products such as piping-related products, start with the index for volume 01.01 (formerly part 1), which addresses materials commonly used for pressure piping construction in a number of industries. Similarly, for plate materials you would begin with the index for volume 01.04 (formerly part 4), and so on. These indexes are easier to find things in. Once you have found a likely specification, turn to it and examine the chemical and physical properties it describes. You may find that this process takes time, especially if you are looking for products such as nickel alloys that are spread out over a number of different specifications.

Once you have narrowed down a material specification, how do you decide which standards apply? For one thing, you already know what product you want, say a valve for water service at X pressure and X temperature. You know that this valve is in a power plant, so you know that the ASME B31.1 Code for Power Piping applies. This code will reference applicable specifications. You know, from the rating charts that are referenced in the code, what the rating of the valve has to be. You can then look at manufacturers' catalogs and find several valves that meet these criteria. You can then decide which of them (speaking in terms of material and type of construction) you can use. Let's say that you have decided this valve has to be flanged end. You then can choose from bronze, iron, or steel body material. (You know from the conditions that an alloy steel is not required, and you have ruled out a nonmetallic material.) Then you can look at other criteria to arrive at one choice. The code will then tell you, if you have chosen a steel body, that the applicable specifications for the body and bonnet material are ASTM-A216 for cast components and ASTM-A105 for forged components. You may have more than one choice, or in this case you may have several different grades of cast-steel material available within the specification. By this time you may need to seek advice from the manufacturer or an experienced specification engineer, or consult past history that may be available to you. In this case, though, you will find that the only casting grade that is available for normal delivery is grade WCB. This, then, becomes your choice.

Can you get by, though, by saying something as simple as "carbon steel" and letting it go at that? Perhaps—if you are not the specification engineer, and if you are looking for a shorthand way of describing the sort of material you want, and if the person you are communicating with understands what you mean. But, it certainly is not all right to skip lightly over material or performance specification, or anything else that helps you to describe *exactly* what you want, when you are at a critical point where exact meanings are important. One place where this certainly applies is on data sheets, or descriptions, or narrative specifications that are transmitted from you to a person outside of your company for pricing or for construction. Anytime you perform any of these functions you are under the perview of commercial law, and this is certainly true when you have entered an order, which is a contract, and you are a party to the actions described in that contract. There is an old saying, "Be careful what you ask for, because you just might get it."

The rather lengthy procedure described above sounds like a daunting task. It really is not, for a number of reasons. One is that, after you spend a little time doing this work, many of the sources and the research become second nature. You quickly develop a set of data, or a feel for the data available to you in files or libraries, and then you know your way around. Another reason why this process is not daunting to experienced engineers is that they do not spend much time hunting new alloys and determining which specifications might be suitable, unless they are researchers or consultants. Most of the material selection that goes on consists of validating existing choices or extrapolating somewhat from existing practices in your plant or some other prototype.

In the engineering/construction business, also known as the architect/engineer business, much of the material specification work that goes on consists of lifting specifications intact from one format and converting them to some other format with most of the original requirements or engineering content intact. Although this seems pointless at first thought, in fact it is a very real requirement. This is especially true now, with heavy use of computerized engineering and design such as computer-aided design (CAD) systems. These systems do save tremendous amounts of repetitive drafting, aid in reaching design solutions faster, and in some cases allow engineering solutions that were once too labor-intensive to bother performing. These systems require a set of specifications in a form that the computer can understand. In addition, depending on the source of the engineering specifications, they may be geared toward plant maintenance and light construction, and not adequately address the concerns of heavy construction, or vice versa. Another reason for recreating specifications could be that a set of engineering data has accumulated in a company whose personnel are aware of where everything is, even if it may be a little disjointed, but a new organization working from the same set of specifications may not be able to follow or find all the requirements without reworking them and reassembling the data.

When you select a given material specification, you also will get a cross-reference to other specifications that apply, such as general specifications that define how to conduct tests on the material and specifications that contain definitions of terms. What you will not get, in general, is guidance on what to do with the material after it is produced. For example, there are no ASTM specifications that tell the manufacturer how to build a valve, only ones that get up to the point of producing the raw material itself. At that, many times in ASTM specifications there will appear a phrase to the effect that such and such is subject to agreement between the manufacturer and purchaser. Here is where the next level of specifications, or standards, comes in.

### *Ratings*

There are many specifications and standards that describe requirements for physical attributes, such as ratings and dimensions. Sometimes these apply to one narrowly defined class of product, or they can apply to practically anything. Sometimes a large number of standards and specifications apply to a given product, which is true of valves. In any given valve specification, references are made to all manner of other specifications and standards. A valve has so much detail, and there are so many possible variations, that no one standard can possibly address all combinations. If nothing else, at least the details of the materials of construction are specified in detail somewhere else, and the valve specification itself simply refers to the material specification by its designation, such as ASTM-A105.

In the early days, valve specifications were not often published separately, since there was less detail then, and it was considered sufficient to specify general requirements for the materials in the overall piping system. Dimensional standards were one of the first individual details addressed, and after the dimensions of pipe sizes and their conventional designations were agreed on, the thread and flange standards were next. This was even more significant then than it would be today, since the mechanical means of threading and bolting were the only joining techniques available in the late years of the nineteenth century, when the standardization process was beginning to take place in earnest.

During this period, which was one of rapid industrialization all over the world, there were several common types of industrial piping construction. Besides the wooden piping built from bored-out tree trunks or stave-and-strap construction like an elongated barrel, there was seamless extruded copper and brass piping with cast-brass fittings and valves, cast gray iron piping for both nonpressure and pressure applications with cast-iron fittings, wrought-iron pipe, riveted pipe and fittings made from wrought iron or steel plate, and beginning in the late 1880s, the first seamless steel pipe. The choice of end connections varied with material. Copper and brass pipe and fittings could be threaded or soldered. Iron pipe could be threaded, and various types of caulked joints using oakum and molten lead were common (but not very suitable for pressure unless absolutely immobile), and steel pipe could be threaded, lapped, and riveted.

Any of these materials could be flanged also, although for years the primary method of building flanged pipe was to cast the pipe and flanges in one piece. Foundry technology was available to cast such pieces, using pits with vertical molds for many years and later using centrifugal casting. Many miles of pit cast pipe is still present in many cities, in service in municipal water lines and in building drains. It suffered from the disadvantage of being thin on one side sometimes, if the core shifted during pouring, so higher pressures and longer lengths were difficult. The cast pipe with integral flanges tended to be in lengths of 5 or 10 feet, and in industrial installations with closely spaced equipment the lengths of straight pipe were often fairly short. These cast pieces of pipe with a flange at each end resembled wooden thread spools, and so to this day a small fabricated piping subassembly is referred to as a spool or spool piece. Eventually the cutting of large-diameter threads became more standardized and more reliable, so the dominant technology changed to straight pipe with threaded ends, with separate flanges threaded on. The pipe could be seamless steel, and the flanges could be forged steel, or any combination of the other materials then in use. Fittings, and this included valves, were still primarily cast because of their complex shape.

To some extent the valve standards evolved from the flange standards. Once there were four ANSI standards for flanges and flange ratings, one for iron flanges (now known as ANSI B16.1), one for bronze flanges (now ANSI B16.24), one for steel and steel alloy flanges (now ANSI B16.5), and one for high alloy and nonferrous flanges (was ANSI B16.31). The first three also covered flanged components, such as flanged fittings, and the standard for steel flanges also once covered flanged valves. Eventually, and gradually, over a period of years, the coverage of steel valves was dropped from the steel flange standard and incorporated into a separate standard for steel and steel alloy flanged valves (ANSI B16.34), while the MSS standard (MSS-SP-66) that had also covered steel valves was gradually incorporated. All along, there had been a standard for face-to-face or end-to-end dimensions for flanged and butt-weld end valves (now ANSI B16.10). Since then, the standard for high alloy and nonferrous flanges has been discontinued and most, but not all, of its ratings were incorporated into the standard for steel and alloy flanges. The ratings for high alloy flanged valves have, likewise, mostly been incorporated into the standard for flanged valves.

A number of other ANSI standards also evolved from these first standards or were created separately to parallel the coverage of valves for other industries. The utility gas industry developed ANSI B16.33, B16.38, and B16.40 for their valve requirements. Other standards for flanges and fittings also were derived from these over the years. However, the majority of user industries were never or have not remained within the ANSI framework for their valve specifications. In many cases, these users already had their own specifications by the 1920s when much of the first ANSI standard work was begun.

Although valve specifiers in the petroleum and refining industries adopted the ANSI flange dimensions and ratings, the ANSI valve standards were never of much interest for them because there was an API standard that covered steel flanged valves (API 600), and a standard that covered face-to-face and end-to-end dimensions (API 6D). This was true even though there was some cooperation between the two organizations, such that similar valves matched up quite well from one standard to the other. API had several other standards that covered specific types of valves that ANSI never has addressed, or had not until the 1988 revision of B16.34, such as small forged body valves (API 602 and API 606) (Fig. 1-1). The use of ANSI B16.34 by the refining industry was largely confined to types of valves that no API standard addressed, such as pressure-seal bonnet valves for high-pressure service. Control valve dimensioning is to ISA S4.01.1 for globe-type valves (Fig. 1-2) and ISA S4.01.2 for flangeless ball valves, although flange dimensions are ANSI.

Similarly, the municipal water industry prefers to use the standards that they developed, such as the one for iron gate valves (AWWA C500) (Fig. 1-3) and for butterfly valves (AWWA C504). For general govern-

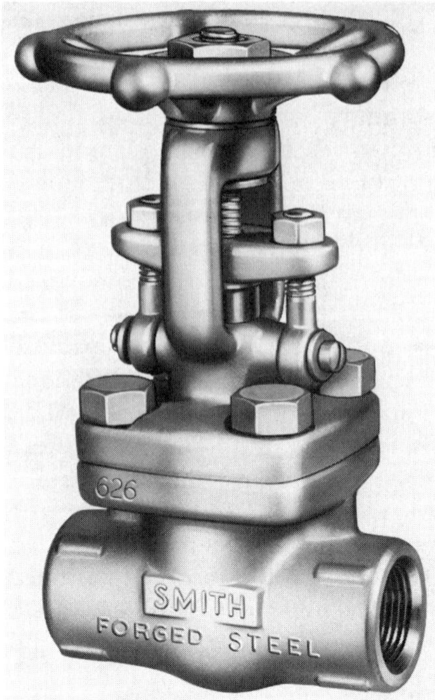


**Fig. 1-1.** Threaded-end, bolted-bonnet, rising stem, bolted gland forged steel API 602 gate valve. (*Courtesy of Smith Valve Corporation.*)

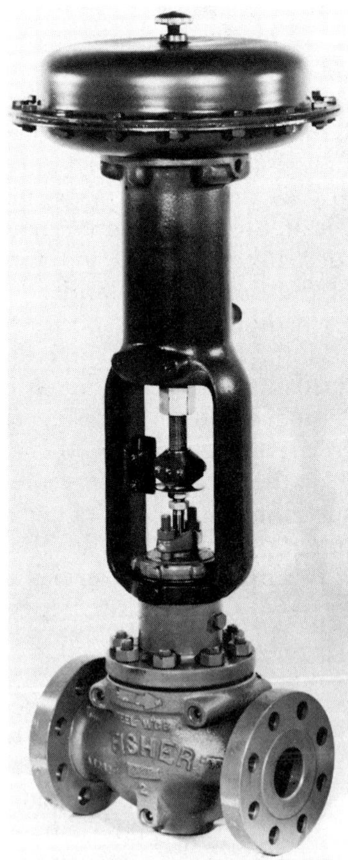

**Fig. 1-2.** Air-actuated globe-style control valve. (*Courtesy of Fisher Controls International Inc.*)

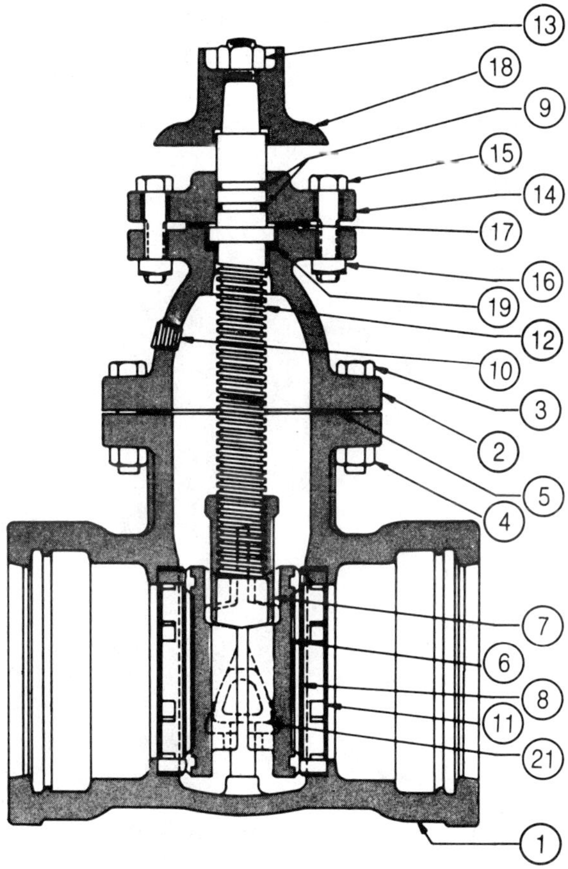


**FIG. 1-3.** AWWA standard nonrising-stem gate valve with parallel seats and operating nut. Ends are designed to mate with bell and spigot pipe. (*Courtesy of Stockham Valves and Fittings.*)

ment contract work, the applicable standard for similar iron gate valves is WW-V-58, and for military installations, the standard for steel gate valves is MIL-V-18434. In the pulp and paper industry there is TAPPI TI 0420-07 for stainless wedge gate valves and TI 0420-04 for knife gates. Yet another series of valves are the ones approved by UL, such as UL 262 for iron gate valves (Fig. 1-4). They may be very similar to nonapproved valves, but since the specific valve model is approved by UL, or in a similar manner by FM, the approved valves will be cataloged and sold separately.

This plethora of standards is a fact of life. It is possible that in future years some attempts to consolidate standards may occur. Even for specification engineers who confine their activities to one industry, not all applicable standards come in one nice neat package. Most engineers still find themselves making reference to standards from a host of different organizations, and probably will for a long time to come even if some rationalization and streamlining occurs in the standardization

**Fig. 1-4.** Gate valve with post indicator designed for buried service on fire water line. (*Photograph by R. C. Merrick.*)

world. For this reason, specification engineers need to be aware of these different organizations and how they and their standards interact. This book provides a list of standards that, while perhaps not covering everything that has a possible relevance to valves or valve components, should give the specifier an idea of where to start in the search for standards.

The standards themselves help in the cross-referencing, as mentioned before. Any standard that invokes another standard as an additional requirement will so state, and may list a number of other standards as references, although they may or may not be a requirement in any particular case. However, many other combinations, especially those that involve material specifications, may not be stated, and it will be up to the engineer to discuss with the manufacturer whether or not a particular combination of requirements is feasible. Again, although various code requirements and various standards that are invoked by the codes are mandatory, it is in the end the specification engineer's responsibility to be sure that all the proper specifications and standards are invoked for the construction of an item.

Are all the possible specifications and standards in agreement with one another? Do similar specifications, covering basically the same product, require the same things? Do two similar specifications conform to a particular code they might be used for? The answer to all these questions, as you may suspect by now, is no, but the situation is better than it used to be. There will always be "parallel" specifications—for example, the ones covering flanged steel gate valves—that will be different in details because of the differing emphasis and needs of different industries. But, first, some of these specifications are in use in various places because of their usefulness or because many manufacturers would rather not make products to so many different end-user specifications and would rather see consolidation. Second, many engineering societies, standards organizations, and owners are thinking more about cooperation and consolidation because of the cost of maintaining and updating so many standards. It seems likely, though, that the number of standards currently in use will shrink by only a small percentage in the foreseeable future, while the number of new standards that are added to cover new products and new ways of doing things will grow by a greater number.

The flange and body rating standards described above are the product of many years of evolution. As a result, they have grown into a pattern, which is not entirely unrecognizable in terms of design basis, but is puzzling to the outsider. The ratings of steel flanges were originally determined by the first services they were used in. In the era that steel flanges were first used, steam was the most severe service, so the maximum conditions of steam were used. Steam had useful pressure only when the temperature was elevated, so the ratings were based on a given working pressure at an elevated temperature. This fact gave rise early on to the concept of the pressure-temperature curve, in which the maximum allowable working pressure varies inversely with temperature. The first issue of American Standards Association (ASA) B16e in 1927 (the predecessor of ANSI B16.5) had ratings for flanges of 150 lb, 300 lb, 600 lb, and 1200 lb working pressure, meaning pressures that were expressed in pounds of force per square inch of surface area. Over the years the improvements in metallurgy allowed the ratings to creep upward, since the dimensional interchangeability principle was established. The original numeric values of the ratings soon lost any relationship with the actual pressures allowed, and in recognition of this fact ANSI redesignated the ratings as dimensionless numbers. The proper form of these ratings now is "class 150," not "150#" or "150 lb" or "150 psi," since any rating expressed in pounds or psi could be confused with other ratings, such as those for rubber-lined butterfly valves and other non-class-rated items that are still rated in terms of their *actual working pressure* expressed in psi. The ratings defined by ISO are similar, being expressed as dimensionless numbers (PN, standing for "nominal pressure" in most European languages) that are based on the allowable pressure in bar. The ISO ratings overlap, since certain of these ratings have been defined to coincide with ANSI ratings while others are in common use only in certain industries or countries and not in others.

The same evolution has occurred in other product forms. Steel pipe dimensions, which are probably the oldest current standard data in use today, are based on standardization work that began in the 1880s. The original concept was based, like the flanges, on a maximum common industrial pressure, which was recognized as requiring a heavier wall thickness in proportion to increased diameters. The inside diameter was at first considered the most important, so inside diameters were established at even inch and fractional inch sizes and wall thicknesses were calculated as a ratio of the diameter. Wall thickness to diameter ratios of 1:10, 1:20, 1:30, and 1:40 were calculated, and the wall thicknesses of each one for a given size were referred to as a schedule. Thus the terminology, still in use today, of "schedule 40" or "schedule 160." Since the outside diameters were at first calculated from the inside diameters and then used as the basis for other pipe wall thicknesses, like the standard and extra strong walls, the outside diameters became "nominal" and finally ANSI redesignated the sizes as dimensionless using the term NPS, for "nominal pipe size," in place of inch designations. Here the correct designation now is "NPS 2," not "2 in." The ISO equivalent is DN, standing for "nominal diameter" in most European languages.

Although these dimensionless rating and size designators are used by ANSI, although most specification-writing organizations use these terms in conformance with ANSI, and although ISO and other organizations outside the United States make use of dimensionless units, common usage has never followed. Try asking a material picker to get you a joint of "NPS 2" pipe sometime.

## *Performance*

All of the material and dimensional specifications in the world, and for that matter all the good intentions of the specification engineer, the designer of the product, and the people who machine and assemble the product, are no guarantee that the item will work as it was intended to work. This is where performance specifications come in.

Simply stated, *performance specifications* are those that can be used to measure the functions or performance of the complete valve or valve package, including not only closure time, leakage rate, and flow capacity, but also the dimensional checks, material analyses, and other examination and inspection functions that can be undertaken. In fact, for manual valves, the dimensional checks, material confirmations, and pressure testing are of primary importance, and most of the time there is an unspoken assumption that the valve will actually operate successfully. You could build a case for the argument that nowhere in most specifications does it say that the valve actually has to work.

Performance specifications are written to cover the expected performance of an item, without much regard for what the item is. That is, a specification that describes the hydrostatic pressure test of a valve body is not especially concerned with rating or material, except insofar as they need to be known in order to determine the test pressure, or the mechanical details of the valve or how it works or how fast it is supposed to close, just that the body is required to hold *X* psi of internal pressure for *Z* seconds without losing pressure or showing signs of leakage.

In some cases, the performance specification is simply part of the design or material specification. This is more true of products like pipe than it is of valves. In recent years, where test methods and test pressures had been described in the ANSI valve standards, they have generally been deleted and replaced with a reference to one of the common test standards. Most ASTM material specifications make reference to a general or "background" ASTM specification covering, for instance, all types and materials of plate products, which covers general test methods, packaging, referee methods in case of disagreement on test results, and other things less related to any specific material. In this specification, and sometimes also in the primary specification, are the references to the test methods that are required and the ones that ASTM considers desirable as additional testing. The circumstances under which this additional testing is desirable are sometimes explained in the specification, but often not. Invariably, though, reference will be made to the testing methodology standards, such as ASTM-E165 for dye penetrant examination and ASTM-E446 for casting reference radiographs.

For other things beyond material testing, such as pressure test (commonly referred to as hydrotest), the higher level specifications covering complete valves also commonly refer to other specifications, rather than attempt to describe testing requirements. Partly, this is for the purpose of paper control, in that if the testing specification is published separately, there are fewer pages to update every time a standard is reissued. Also, and perhaps more importantly from an engineering point of view, this more neatly separates the coverage of requirements since testing specifications can be applicable to many different items; and, conversely, but to a lesser extent, valve products can be tested in many different ways depending on the end use and the industry using them. For instance, API valve standards have always referred to API 598 as the test standard, while very similar valves built to ANSI standards tend to use MSS-SP-61 as the test standard, while seat leakage of actuated valves is defined by ANSI B16.104, which is under the sponsorship of the Fluid Controls Institute (FCI).

In some cases, the performance specifications can be quite elaborate.

This is generally true of individual project specifications written for a particular item when the performance requirements are quite critical. National standards on performance, on the other hand, are written to the same level as the standards and specifications for material and fabrication. That is, they are the minimum required to assure basic functionality, including the safety factors that experience has demonstrated to be adequate. This is the same level as national standards on valve construction, with the minimum requirements to assure design integrity. Material, fabrication, and performance standards all have stood the test of time, that is, where problems or incomplete coverage in standards has surfaced, over time most of these discrepancies have been addressed in the standards maintenance process. Either the standards in question have been updated, or where they have become inadequate to cope with evolving requirements, they have been discontinued and replaced with new ones.

The point is that you can rely on national standards and specifications for all, most, or some of your work, depending on where you are and what you are doing. In common industrial work the national standards were written with you in mind, and you need not worry about how good they are. But you will have to be the ultimate judge of available standards. The following observation may help you in determining how you stand in this respect.

There is a normal distribution of results from any investigative activity: Any testing that is done on any number of items at any time may, when repeated, find some that fail where they have passed before. Given that this is true, the minimum acceptable standard is set so that some time after testing you will still have a group of items that performs on the aggregate in a good enough manner. However, you may not be comfortable with operating that way. You may need to guarantee that at no time and under no circumstance will your performance values ever fall below *X* point. You can achieve this result in one of two ways: One way is to monitor, through constant inspection and review, the performance of the item, which is the way it is normally done for existing systems in existing plants. The other way is to make your minimum acceptable value *higher* than the value you really need, to provide some margin of safety over and above those provided by national standards.

No amount of specification writing can change physical reality or test results. But you do have the ability to do more rigorous examination and more thorough testing, in order to weed out the cases where the results fall below the minima. In effect, you can move the minimum standard higher so that, when the in-service performance deteriorates for reasons such as age, contamination, and wear, the likely minimum performance will still be higher than you would have gotten had you set that minimum value in the first place. The first question to ask, as noted before, is whether you have enough coverage with the standards and specifications already available. Only if the answer is "no" or "maybe" do you need to concern yourself with developing your own new standards.

In order to get a full set of performance specifications, the specification engineer normally has to pull together separate requirements. This extra nonprepackaged work is mostly a matter of customizing. It applies most often to material specifications, since the choice of test methods and individual options are not that well spelled out compared to, say, an API specification that is almost a checklist of available options. Many other valve specifications are pretty much silent on the subject of options, leaving it to the engineer's ingenuity to come up with extras or additions. Valve manufacturers, knowing that this is true, often will catalog an extensive list of options for their product, sometimes tailoring a valve for this or that market. Given that not many engineering organizations work extensively in more than one industry, you often have to page through many brochures or separate catalogs to come up with a specific variation that you might be able to make use of.

Performance examination of materials—which might include radiography or other internal examination, external examination, or special chemical resistance tests—is often considered to be "a matter of agreement between seller and purchaser," to use the phrase that ASTM specifications use so often. This is because the ASTM specifications are so often written in an attempt to make them "nondenominational," or capable of being used by as many different user groups as possible. The minimum requirements are spelled out in the specification, with the requirements sometimes depending on the exact grade of material being specified. In general, however, a number of optional requirements are included in the "supplementary requirements" section at the end of many ASTM specifications. The individual users are then supposed to choose the additional testing and examination required. In some cases, the codes will require certain minimum levels of examination, or give additional credit in the form of higher allowable stresses when additional examination is performed. In addition, of course, the user always has the option of adding any more requirements that might be considered necessary or desirable.

What this means in practice is that those users who operate under the requirements of a given code, such as the Boiler Code, are sometimes required to add at least a minimal examination that no product specification specifically calls for. In most cases, the manufacturer expects that the product you specify will be used in code situations, and the necessary inspection procedures and practices are already built into the manufacturer's standard work plan (Fig. 1-5).

Here it might be useful to conduct a brief review of the different examination methods available to the specification engineer, and look at the circumstances under which each one is most useful. These methods are formally defined in a number of places, sometimes with a wide

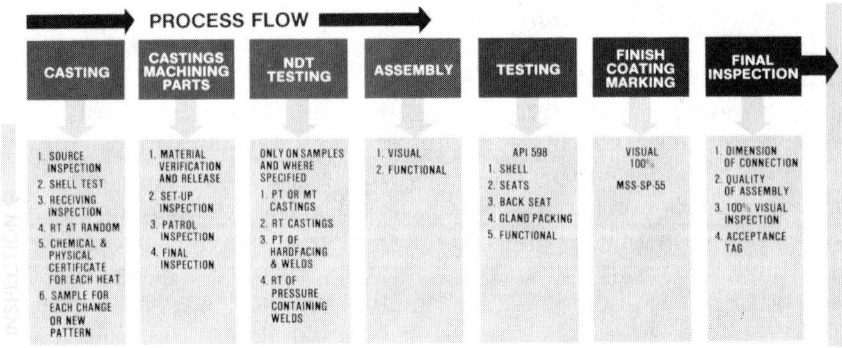


**FIG. 1-5.** Typical valve manufacturer inspection flow. (*Courtesy of Velan Valve Corporation.*)

variety of acceptance criteria depending on end use. The organizations whose sets of examination procedures are the most widely used are ASME and AWS.

A distinction should also be drawn between examination and inspection. Examination is the practice of looking at an object or at a test being conducted on that object and comparing the results with a set of acceptance criteria. Examination can be performed by anyone—the person who makes the part, the manufacturer's quality assurance personnel, an outside person like a contractor's engineer, or any of the owner's personnel. Inspection is the act of performing that same work, witnessing the performance of that work, or reviewing the results of that work, by a person authorized by the owner or the owner's insurance carrier to officially accept or reject the examination results.

The first, and most easily performed, examination technique is *visual,* or VT (the two-letter codes are AWS designations). This can include looking to see that all the parts are there, looking at surface finish, looking at workmanship, or looking at casting quality compared to the standard set of reference photographs of acceptable and unacceptable surface defects. Other examination methods that are suitable for finding surface defects are *magnetic particle examination,* or MT, and *liquid penetrant examination,* or PT. Magnetic particle examination is faster, and can actually detect defects down to about 1/16 in. (1 to 2 mm) into the surface, but can be used only on magnetizable materials such as ferritic steels. Liquid penetrant, or dye penetrant, can be used on almost any surface, but the examination requires several minutes to perform. Magnetic particle techniques involve passing a current through a part of the object to be examined using prods, and sprinkling iron filings or spreading an iron filing paste over the surface. Defects such as cracks and pits will interrupt the smooth flow of magnetic lines of force and cause the iron filings to accumulate around the defect. Dye penetrant examination is done by spraying a deep red or fluorescent dye onto a surface, letting it dry, and then wiping the excess off the surface. The examiner then sprays a developer over the surface. The developer dissolves and absorbs any remaining dye on the surface, leaving a whitish powder, but any defects will have collected a larger amount of the dye, which then bleeds out and becomes visible.

There are several techniques for finding defects below the surface of a part. One is *radiography,* or RT, commonly known as X-ray but usually using gamma radiation, in which the radiation source is placed on one side of the region to be examined and the film is placed on the other side (Fig. 1-6). The source is moved out of its lead housing for a specific exposure time, then retracted. The film is developed, sometimes on site, and any defects in the part can then be seen. The exposure is calibrated by placing a reference object known as a penetrameter, or penny, under the film. The penetrameter has holes of a standard size drilled in it. The appearance of those holes on the X-ray film provides a reference for the actual size and magnitude of the defects compared to the way they show up on the film. Certain defects will not show up on the film, including cracks or laminations that run substantially parallel to the plane of the film.

Another method is *ultrasonic examination,* or UT, which is performed

**Fig. 1-6.** Heavy-wall gate valve body casting being radiographed. Radiographic energy source is a linear accelerator. (*Courtesy of Pacific Valves, a unit of Mark Controls.*)

by sending a high-frequency wavelength into the material and looking at its reflection on an oscilloscope. The reflection of the wave off the near surface of the part shows up on the screen at the bottom, and the reflection off the far or back surface of the part shows up at the top. Any laminations, cracks, or inclusions in between will also show up. Ultrasonic examination has difficulty detecting small rounded indications that radiography can see clearly. The two methods thus complement each other.

These methods are applicable to valves depending on the type of material to be examined. Generally, larger and higher pressure valves warrant more examination. But, although radiography of castings significantly improves the ability to locate defects, there is not much point in radiographing a forged part because the type of defects that radiography sees most clearly do not often show up in forgings. Either cast or forged parts will profit from a magnetic particle or dye penetrant examination of finished machined surfaces. The codes recognize this by assigning higher allowable stresses to components that receive various combinations of the examination techniques. But most users also take advantage of these techniques to increase their confidence in the material beyond the requirements of the codes.

In many cases, the user can pick and choose among performance specifications as among material and manufacturing specifications and standards. Or you can write your own, as long as you meet the code minimum requirements. If you are not working to code, you can make up your own performance requirements, of course. The only caveat here, as with material requirements, is that you have to be practical. Certainly in the case of general industrial work, or whenever you are working within a budget, you do not ask for fantastic requirements for ordinary services. There are applications that have to be done up in a first-class manner, and there are applications where just plain ordinary is the right choice. For military and space work, where the risks are uncertain and the budget does not have any relation to direct payback, there is a little more room to take precautions.

Common sense also should be applied to the selection of requirements, not just to the overall mass of extra requirements imposed. For instance, it helps to know the industry conventions on examination methods and exactly what each examination method will tell about the part being examined. As an example, both dye penetrant examination and magnetic particle examination are reliable methods to detect surface flaws in a metal part. But MT is useless on nonmagnetic materials like stainless steels. Conversely, because MT is faster, it is the preferred method for steel and ferritic steel alloy parts such as low chrome steels. For small areas and quick checking, PT is easier because the cans of dye and developer and the cleaning rags are easier to carry around than the MT apparatus, which requires electricity.

Manufacturers can help out here, since they are willing to tell you what methods they commonly perform and what they are capable of performing, as well as what they can do for extra charges and additional ways that they can recommend to increase the expected reliability of their product. It should be noted, of course, that just as manufacturing competence and quality levels can vary between competing manufacturers, so can examination competence. Your own inspectors or third party inspectors are one solution to this problem, and, again, you should be able to discuss with the inspectors their abilities and their opinions about the most reliable methods. You may choose to disbelieve or have skepticism about the information you receive from manufacturers, since it is true that their primary interest is in selling you an item, but an adversary relationship is never helpful and it is in the best interest of a reliable manufacturer to be honest and straightforward as a means of assuring a long-term relationship and future sales. Your knowledge of what to believe, and who to believe, will come with experience.

Another consideration is that material performance generally cannot be "tweaked" the way that valve operational performance can. Tweaking is the practice of making minor adjustments or playing around with the item being tested until it can pass the tests. Most manufacturers do this before the inspector shows up. There is probably a thin line, at best, between tweaking and rework. The person witnessing the tests should bear in mind, though, that the manufacturer is entitled to do at least as much tweaking as any operator would do in the field to get a valve to shut off.

The real point here is that materials, especially as concerns chemistry requirements, either meet specification or they do not. No amount of rework will change the chemical composition of a casting. Mechanical properties, such as hardness, can sometimes be adjusted by heat treatment, but a product like a valve that is made up of highly machined parts and then assembled is not really amenable to being heat treated if a problem is discovered at the time of final inspection. The entire valve would have to be disassembled, the offending components re-heat-treated, and then the machined surfaces would need at least a clean-up and probably a complete remachining since a fair amount of distortion would have occurred. In other words, when such properties are out of spec it is a pretty significant matter no matter what it is. It then becomes a matter of the engineer's judgment (not the inspector's) to consider how significant such an out-of-spec occurrence really is. Most of these problems constitute cause for rejection, but some are not necessarily so bad that they would not work just as well anyway.

### *Controls*

The standards on valve control and actuation elements are somewhat different from the standards covering any of the valve elements exposed to service conditions. The standards tend to be not as specific or all-encompassing, and the applicable standards tend to be larger ones written for, say, electrical components in general, or other generalized equipment that happens to find its application as part of a valve automation package. There are specific material standards for some few of these items, such as wire and cable, and many of the standards that apply are application standards that discuss interfaces, and standard signals to be sent across these interfaces. The discipline with the most comprehensive set of standards is electrical and electronics, but there are also standards for pneumatic signaling and some of the lesser-used systems.

Not only are there standards, but also codes, primarily the National Electrical Code. Other codes like the Boiler Code discuss controls only in a limited fashion, with certain performance or location requirements for instrumentation and signaling, but they do not neglect the subject, and the controls engineer needs to be cognizant of the fact that there are at least a few requirements.

## Valve Material Selection

### *Chemical Resistance*

Regardless of how valves are designed or what type of valve is selected, the valves must be suitable for the services they will be used in. This section will discuss suitability for the intended service, first from the standpoint of corrosion resistance and then considering the other factors that make one valve more suitable than another. Very definitely, the intended service has a profound influence, first on the types of valve that are suitable, and then on such details as the internal construction and the combination of materials chosen. There are many specialized services that require some, or a lot of, design consideration, and this section will attempt to list some of the ones that are commonly encountered. Very few individual users will ever encounter even half of the services listed here.

There are many individual corrosion resistance or chemical resistance charts available, many of which are published by individual manufacturers based on their tests of their own product. Some, such as the National Association of Corrosion Engineers (NACE) corrosion resistance tables, are compendiums of data from many sources. Most are quite comprehensive, yet none could cover all possible situations. This is especially true for the many services whose corrosivity varies with temperature or with combinations of other chemicals. The charts published by manufacturers are concerned with a select group of materials, either with metals, nonmetallic body materials, elastomers, or some specific material, as long as it is produced by that company. In general, these are based on actual tests, but you may notice that all charts carry a disclaimer to the effect that the data presented therein represents the best efforts of the data gatherer to be representative of the chemicals that you encounter, but that there is so much variation in terms of other compounds in the stream, temperature, or other physical property, that the publisher of the chart cannot guarantee that your chemical will react the same way as the ones that they tested. That is a fact of life, but most of the chart data is pretty realistic. If you are at all concerned that you might not be paralleling the published data, or if you are using something for which no data exist, you would be well advised to do your own testing or provide the manufacturers some samples of your material for testing. Many manufacturers, especially the specialty product ones, will be happy to do this.

Most charts will list anywhere from fifty to several hundred services that the product in question might be used with. Sometimes a fairly specific set of conditions is included, such as temperature or concentration; sometimes a set of data is shown for a spread of conditions; and sometimes no other data are provided. Chemical resistance or service life is often shown in the categories "excellent," "good," "fair," and "poor" or "not recommended." Depending on the choice you have, it is usually a good idea to stay away from anything that is not rated "excellent."

It is wise to remember that optimism may play a role in the services listed for a particular material. Laboratory tests may or may not reflect the performance of a piping material in the field, whether or not the chemicals involved are accurately represented in the testing. There are many variations resulting from production work, ranging from misoperation to inaccurate readings to accidents, that cannot be extrapolated from laboratory data. Time is a factor, also. Although labs can simulate long-term service by changing other factors such as temperature or concentration, not all natural phenomena can really be simulated. What works fine for a month under laboratory conditions may or may not work for 10 years in a plant. It is sometimes possible to get actual service results of a product that has seen years of service in test installations, and those are generally the most reliable indicators. Some licensees of hazardous or critical processes will recommend materials or valve designs for their process only after the item in question has survived 2 or 5 years in a test installation and then been removed and examined for evidence of deterioration.

## *Valve Selections for Common Services*

There are an almost infinite number of fluids that can pass through a valve, as any set of corrosion charts will demonstrate. But simply determining whether a body material is acceptable for a given service is nowhere near all that is required. Not only do all the components that come into contact with the process (known as wetted parts) need to be considered, but the combinations of materials are sometimes important. In many cases the material and design of nonwetted parts are significant. In some situations, such as offshore, underground, or in large chemical plants, the externals of valves may be exposed to as much or more chemical action than the internals.

Many other factors must be considered. Temperature is almost always significant, not only in how the corrosion rates are affected by the fluid temperature, but in how the material itself is affected by temperature, through allowable stress levels or notch toughness or many other mechanical properties. Pressures have an impact on material choices, and on whether a given material is even feasible. Adding an additional dimension to all the other factors is cost. Many materials and designs are possible, but unfeasible or a less desirable choice because of cost.

The services listed below are some that the specification engineer is likely to encounter. Not all industries solve the same problem in the same way, but the listings include the normal materials and design solutions that are applied. These thumbnail sketches will describe briefly the valve types, valve and piping material commonly used, end connections, and other relevant useful information. Remember that in some cases only a few choices are possible or feasible, while in other cases almost anything will work and the constraints are economics or common custom.

If you need more information on these specific services or services not listed here, especially for specialty products, the producer of the product will generally be able to give adequate information about acceptable materials and design restraints. For some hazardous chemicals, the producer or a few users are almost the only source of data. Sometimes a trade or technical association such as the Chlorine Institute or the Compressed Gas Association fulfills the role of information provider. What these organizations provide, though, are general guidelines similar to those shown below. Specific cautions, prohibitions, or recommendations will generally be stated, but most organizations prefer to stay generic, to give the specification engineer latitude to do the detail work on the piping or equipment specifications.

**Acids (general).** Corrosivity varies widely with temperature and with coreactants or buffering agents. In general stainless steels are required, sometimes high alloys, and valves are generally flanged because line replacement is sometimes needed and because threading or socket-welding is prone to crevice corrosion. Block valves are often sleeved plugs because of their large sealing area. In some cases, nonmetallic piping such as PVC (polyvinyl chloride) can be employed, and there the valve of choice is the all-plastic ball valve.

**Acid, acetic.** Depending on concentration, 316 stainless is the usual choice. If 316 is not sufficient, the next choice is Hastelloy C or C276. Valves are usually flanged, the same material as the pipe, either sleeved plugs or a ball valve that does not have a lot of internal parts, such as a top entry type. Sometimes the springs and other small wetted parts in the ball valves are titanium, and in the plug valves a zirconium plug is sometimes used.

**Acid, hydrochloric.** Carbon steel piping is sometimes adequate but lined pipe, using polypropylene, PVDF (polyvinylidene fluoride), or Saran lining, is more common. Valves are generally plastic lined but 316 can also be used.

**Acid, hydrofluoric.** Hydrofluoric (HF) acid as used in alkylation units is extremely corrosive. For predominantly HF acid streams, Monel piping is required, with Monel gate and plug valves. For hydrocarbon streams with small amounts of HF acid, carbon steel is adequate. The valves are carbon steel with Monel trim and extra internal clearance because the hydrofluorine reacts with the steel to form iron fluoride deposits that can bind the moving parts. Valves are gates where fire safety is required and sleeved plugs where tight shutoff is required. It is common to use a gate and a plug in series. HF acid is corrosive to glass, so level "gauges" consist of a standpipe with a valve every 6 in. (150 mm). Level determination involves climbing down the standpipe and opening valves until liquid comes out. These valves are known as trycocks and are special globe valves with a Teflon insert in the disc.

**Acid, sulfuric.** Sulfuric acid at low velocities can be piped in carbon steel, since the acid forms a protective iron sulfide film on the inside surface of the pipe, but above about 7 ft/sec (2m/sec) the turbulence causes the film to erode and corrosion proceeds continuously. Alloy 20 or Hastelloy B is then required. Sometimes Alloy 20 is used immediately downstream of pumps and other turbulence-producing equipment, with the balance of the system in carbon steel. Valves can be carbon steel with Alloy 20 trim, or solid Alloy 20. Solid alloy valves are normally flanged, but small steel valves can be socket-welded. Vent and drain valves can be 316, since they can be cheaper than Alloy 20 and are not subject to erosion.

**Air, utility, or plant.** Threaded steel is perfectly adequate, but is often galvanized to minimize rust carryover into equipment. Valves are usually threaded brass or bronze, either gates or balls although ball valves are better since they operate faster and may be cheaper. Larger valves may be flanged bronze, but flanged cast iron is cheaper.

**Amines.** Carbon steel is adequate, but corrosion rates are sometimes high, especially with rich (sulfur-bearing) amine solutions on the way to the stripper. Valves are normally socket-welded or flanged, usually carbon steel with 316 trim. Welds are stress-relieved. Control valves that will see a flashing condition, and sometimes the piping for a few feet downstream, may all be 316 due to the high corrosivity of vapors with free water and sulfides.

**Ammonia.** Ammonia is very hazardous to health when it leaks, but is not especially corrosive. Old technology used globe valves with lead inserts in the seat for tightness, but Teflon seats are commonly used now. Globe valves are used because of their superior shutoff ability. Piping is carbon steel, but special cleaning is required to eliminate scale and water. Formerly, special square ammonia service flanges were used, sometimes with lead seals, but newer designs use conventional flanges.

**Boiler blowdown.** Blowdowns are high-pressure drop, flashing services. Intermittent blowdown (from the bottom of the boiler) is high in suspended solids.

**Boiler feedwater.** Boiler feedwater (BFW) is a much less severe service than steam service, although a 600 psig (42 bar) feedwater line may still be operating at 400°F (200°C) or more because of preheating. Often the same pipe spec is used for both steam and BFW, normally carbon steel with gate valves, often butt-weld ends with pressure-seal bonnets. Throttling boiler feedwater can present cavitation problems, since the hot water can flash into steam in the throttling region of the valve.

**Brine.** Depending on the concentration and temperature, brine solutions can be very corrosive. Shipboard and offshore platform seawater and brine piping is often a nickel-copper alloy, sometimes bronze. Carbon steel can be used in some applications, but the corrosion rate is high. In shipboard piping, sometimes carbon steel valves with bronze trim are used for a combination of fire safety and mechanical strength with corrosion resistance.

**Butadiene.** Because butadiene tends to polymerize, valves should be of designs that contain as little dead volume as possible, such as plug valves or ball valves with cavity filler seats. Butadiene permeates sintered TFE, so seats and seals should be melt-processed FEP. Carbon steel bodies and piping are acceptable.

**Carbon dioxide.** Processes that produce $CO_2$ may operate in the range of −40°F, where normalized steels are required. Trim should be 316 stainless. Where low temperatures are not a factor, carbon steel is adequate for dry $CO_2$ but wet $CO_2$ can be very corrosive, generally requiring stainless steel. Production services where wet $CO_2$ and wet $H_2S$ are both present are so highly corrosive that duplex steels or Inconels are sometimes required.

**Catalyst, solid.** Catalysts are usually transported either suspended in air or in a liquid slurry. Depending on the hardness of the catalyst and whether there is any process fluid present (catalyst withdrawal lines), the piping can be aluminum, steel, or stainless steel. Valves are often knife gates, sometimes with hardfaced edges and air or steam ports to blow the catalyst away from the seating surfaces as the valve is closed. Another common choice is a full port ball valve, usually a high-performance type with hard coatings such as chromium carbide or tungsten carbide. For hard catalysts, sometimes a rubber lining is used.

**Caustic.** Carbon steel is adequate, but steels are prone to stress corrosion cracking so stress relieving of welds is generally used. Sometimes stress relieving is employed only for services above a certain operating temperature. Valves can be sleeved plugs or gates, either flanged or socket-welded. Spray shields over bonnet and end flanges and threaded connections are normally used for personnel protection. Carbon steel valves can have F6 trim where caustic concentrations are not high, but Monel trim is more suitable for high concentrations or higher temperatures.

**Chlorine.** Wet chlorine is extremely corrosive, and any chlorine leakage is very hazardous. Absolutely dry chlorine is not corrosive, but even the humidity in the air is enough to cause extreme corrosion. Carbon steel bodies are adequate, but trim is Monel where water is never present and Hastelloy C where wetness is possible. In-line valves may have Monel ball or disc with Hastelloy C stem, and drain valves or end-of-line valves are all Hastelloy C trim. Ball valves must have a vent hole in the ball. The Chlorine Institute pamphlet No. 6 provides full details of valve requirements.

**Coal slurry.** Coal slurries are fairly abrasive, but not especially corrosive, and coal falls out of suspension readily so velocities are often fairly high. Full port ball valves are the best choice, usually metal-seated or high-performance types. Flanged valves and flanged piping, for ease of disassembly, are the most common.

**Crude oil, sour.** Depending on the concentration of $H_2S$, and the presence of water, the corrosivity may be somewhat high. The risk of $H_2S$ leakage is more significant, though, so tight shutoff valves are important. In very sour services, corrosivity may be high enough to require duplex stainless steels or Inconel.

**Crude oil, sweet.** Standard carbon steel piping and valves are adequate. In production areas, there may be sand and brine with the crude, which are a problem for corrosion and erosion.

**Cryogenic services.** Stainless steel is almost always required below −50°F (−45°C). Most codes allow stainless steels to be used to −325°F (−198°C) without impact testing, except for welds and components that contain welds. If stresses and conditions allow it, aluminum alloys, nickel steels, bronze, or bronze alloys such as aluminum bronze can be used, but in the United States stainless steel may be cheaper. Liquid services where boil off or vaporization might be a problem (which is usually the case) must have some provision for venting boiled-off gases to prevent overpressuring damage. Vent holes in gate valve discs, or the use of globe valves that are inherently free from trapped volumes, are the common solutions. In most services, corrosion is not a problem, but lubricants or anything that hardens or becomes brittle at the intended service temperature cannot be used. The most common resilient material used is Kel-F. Vents may have double valves, so if the outer valve ices up, the inner valve can still be used to shut off.

**Demineralized water.** Demineralized water is, in effect, very corrosive because it tends to pull ions from anything it comes in contact with. Stainless steel is normally used, but any grade is suitable. 304 piping is the cheapest grade, with 316 valves. End connections can be anything, screwed, socket-weld, or flanged.

**Ethylene.** Although pure ethylene is not corrosive, it is hazardous even without leaking. Welding near ethylene lines can be dangerous since the ethylene can be ignited by a fairly low heat. Normally standard steel valves are adequate, but often tight shutoff valves such as globes or soft seated gates are used. Ball valves may be used, but a grounding spring or other means of electrically grounding the ball is required.

**Fire water.** This is water used for fire fighting. Such water is intended to be pulled from any source, just so it is nearly inexhaustible. River water and seawater are common, and the lines are usually stagnant, with no flow for months at a time. Corrosivity is thus high, and since the lines are buried, it is almost impossible to find leaks. Cement lined piping has been the most common, but in recent years plastic piping, either filament-wound epoxy, PVC, or polyethylene, has been used. Nonmetallic piping may have a long-term relaxation problem, though. Valves are gates or rubber-lined butterflies, normally equipped with indicator posts for direct burial. Some users have had problems with the shutoff ability of rubber-lined butterfly valves when sections of the fire water loops had to be isolated.

**Freon.** Although the Freons are noncorrosive by themselves, a chemically cleaned and dewatered system is required. Piping may be copper, as in home air-conditioning systems, or carbon steel.

**Gas, sour.** Sour gas is even more prone to causing sulfide stress cracking than sour liquids are. The NACE standard MR-01-75 is used as a guideline for material selection and hardness requirements. Depending on the concentration, carbon steel is often adequate, but stress relieving may be required. Valve components may be standard 410 stainless, but Monel or precipitation-hardening steels like 17-4PH may be used. Very sour gas or high-pressure hot services may require duplex steels or Inconels. Since liquid water with sulfides dissolved in it is a prime culprit, sometimes the piping can be heat traced to keep condensation from occurring. Heavier walls are usually ineffective, since stress cracking occurs within the metal rather than on its surface.

**Gas, sweet.** This is an easy service, with carbon steel being perfectly adequate. Gas utilities have their primary corrosion problems from outside the pipe, from corrodents or currents in the soil. Cast-iron pipe was once used for utility gas piping, but many locales allow polyethylene piping for gas. It is usually color coded orange. Valves are a cylindrical plug type, all polyethylene, with heat fusion end connections.

**Glycols.** Glycols are generally not corrosive, but they and their intended uses are sensitive to contamination. For this reason almost all glycol piping is stainless steel, commonly 304 with 316 valves. End connections are flanged, threaded, or socket-weld.

**High purity gases.** Examples of these services would be medical oxygen, reagent gases, reference gases for chromatography, and the like. Here the piping is generally small-diameter tubing, brass, stainless, or high alloy, normally specially made for these services and for refrigeration service, usually purchased precleaned. Valves are often special diaphragm valves for tight shutoff or small ball valves, being in both cases made from small forgings or barstock. End connections are usually special ultra-tight tube connectors, capable of pressures from fairly high down to high vacuum conditions.

**Hydrocarbons, general.** As used in processing plants, carbon steel piping with gate valves is acceptable for the majority of services regardless of molecular weight. However, much processing is done at elevated temperatures, which simultaneously lowers the strength of the pipe material and (often) increases the corrosivity or reactivity or permeability of the fluid. For oils that contain any substantial sulfur, at about 500° F (260°C) and above the preferred material is a chrome steel, either 1¼ chrome, ½ moly or 5 chrome, ½ moly. At temperatures of 800°F (425°C) and over, the preferred material is 9 chrome, 1 moly. For processing at temperatures of 900°F (480°C) and higher, an austenitic stainless is normally required, sometimes 316 stainless but more often a stabilized grade like 347 stainless. For hydrogen services, a very similar progression occurs but it is more formally described in the Nelson Curve (API RP 942). For noncorrosive processes at high temperature, carbon steel is adequate to about 750°F (400°C) long-term and 850°F (450°C) for short-term use. Chrome steels are good to 1000 to 1100°F (535 to 593°C), depending on the grade; and stainless steels, preferably high-carbon grades like 304H, are good to about 1200°F (650°C). Above that temperature the Inconels and Incoloys can be used up to around 1500°F (815°C). Very few processes have ever been intended to operate beyond that point, although some exothermic reactions can reach 1800°F (1000°C) in a runaway case. Where such conditions occur, the normal recourse is the use of refractory linings.

**Hydrogen.** Hydrogen by itself is noncorrosive but very permeable, especially at elevated temperatures. Hydrogen autoignites at about 550°F (285°C) in air, and burns with a flame invisible in daylight. Some techniques used in detecting hydrogen leaks include walking through the plant with a broom or a piece of rolled-up newspaper to hold over any place where a leak is suspected, and if the broom ignites, you have found a leak. The Nelson Curve (API RP 942) relates the partial pressure of hydrogen at a given temperature to the steels required to contain it. But, in addition to permeating through steels, anything such as valve seats and packing intended to retain pressure will be more sensitive to hydrogen leakage. For packing, graphite rings and braided graphite are the primary choice. Often the packing chamber and the stem need to have a smoother than normal finish (down to 16 RMS) (0.6 $\mu$m) to effectively seal. Seats need to be lapped very thoroughly. Double valving is helpful, especially at elevated pressures.

**Nitrogen.** Nitrogen is one of the best possible services, at utility conditions, since it is inert. But if it is purchased in bottles or truckloads, it can be expensive. So, generally, nitrogen lines are all welded with socket-weld or flanged valves to cut down on leakage. For safety reasons, also, it is a good idea to make the nitrogen lines impossible to connect air hoses to (by using different couplings, for instance).

**Oil, coking or heavy.** These services are generally hot, by nature of the processing they undergo and, if for no other reason, because their pour points are so high. Piping is either carbon steel or a chrome steel, depending on temperature. Corrosion allowances range from 1⁄16 to 3⁄16 in. (1.6 to 5 mm), but sometimes it is cheaper to change metallurgy to a more corrosion-resistant chrome steel than to add extra wall thickness required for the corrosion allowance. Because of the viscosity and the necessity for cleanout if the oil sets up, the minimum size for drain valves is generally 1½ in. (DN 40) instead of the normal ¾ in. (DN 18). Other block valves are standard flanged or socket-weld. For coking services, sometimes purge ports are provided on gate valves for injection of flushing oil to help clear deposits prior to operating the valve.

**Oil, lube and seal.** The requirement is for cleanliness or for the elimination of particles that the oil could pick up and carry into machinery bearings. Stainless steel often is used, since after assembly only a flushing of the lines is required. Flanged carbon steel valves are often used with the stainless steel piping since they can be inspected for cleanliness before installation. Carbon steel piping can be used, but it requires a chemical cleaning to remove the scale and rust, which may make it more expensive than stainless steel in terms of installed cost.

**Oxygen.** Oxygen is highly reactive, and vigorously promotes combustion. A loose rust particle inside an oxygen line can start a fire that will burn the pipe at several hundred feet per minute. Most oxygen piping is stainless steel, which also burns but is more resistant to sparking. Carbon steel can be used, but it requires much more extensive cleaning to remove scale and rust. Monel is also used, and is recommended for high velocities and at places where impingement onto the pipe wall could occur, such as flowing into the branch of a tee, since it is more resistant to sparking than steels are. Valves are the same material as the pipe, preferably with Monel trim. Bronze can also be used, and is the easiest material for low-pressure utility oxygen systems. Regardless of material, the interior surface of every component must be cleaned with a degreasing agent such as trichloroethylene and inspected under ultraviolet light to ensure that no hydrocarbons, down to the level of a fingerprint, are present. The Compressed Gas Association standard G4.4 gives guidelines for oxygen services.

**Paint.** Valves used in paint services, like spray booths, can be conventional designs except that they must be assembled dry, without lubricant. Ball valves are the normal choice.

**Plastics, solid or pellet.** Many solid materials can be conveyed in air. Piping designed for conveying normally has large radius bends, often ten diameters or more, and may require a specially roughened interior surface so that the pellets bounce off instead of wiping along the surface and forming streamers. Often the piping is aluminum, usually with rod-type couplings rather than welded construction. Valves are special designs, such as knife gates, pinch valves, and diverter valves, almost always flanged. If small drain or purge valves occur, they are sometimes threaded stainless.

**Polyethylene.** Older polyethylene plants used a high-pressure reaction operating at 30,000 to 50,000 psi (2000 to 3500 bar). Piping is special high-strength alloy tubing, usually unweldable, usually autofrettaged or deliberately overpressured to produce plastic deformation and compression in the inner regions of the pipe wall while still in tension in the outer regions of the pipe wall. Sections of tubing are joined with cone ring or lens ring gaskets that seal by means of plastic deformation, with the seating surfaces being on the ends of the tube. Valves and fittings are block forgings. Surface finish of the tubing is very carefully controlled to eliminate any discontinuities that could lead to cracking. The entire high-pressure portion of the unit is normally surrounded by a concrete blast wall. Newer processes have made it possible to produce polyethylene at much lower pressures, but many of the high-pressure plants still exist.

**Polymers, molten.** Many polymers and heavy compounds are formed by reactions at somewhat elevated temperatures, 200 to 400°F (100 to 200°C), and either after formation or during further processing are often conveyed in a molten state. The viscosity of these polymers is very high and they solidify very easily, requiring constant heat input over virtually every square inch of piping surface. If only moderate heat input is required, a tracing system can be used with either electric tracing, which uses resistance wire wrapped around the pipe, or steam tracing, which uses tubing wrapped around the pipe. The insulation is then applied over the heat sources. If more heat input is needed, the usual recourse is to jacket the lines so that the process piping is completely surrounded by steam. Valves are sometimes wrapped with tracing, or they may have special clamp-on aluminum jackets with steam chambers inside, where the aluminum acts as a heat-transfer medium, or they may be fitted with welded-on jackets, which are the most expensive but the most effective. Sometimes standard flanges are adequate, but if a very thorough heat coverage is needed, the flanges are oversized so that the jacketing can be brought up to the very edges of the flanges.

**Process sewers.** The requirements for process sewers vary with the materials that are expected to go down the sewer. A standard oily water sewer is usually cast iron, either service weight or extra heavy, now assembled with neoprene gaskets at the joints. Acid sewers were formerly built with vitrified clay pipe, as were many oily water sewers, but vitrified clay is extremely fragile which, for environmental reasons, makes it less desirable. More common materials now are high-silicon iron or plastic pipe, often polypropylene or polyethylene. Steel sewers can be used if very hot discharges can be expected, or if the entire system must be pressurized or closed, such as with caustic or amine sewer systems. Occasionally part or all of the sewer is stainless steel, if very acidic or cold discharges are present.

**Sanitary services.** There are two degrees of sanitary service requirements, the first being a service that requires validation by the FDA. These services include all processing of food for human consumption and all chemicals that are part of a product for bodily use, such as mouthwashes and contact lens products. The second set of requirements are for nonfood products that would be affected by the presence of unwanted bacteria, such as chemicals that are made by fermentation. The FDA has more stringent rules about the process employed to ensure that products are uncontaminated, but the process requirements are sometimes more stringent for biotechnology processes. For agricultural food processing, the USDA also has a set of regulations. The dairy industry and some other food processing industries use 3A standards in addition. For all cases, the piping is usually stainless steel tube (slightly different in size from pipe), with a polished interior surface and often a polished exterior surface to aid in washdown. The system must be capable of either complete disassembly for cleaning or cleaning in place (known as CIP, or SIP for steam in place). Mechanical design of components is with a minimum of entrapped volume or crevices. Valves are generally diaphragm type, for ease of cleaning and tight shutoff, or three-piece ball valves with special seat designs that fill the entire body cavity. In true sanitary services, gate and globe valves are virtually never used. There are some special plug designs that were common once, but are now largely obsolete. Actuators are usually an all-stainless diaphragm type capable of being washed down, and air lines are usually transparent plastic tubing. Control valve bodies are often angle type with integral seat and a solid Teflon plug, or else diaphragm valves with a special two-part compressor capable of considerable turndown. Flanges are rarely used. End connections are either a special butt-weld design that can be made by automatic welding machines, for surface uniformity, or any one of a dozen or so proprietary, noninterchangeable, clamped end gasketed joint designs.

**Sewage.** Sewage varies in nature, depending on its source, but municipal sewage normally is heavily laden with solids, anything that people can put into it, plus some infiltration that adds sand, mud, and excess water. Piping can be cast iron, PVC, polyethylene, or concrete. Plastic piping stands up best but is limited in size. Polyethylene liners are often inserted into older cast-iron mains instead of replacing in kind, which is both cheaper and an improvement in flow characteristics. Concrete is best for large mains, but it is expensive because of its weight and the labor-intensive nature of its installation. Valves in sewer systems are best avoided, but if required, they should be knife gates or pinch valves to cut through all the solids present. Cast-iron wedge gates are often used, but they can be provided with small flanged openings to gain access to the area around the bottom of the disc to clean it out if necessary. Where sewer lines also must carry storm runoff in flat terrain with heavy rains, it may be necessary to provide check valves, also known as flap valves, to prevent overflowing streams from backing up into nearby low areas.

**Steam, utility or low pressure.** Depending on the industry, utility steam may range from near-atmospheric to 150 psi (10 bar) or so. Piping is carbon steel, always insulated to keep heat loss and condensation down. In a few applications such as laundries where substantial amounts of copper piping are used, low-pressure saturated steam can be piped in copper also, but joints must be threaded because the temperatures exceed the safe limits of solder joints. Valving for steam was always gates and globe valves, but in the last few years a few resilient seat materials such as bronze-filled Teflon and PEEK have made it possible to use ball valves in small sizes. Larger sizes use flanged valves. Butt-weld end valves are available but are more expensive and not really needed for utility steam services.

**Steam, medium pressure.** Medium pressure steam might range from 200 psi (14 bar) up to about 500 psi (35 bar), sometimes saturated but usually superheated. Carbon steel piping is used, normally ANSI class 300 (PN 25). Larger valving is usually butt-welded, and as few flanged joints are used as possible since steam does leak quite aggressively and steam systems are hard to shut down for repairs. Valves are otherwise conventional bolted bonnet designs, with F6 trim and hardfaced seats. Smaller valves are socket-welded, often with welded bonnet.

**Steam, high-pressure.** Steam systems of 600 psi (40 bar) and above are usually considered high pressure. Carbon steel is acceptable for class 600 (PN 40) systems, depending on the degree of superheat present. The temperature limit for carbon steel is usually considered to be 750°F (400°C) for steam. Conditions slightly greater than that call for 1¼ chrome, ½ moly, and conditions in excess of 900°F (480°C) usually use 2¼ chrome, 1 moly because of its greater strength at temperature. Very hot high-pressure steam services in central stations may use an austenitic stainless, usually 316 but sometimes 347 or other, for temperatures up to 1200°F (650°C) and pressures up to and beyond the critical point (3206 psia) (225.7 bara). Valving for 600 psi steam systems is normally butt-weld end. In plant steam systems bolted bonnet valves are usually used. Pressure-seal bonnet designs can be used, but their real advantages over bolted bonnet designs are in class 900 (PN 60) and higher. Small valves are almost always socket-weld, with welded bonnet, and sometimes with a bellows seal bonnet design. Large block valves can be wedge gates or parallel seat gates, with a bypass either integral to the valve or in the piping around the valve. Smaller valves, letdown valves, vent and drain valves, and any atmospheric connections such as instruments are normally globes because of their tighter shutoff ability, but it is common practice to provide two valves in series anyway. Valves directly connected to boilers are special, since ASME Boiler Code rules apply.

**Steam condensate.** Steam condensate would seem to be a fairly easy service, but it has flashing and two-phase flow to make it difficult. The corrosion allowance may have to be greater in a condensate line than in a steam line. Piping is carbon steel, usually insulated for heat retention. Where erosion is a significant problem, stainless steel is used. Valves are conventional socket-weld or flanged.

**Sulfur.** Sulfur is usually piped in the molten state, at a little over 400°F (200°C). Piping is carbon steel. The difficulty with sulfur is that when it freezes, it becomes like yellow concrete. It has poor heat-transfer qualities, so it is very difficult to remelt sulfur that has solidified in a line. Virtually all molten sulfur piping is fully steam jacketed, so that continuous heat input is available. Valves in a sulfur line are normally also fully jacketed, because merely wrapping the valve with steam tracing tubing cannot provide enough heat input. Jacketing can take the form of a steel jacket directly welded to the valve body or a special aluminum casting that encapsulates a steam chamber and is then bolted around the valve. Some special valves, such as butterfly valves in sulfur vapor service, also have the shaft drilled for a steam flow through the inside of the shaft and disc. Piping normally has flanges every 20 feet or so, because of the jacketing and to permit disassembly and rodout to remove sulfur buildup on the pipe walls. Changes in direction are usually made with flanged crosses, to permit easy access for rodout.

**Vacuum, hazardous.** Hazardous vacuum services are those where a reaction or combustion could occur inside a line if atmospheric air were pulled inside it. For these services, valve packing must be designed to seal against pressure differentials in both directions. The most common practice is to use an extradeep stuffing box with two sets of Teflon chevron rings, with the one on the bottom conventionally oriented and the one on the top, or toward the atmosphere, oriented upside down. Bellows seal valve bonnets could also be used. Gasketing is subject to the same problem, although perhaps not as severe as valve packing. Flat gaskets are bidirectional, but if the superior sealing ability of spiral-wound gaskets is desired, two sets of windings must be used with the outer one being backward or with the V-shaped windings facing outward.

**Vacuum, nonhazardous.** If small amounts of air pulled into a line are tolerable, such as in a below-atmospheric steam condensing line, standard braided or continuous ring packing is acceptable. Chevron ring packing is still not acceptable, since it offers very little resistance to differential pressures in the opposite direction to its orientation.

**Water, municipal.** Historically, large cast-iron gate valves have been used for pumping stations and distribution. Valves are available with ANSI flat-face flanges, with special ends for iron pipe bell and spigot joints or mechanical joints with bolted gland followers, with ends for asbestos cement or for PVC bell and spigot joints. The valves themselves are either wedge gates or parallel seat types, sometimes with integrally cast flanges on the sides for bypasses. Butterfly valves are also used, normally a standard design with ANSI flat-face flanged ends that is longer face to face than a wafer butterfly but not as long as a gate valve. All of these valves are normally built to AWWA standards for utility work. Occasionally large plug valves have been used in pumping stations and elsewhere. Where tapping into an existing main is required, a special oversize valve is used. Topworks are either conventional yokes with handwheels for valves in pumping stations or elsewhere aboveground, or with a standard 2 in. (50 mm) square-tapered operating nut for direct burial where the top of the valve is located in a road box, or with a post indicator for direct burial or installation in a wall.

**Water, potable inplant.** Depending on local codes, copper, galvanized steel, PVC, polyethylene, or stainless steel is used. Any potable water line in a plant must be designed so that process and contaminated water systems cannot be connected to it. When connections are required for makeup or fire fighting, a backflow preventer should be installed and is required by code in many jurisdictions. Sometimes two degrees of hazard are recognized, which influences the separation ability of the backflow preventer.

**Water, utility or plant.** Piping material is usually whatever the lowest piping specification permits in a given plant, either copper, galvanized steel, or black carbon steel. For consolidation of warehouse stock and avoidance of substituting lower grade material in a more severe service, plants may pipe water in exactly the same material as used for hydrocarbon services. Even if pipe and fittings are the same, though, it is still common to use brass ball valves or gate valves for water lines.

## Valve Specifications and Requirements

### *Valve Specifications*

Much of the material discussed so far in this chapter is a description of what specifications mean to the engineer and user, and a description of the different documents and organizations involved in this work. Now the discussion will turn to how to produce a valve specification and what it should look like. Then the following two chapters will describe the nuts-and-bolts details of the different types of valves and actuators available and commonly used today. Chapter 4 will discuss the next step, putting the specifications to work in buying the material that has been specified.

Valve specifications can take many forms, depending on the complexity required and the number of implicit assumptions. Often the complete valve specification will be in two sections. One will be a general specification describing basic requirements for all valves, and the other will be an individual specification that describes the specific requirements for one valve or one group of valves. For very basic work, the general specification is often very short or sometimes absent altogether. Caution is warranted here, though, because sloppy specification procedures can make it easy to lose track of the significance of this general specification, or they can make it seem as though "everybody ought to know that." When it turns out that everybody did *not* know whatever it was that you implied, you as the valve specifier are left with a problem or a useless piece of expensive equipment.

The underlying requirement is that the valve specifier must supply every piece of information necessary for all of the other parties involved with the valves to do the job that they are responsible for. This includes both other people in your own organization and the people involved in building and selling the valves you need.

How much is enough, and what do you put in? Any given industry has common specifications that are important in regard to the valves to be used. Make sure all of those have been covered, and you probably have done well enough. Most of the time, you will not have to make up data literally from scratch, and there will be examples from other work you have seen. They might be good examples or they might not be—but think of them in terms of conveying all the information needed for others to do their part. In a large company there are probably standard specifications, and for that matter, standard methods of developing specifications.

Over the years, there have been about as many ways of writing specifications as there have been people to write them. In most cases of corporate, military, and government specifications, the valve specifications tend to look like the specifications for any other piece of mechanical equipment, because of the standards for specification writing that these organizations have. Smaller organizations are often not as organized, so there the specifications done by one particular engineer tend to look the same, regardless of what is being specified.

As described earlier in this chapter, it is possible to make use of any one of a number of valve standards right off the shelf, if you are in an industry that has developed any, and use that as the basis for your specification. For instance, the API standards for everything from valves to compressors are designed such that the user can specify an API valve and add only a few choices, like body and trim material and end connections, plus, of course, the size and quantity. Some industry and government standards are even more all-encompassing, or specific, such that the user needs to add only the size and quantity. Other standards are more in the nature of general guidelines, and it is up to the user to invoke all the related requirements.

In some cases there really are not any standards that directly apply. Steel globe and check valves for the refining industry were not covered by the API standards, for instance, which specifically addressed *gate* valves. For years, engineers have been lumping all these valves together, though, and sometimes had invoked, say, "API 602 insofar as applicable," as a standard reference. This is a valid approach for anything that is not directly covered by a standard. For several years in the mid-1980s the scope of the cumulative body of standards did not change very much, but then in the last years of the decade standards for additional categories of valves appeared for the first time, primarily by MSS, but by other standards organizations also. Perhaps in the future everything imaginable, in every industry, will be covered by some organization's standard. At present, though, there is an uneven depth of coverage, with several similar and, in a sense, "competing" standards for some types of valves and very sparse coverage of others.

Many valve companies have been willing to write sample specifications for their products. In general, those specifications have been aimed at the general industrial market, where specification efforts have been more limited than in the refining and power industries, and the specifications have been written in the narrative style used for equipment specifications. Most of these specifications, which you can get free from some valve catalogs, are quite self-serving and are designed to give the impression that nothing else is an equal or equivalent to their product. Some are written fairly, though, and give you an idea of what items the manufacturer considers important.

Most standards organizations have not seriously addressed the mechanics of writing specifications. In general, the engineering organizations have concerned themselves with establishing minimum *requirements* for material, rather than with the ways of specifying, on the theory that the writing of specifications for purchase is such an individualistic process and so dependent on individual tastes that no such work would be feasible. Very few organizations beyond the corporate level have ever tried to propose sample specifications for all types of material, or until recently, ever tried to make specifications look more alike regardless of origin.

The one area where standardization has been in evidence for many years has been in instrumentation. There are standard ISA data sheets for almost every major type of instrument and control item, including a number of data sheets for control valves (Fig. 1-7a–g). The ISA data sheets are suggested formats that contain spaces for all the appropriate data, very similar to the ASME welding procedure forms, and the user can adopt the data sheet and modify it to suit specific needs. Since the data sheet is suitable for describing only the requirements for the individual item, the general requirements for all valves still need to be stated somewhere.

The organization that has begun, over the last few years, to promote standard specifications for various manufactured products and various craft activities such as welding is the Construction Standards Institute (CSI) and Construction Standards Canada (CSC). This group is oriented primarily toward general construction, meaning buildings and light industrial construction, and their specification breakdowns do not in-

| **FLUOR DANIEL** INSTRUMENT SPECIFICATIONS CONTROL VALVES | NO. | BY DATE | REVISION | SHEET NO. | REV. |
|---|---|---|---|---|---|
| | | | | DATE | CONTRACT |
| | | | | MANUFACTURER | |
| | | | | FLUOR P.O. | |
| | | | | BY | CHK'D / APPR'D |

UNLESS OTHERWISE NOTED BELOW ALL VALVES SHALL COMPLY WITH GENERAL SPECIFICATION SHEETS:

| | | | | | | REV |
|---|---|---|---|---|---|---|
| | 1 | | | | | REV |
| GENERAL | 2 | TAG NUMBER | | | | |
| | 3 | FLOW SHEET & FRAME NO. | | | | |
| | 4 | SERVICE | | | | |
| | 5 | | | | | |
| | 6 | LINE SIZE | | | | |
| | 7 | LINE SPEC | | | | |
| | 8 | PIPE SCHEDULE | | | | |
| BODY | 9 | BODY SIZE | | | | |
| | 10 | PORT SIZE | | | | |
| | 11 | CALCULATED C | | | | |
| | 12 | SELECTED C | | | | |
| | 13 | VALVE TYPE | | | | |
| | 14 | MATERIAL | | | | |
| | 15 | END CONNECTIONS | | | | |
| | 16 | BONNET | | | | |
| | 17 | PACKING OR SEAL | | | | |
| | 18 | GUIDE MATERIAL | | | | |
| | 19 | | | | | |
| | 20 | | | | | |
| TRIM | 21 | NO. OF PORTS | | | | |
| | 22 | PLUG FORM | | | | |
| | 23 | PLUG MATERIAL | | | | |
| | 24 | SEAT MATERIAL | | | | |
| | 25 | | | | | |
| | 26 | | | | | |
| ACTUATOR | 27 | TYPE | | | | |
| | 28 | SIZE SELECTED | | | | |
| | 29 | OPEN AT | | | | |
| | 30 | CLOSE AT | | | | |
| | 31 | FAILURE POSITION | | | | |
| | 32 | HANDWHEEL | | | | |
| | 33 | | | | | |
| | 34 | | | | | |
| | 35 | | | | | |
| POSITIONER | 36 | POSITIONER REQ'D. | | | | |
| | 37 | TYPE | | | | |
| | 38 | OPEN AT | | | | |
| | 39 | CLOSE AT | | | | |
| | 40 | BYPASS | | | | |
| | 41 | | | | | |
| ACCESSORIES | 42 | I/P CONVERTER | | | | |
| | 43 | MA RANGE | | | | |
| | 44 | CONVERTER OUTPUT | | | | |
| | 45 | | | | | |
| | 46 | | | | | |
| MFG. | 47 | OUTLINE DIMENSIONS | | | | |
| | 48 | MFG. MODEL NUMBER | | | | |
| | 49 | | | | | |
| | 50 | | | | | |
| | 51 | | | | | |

| | | | | | | REV |
|---|---|---|---|---|---|---|
| | 52 | | | | | REV |
| SERVICE | 53 | FLUID TYPE | | | | |
| | 54 | | | | | |
| | 55 | FLUID PHASE | | | | |
| | 56 | FLOW UNITS | | | | |
| | 57 | Quan. Sizing | | | | |
| | 58 | QUAN. MAX. ANTICIPATED | | | | |
| | 59 | QUAN. NORMAL @ F.T. | | | | |
| | 60 | QUAN. MINIMUM | | | | |
| | 61 | PRESS. MAX. IN. | | | | |
| | 62 | PRESS. NORMAL IN. | | | | |
| | 63 | $\Delta$P MAX (SHUTOFF) | | | | |
| | 64 | $\Delta$P @ MAX ANTICIPATED FLOW | | | | |
| | 65 | $\Delta$P Normal (Sizing) | | | | |
| | 66 | $\Delta$P @ QUAN. MIN. | | | | |
| | 67 | TEMP. MAX. | | | | |
| | 68 | TEMP. NORMAL | | | | |
| | 69 | TEMP. MIN. | | | | |
| | 70L | SP. GR. Liq @ F.T. | | | | |
| | 70G | Sp.Gr. Gas $= \frac{\text{mol.wt.gas}}{\text{mol.wt. air}}$ | | | | |
| | 71 | VISCOSITY @ F.T. CP | | | | |
| | 72 | VAPOR PRESS. @ F.T. | | | | |
| | 73 | CRITICAL PRESS. PSIA | | | | |
| | 74 | WEIGHT % FLASH | | | | |
| | 75 | $\Delta$P SYSTEM (W/O VALVE) | | | | |
| | 76 | TIGHT SHUTOFF REQ'D. | | | | |
| | 77 | COMPRESSIBILITY @ F.T. | | | | |
| | 78 | SPECIFIC HEAT RATIO $C_P/C_V$ | | | | |
| | 79 | NOISE @ MIN. FLOW | | | | |
| | 80 | NOISE @ NORMAL FLOW | | | | |
| | 81 | NOISE @ MAX ANTICIPATED FLOW | | | | |

NOTES:

(a)

**Fig. 1-7.** Data sheets for actuated and control valves, based on ISA standard data sheets, and dimensional data sheets. (*Courtesy of Fluor Daniel, Inc.*)

ELECTRIC MOTOR OPERATED VALVES
GROUP 9

| **FLUOR DANIEL** INSTRUMENT SPECIFICATIONS | NO. | BY / DATE | REVISION | SHEET NO. | REV. |
|---|---|---|---|---|---|
| | | | | DATE | CONTRACT |
| | | | | MANUFACTURER | |
| | | | | FLUOR P.O. | |
| ELECTRIC MOTOR OPERATED VALVES | | | | BY / CHK'D / APPR'D | |

ALL UNITS SHALL COMPLY WITH GENERAL SPECIFICATION SHEETS

| | | | | | | | | |
|---|---|---|---|---|---|---|---|---|
| GENERAL | 1 | TAG NUMBER | | | | | | |
| | 2 | FLOWSHEET & FRAME NO. | | | | | | |
| | 3 | LINE NUMBER | PIPE SCHEDULE | | | | | |
| | 4 | SERVICE | | | | | | |
| PROCESS | 5 | LINE PRESS | MIN. | MAX. | | | | |
| | 6 | LINE PRESS NORMAL | | | | | | |
| | 7 | DIFF. PRESS | MIN. | MAX. | | | | |
| | 8 | DIFF. PRESS NORMAL | | | | | | |
| | 9 | TEMP. | MIN. | MAX. | | | | |
| | 10 | TEMP. NORMAL | | | | | | |
| | 11 | FLOW QTY. | MIN. | MAX. | | | | |
| | 12 | FLOW QTY. NORMAL @ F.T. | | | | | | |
| | 13 | SP. GR. @ | 60° | F.T. | | | | |
| | 14 | VISCOSITY @ F.T. | | | | | | |
| | 15 | FLUID | | | | | | |
| VALVE | 16 | SIZE | | | | | | |
| | 17 | TYPE | | | | | | |
| | 18 | PACKING | | | | | | |
| | 19 | TESTING | | | | | | |
| | 20 | FIG. (A OR 3) SHEET 8 | | | | | | |
| | 21 | BODY MATERIAL | | | | | | |
| | 22 | PRESSURE RATING | | | | | | |
| | 23 | CONNECTIONS | | | | | | |
| | 24 | VALVE MFG. TYPE NO. | | | | | | |
| | 25 | | | | | | | |
| | 26 | | | | | | | |
| | 27 | | | | | | | |
| | 28 | | | | | | | |
| TRIM MATERIAL | 29 | TAG NUMBER | | | | | | |
| | 30 | | | | | | | |
| | 31 | STEM | | | | | | |
| | 32 | STEM PACKING | | | | | | |
| | 33 | PLUG | STELLITED | | | | | |
| | 34 | SEAT | STELLITED | | | | | |
| | 35 | SEAT SEAL | | | | | | |
| | 36 | GUIDES OR BEARINGS | | | | | | |
| | 37 | CALC. CV | VALVE CV | | | | | |
| | 38 | | | | | | | |
| | 39 | | | | | | | |
| | 40 | | | | | | | |
| ACCESSORIES | 41 | POSITION INDICATOR | | | | | | |
| | 42 | POSITION TRANSMITTER | | | | | | |
| | 43 | HANDWHEEL | | | | | | |
| | 44 | | | | | | | |
| | 45 | VALVE STEM PROTECTOR | | | | | | |
| | 46 | | | | | | | |
| | 47 | | | | | | | |
| | 48 | | | | | | | |
| | 49 | | | | | | | |
| | 50 | | | | | | | |
| | 51 | | | | | | | |

(b)

Fig. 1-7. *(continued)*

ELECTRIC MOTOR OPERATED VALVES
GROUP 9

**FLUOR DANIEL**
INSTRUMENT SPECIFICATIONS

| NO. | BY / DATE | REVISION | SHEET NO. | REV. |
|---|---|---|---|---|
| △ | | | DATE | CONTRACT |
| △ | | | MANUFACTURER | |
| △ | | | FLUOR P.O. | |
| △ | | | BY / CHK'D | APPR'D |
| △ | | | | |
| △ | | | | |

## ELECTRIC MOTOR OPERATED VALVES

ALL UNITS SHALL COMPLY WITH GENERAL SPECIFICATION SHEETS

| | | | | | | | | |
|---|---|---|---|---|---|---|---|---|
| ACTUATOR | 1 | MECH. DIAL POS. IND. (YES OR NO) | | | | | | |
| | 2 | AMBIENT TEMP. RANGE | | | | | | |
| | 3 | HAZARDOUS LOC. | | | | | | |
| | 4 | HANDWHEEL W/AUTO RESET TO PWR. | | | | | | |
| | 5 | ELECTRICAL CLASS | | | | | | |
| | 66 | POWER SUPPLY VOLT | PHASE | FREQ. | | | | |
| | 7 | CONTROL VOLTAGE | | | | | | |
| | 8 | STROKE TIME | MIN. | MAX. | | | | |
| | 9 | STARTER LOCATION | | | | | | |
| | 10 | CONTROL, ON-OFF | THROT. | | | | | |
| | 11 | INTEGRAL CONTROL | LIGHTS | | | | | |
| | 12 | REMOTE CONTROL | | | | | | |
| | 13 | TORQUE LIMIT | CLOSE | OPEN | | | | |
| | 14 | LOCAL-REMOTE INTERLOCK | | | | | | |
| | 15 | OPERATOR MANUFACTURER | | | | | | |
| | 16 | OPERATOR TYPE NO. | | | | | | |
| | 17 | OUTLINE DIMENSIONS | | | | | | |
| | 18 | | | | | | | |
| | 19 | | | | | | | |
| | 20 | | | | | | | |
| | 21 | | | | | | | |
| ACCESSORIES | 22 | HEATER | | | | | | |
| | 23 | THERMAL OVERLOAD SW. | | | | | | |
| | 24 | TORQUE SWITCHES | | | | | | |
| | 25 | REVERSING STARTER | | | | | | |
| | 26 | AUX. LIMIT SWITCHES | | | | | | |
| | 27 | | | | | | | |
| | 28 | | | | | | | |
| | 29 | | | | | | | |

***NOTES:***

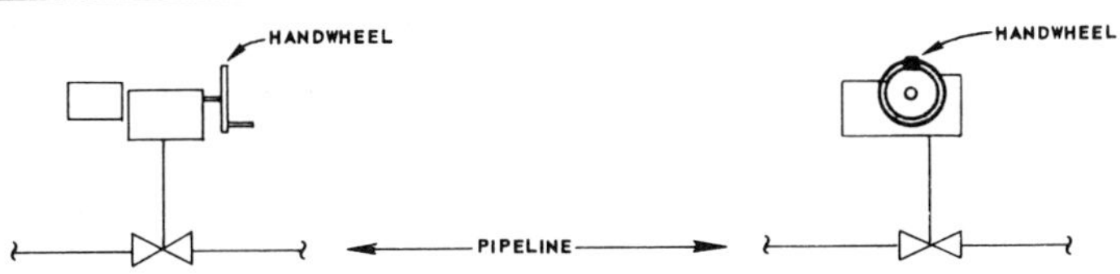


***FIG. A*** ***FIG. B***

(c)

CONTROL VALVES – GROUP 9
GLOBE TYPE
DIMENSIONAL DATA SHEET

| **FLUOR DANIEL**<br>DIMENSIONAL DATA SHEET<br><br>CONTROL VALVES – GLOBE TYPE | NO. | BY / DATE | REVISION | SHEET NO. / REV. / DATE / CONTRACT / MANUFACTURER / FLUOR P.O. / BY / CHK'D / APPR'D |
|---|---|---|---|---|

DIMENSIONS CERTIFIED CORRECT BY

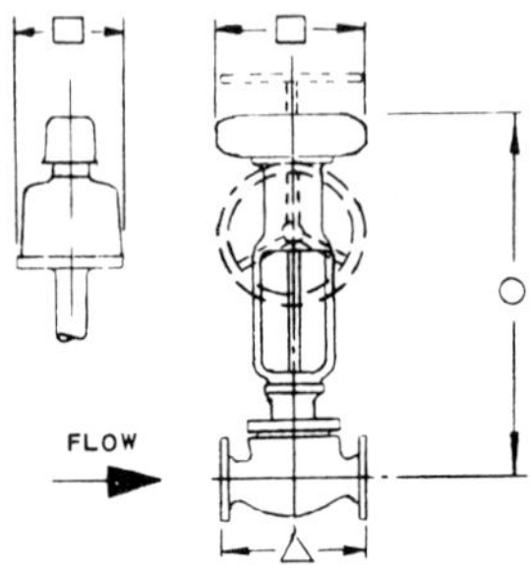


| REV. | TAG NO. | VALVE SIZE | FLANGE RATING | DIMENSIONS IN INCHES | | | HANDWHEEL | | |
|---|---|---|---|---|---|---|---|---|---|
| | | | | △ | ○ | □ | SIDE | TOP | |

(d)

**Fig. 1-7.** *(continued)*

CONTROL VALVES – GROUP 9
BALL TYPE
DIMENSIONAL DATA SHEET

| **FLUOR DANIEL**<br>DIMENSIONAL DATA SHEET<br><br>CONTROL VALVES – BALL TYPE | NO. | BY / DATE | REVISION | SHEET NO. | REV. |
|---|---|---|---|---|---|
| | △ | | | DATE | CONTRACT |
| | △ | | | MANUFACTURER | |
| | △ | | | FLUOR P.O. | |
| | △ | | | BY | CHK'D | APPR'D |
| | △ | | | | |
| | △ | | | | |

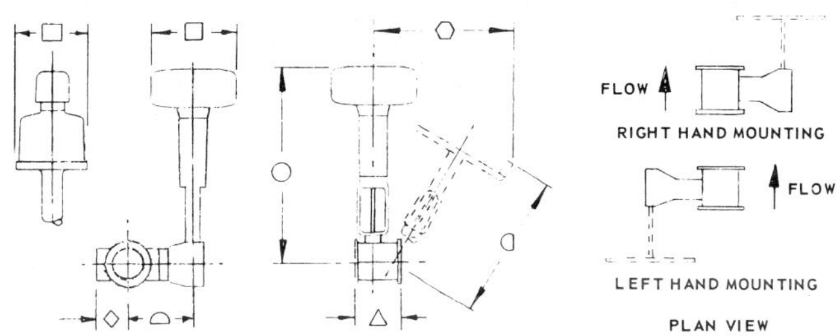


DIMENSIONS CERTIFIED CORRECT BY

| REV. | TAG NO. | VALVE SIZE | FLANGE RATING | DIMENSIONS IN INCHES △ | ○ | □ | ◇ | ⌓ | ⬡ | ◖ | MOUNTING ORIENTATION | HAND-WHEEL |
|---|---|---|---|---|---|---|---|---|---|---|---|---|
| | | | | | | | | | | | | |
| | | | | | | | | | | | | |
| | | | | | | | | | | | | |
| | | | | | | | | | | | | |
| | | | | | | | | | | | | |
| | | | | | | | | | | | | |
| | | | | | | | | | | | | |
| | | | | | | | | | | | | |
| | | | | | | | | | | | | |
| | | | | | | | | | | | | |
| | | | | | | | | | | | | |
| | | | | | | | | | | | | |
| | | | | | | | | | | | | |
| | | | | | | | | | | | | |
| | | | | | | | | | | | | |
| | | | | | | | | | | | | |
| | | | | | | | | | | | | |
| | | | | | | | | | | | | |
| | | | | | | | | | | | | |
| | | | | | | | | | | | | |
| | | | | | | | | | | | | |
| | | | | | | | | | | | | |
| | | | | | | | | | | | | |
| | | | | | | | | | | | | |
| | | | | | | | | | | | | |
| | | | | | | | | | | | | |
| | | | | | | | | | | | | |
| | | | | | | | | | | | | |
| | | | | | | | | | | | | |
| | | | | | | | | | | | | |
| | | | | | | | | | | | | |
| | | | | | | | | | | | | |
| | | | | | | | | | | | | |
| | | | | | | | | | | | | |

(e)

CONTROL VALVES – GROUP 9
BUTTERFLY TYPE
DIMENSIONAL DATA SHEET

| **FLUOR DANIEL** DIMENSIONAL DATA SHEET | NO. | BY / DATE | REVISION | SHEET NO. | REV. |
|---|---|---|---|---|---|
| | △ | | | DATE | CONTRACT |
| | △ | | | MANUFACTURER | |
| | △ | | | FLUOR P.O. | |
| | △ | | | BY | CHK'D / APPR'D |
| CONTROL VALVES–BUTTERFLY TYPE | △ | | | | |

DIMENSIONS CERTIFIED CORRECT BY

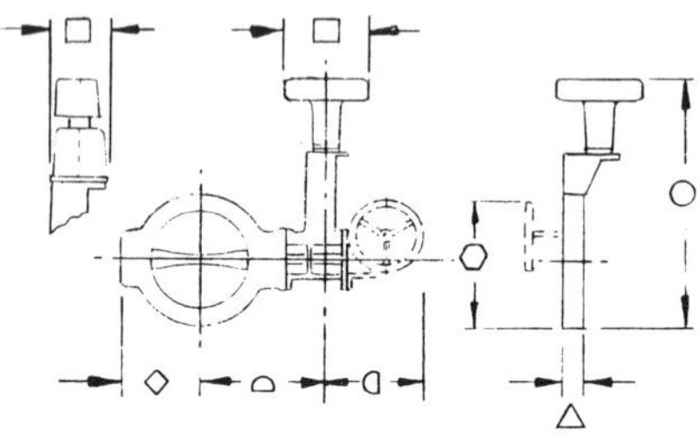

| REV. | TAG NO. | VALVE SIZE | FLANGE RATING | DIMENSIONS IN INCHES | | | | | | | MOUNTING ORIENTATION | HAND-WHEEL |
|---|---|---|---|---|---|---|---|---|---|---|---|---|
| | | | | △ | ○ | □ | ◇ | ⌓ | ⬡ | ◖ | | |
| | | | | | | | | | | | | |
| | | | | | | | | | | | | |
| | | | | | | | | | | | | |
| | | | | | | | | | | | | |
| | | | | | | | | | | | | |
| | | | | | | | | | | | | |
| | | | | | | | | | | | | |
| | | | | | | | | | | | | |
| | | | | | | | | | | | | |
| | | | | | | | | | | | | |
| | | | | | | | | | | | | |
| | | | | | | | | | | | | |
| | | | | | | | | | | | | |
| | | | | | | | | | | | | |
| | | | | | | | | | | | | |
| | | | | | | | | | | | | |
| | | | | | | | | | | | | |
| | | | | | | | | | | | | |
| | | | | | | | | | | | | |
| | | | | | | | | | | | | |
| | | | | | | | | | | | | |
| | | | | | | | | | | | | |
| | | | | | | | | | | | | |
| | | | | | | | | | | | | |
| | | | | | | | | | | | | |
| | | | | | | | | | | | | |
| | | | | | | | | | | | | |
| | | | | | | | | | | | | |
| | | | | | | | | | | | | |
| | | | | | | | | | | | | |
| | | | | | | | | | | | | |
| | | | | | | | | | | | | |
| | | | | | | | | | | | | |
| | | | | | | | | | | | | |

(f)

**Fig. 1-7.** *(continued)*

| FLUOR DANIEL<br>INSTRUMENT SPECIFICATIONS<br><br>SOLENOID VALVES | NO. | BY / DATE | REVISION | SHEET NO. | REV. |
|---|---|---|---|---|---|
| | | | | DATE | CONTRACT |
| | | | | MANUFACTURER | |
| | | | | FLUOR P.O. | |
| | | | | BY | CHK'D / APPR'D |

ALL ITEMS SHALL COMPLY WITH GENERAL SPECIFICATION SHEETS ______

| | | | | | | | | |
|---|---|---|---|---|---|---|---|---|
| GENERAL | 1. | TAG NUMBER | | | | | | |
| | 2. | FLOW SHEET & FRAME NO. | | | | | | |
| | 3. | LINE NO. | CLASS | SIZE | SCH. | | | |
| | 4. | SERVICE | | | | | | |
| | 5. | | | | | | | |
| BODY | 6. | BODY SIZE | PORT SIZE | | | | | |
| | 7. | BODY STYLE | | | | | | |
| | 8. | MATERIAL | | | | | | |
| | 9. | PRESSURE RATING | | | | | | |
| | 10. | END CONNECTIONS | | | | | | |
| | 11. | PACKING OR SEAL | | | | | | |
| | 12. | PLUG MATERIAL | | | | | | |
| | 13. | SEAT MATERIAL | | | | | | |
| | 14. | CALC. Cv | VALVE Cv | | | | | |
| | 15. | CONNECTIONS (SEE LINES 36 & 37) | | | | | | |
| | 16. | | | | | | | |
| ACTUATOR | 17. | COIL VOLTAGE | FREQ. | | | | | |
| | 18. | CURRENT | INRUSH | HOLD | | | | |
| | 19. | COIL ENCLOSURE | | | | | | |
| | 20. | CONDUIT CONN. SIZE | | | | | | |
| | 21. | DIRECT OR PILOT ACTUATED | | | | | | |
| | 22. | NUMBER OF COILS | | | | | | |
| | 23. | MANUAL RESET | | | | | | |
| | 24. | AMBIENT TEMP. RANGE | | | | | | |
| | 25. | | | | | | | |
| PROCESS | 26. | FLUID | | | | | | |
| | 27. | QUANTITY @ F.T. | MAX. | NORM. | | | | |
| | 28. | INLET PRESS. | MAX. | NORM. | | | | |
| | 29. | OUTLET PRESS. | MAX. | NORM. | | | | |
| | 30. | $\Delta$P | MAX. | SIZING | | | | |
| | 31. | TEMPERATURE | MAX. | NORM. | | | | |
| | 32. | SP. GR. @ (STD.) ____° ______ | @ F.T. | | | | | |
| | 33. | VISCOSITY @ F.T. (Cp) | | | | | | |
| MFR. | 34. | OUTLINE DIMENSIONS | | | | | | |
| | 35. | MFR. MODEL NO. | | | | | | |

| NOTES: | CONNECTION DIAGRAMS | | A | B | C | D |
|---|---|---|---|---|---|---|
| | ENERGIZED | 36. | A | B | C | D |
| | DE-ENERGIZED | 37. | A | B | C | D |

(g)

clude as much detail as might be needed for specialized heavy-duty industrial work, such as that covered by the various codes under discussion here. Their emphasis is on areas that up until now were not highly standardized in the way that valving and piping in the process industries were, and on materials involved in general construction like concrete, heating, ventilation, and air-conditioning (HVAC) equipment; and other items that occur in almost all buildings. There is one number series reserved for specialized process equipment, but it is not finely differentiated, as some other areas are. Nevertheless, it appears that this system holds promise for more standardized specification generation and interchange.

The CSI format has three parts. The first part is general conditions, the second part is specific requirements, and the third part, mostly required for work to be performed on site, is execution. Purchase specifications in CSI format need only parts one and two. In applications where much repetitive buying is required, these purchase specifications are then followed by a bill of material or a set of data sheets describing the specific material required. It is possible to write a detailed description in part two of the material to be purchased, which is usually the case with more complex equipment, but bulk and tagged items are handled more simply and uniformly with attachments describing them.

The heart of the valve specification is the description of the specific item required. There are any number of suitable formats for this description. One method is to list every valve on a data sheet. This is, of course, a natural for tagged items (items that have a unique descriptor or designation that relates to one single valve). Valves that have actuators, limit switches, or any other sort of control device to be attached would be best described on an ISA data sheet, but this is not mandatory. It would be possible to write a description for any of these items, but the data sheet is certainly more efficient. Conversely, it would be possible to write a data sheet on which to describe the requirements for manual valves, but in practice this is rarely done.

The distinction between tagged items and bulk items is basically this: For a *tagged item,* some sizing or other specific engineering has been performed that relates to the valve that must be installed at that precise point in the flow scheme, while for a *bulk item,* even though it may be a very heavily engineered product, if there is more than one, any one can be installed in place of any other and no difference in performance will result.

Generally, manual valves are described in one of two ways. One is a series of clauses strung together, listing each point of the specification. The other common way is a series of statements in a standardized order, which lends itself better to editing and modifying when writing new descriptions. The following are some examples:

Gate valve, class 800, socket-weld ends, forged carbon steel, bolted bonnet, outside screw & yoke, rising stem, F6 trim with hardfaced seats, braided nonasbestos packing, in accordance with API 602.

Socket-welded gate valve, class 800 per API 602; OS&Y, standard port, solid wedge, bolted bonnet; carbon steel body (ASTM A105 forging); universal trim, API 600/602 trim No. 8; end preparation per ANSI B16.11; nonasbestos stem packing, handwheel operated; manufacturer's standard construction.

Item—Gate valve
Type—OS&Y—Bolted bonnet
Rating—Class 800
Ends—Socket-weld
Material—Forged carbon steel
  Body—ASTM-A105
  Trim—F6 with hardfaced seats
  Packing—Braided nonasbestos
Standards—API 602

Note that although these three descriptions say fundamentally the same things, the first two are quite compact while the third one is lengthier but perhaps easier to read since it is laid out systematically. Each one describes what is required, point by point. The first and third leave out a number of things that are to be described in the general part of the purchase specification, such as the fact that these valves are to be handwheel operated and the exact end preparation. None of these three mentions that these valves are to be tested prior to shipment in accordance with a recognized testing procedure. Some of these points are implicit in the statements already made in the description, such as the testing. For instance, in API 602 there is a paragraph about testing, which says these valves are to be tested to API 598. It is not a bad idea, however, to say it again so it does not escape somebody's notice.

Descriptions like the ones above can be longer or shorter by a factor of two or three, depending on how much you put into the general portion of the specification and how much you feel it is necessary to reiterate. This is very much a judgment call, depending on the target audience of bidders you are addressing through the medium of specification. There are two points to consider: First, the more things you put into a general specification, or any specification, the less likely it is that any given item will be noticed. Second, the items in the specific portion of the specification are the ones that an inside salesperson or order entry person is most likely to read, since those are the items in front of one who is pricing out the quotation.

In general, the statements that apply to all valves are better located in the general part of the specification. Some of these can be listings of the specific editions of various standards that apply, or a simple statement that "the latest edition of each referenced standard shall apply." Some can be statements along the lines of, "All steel valves shall have graphite packing and bonnet gaskets," which save you the trouble of listing that fact in each individual description. The caveat is, though, that this part of a specification may get read poorly or not at all, or, if it has to be attached by a clerical person, there is a possibility that the intended user may not get it and never know that it was missing. The general rule is to put into the individual descriptions as much detail as possible without making each one too long or repetitive. It is possible to refer in the specific portion of the purchase description to specific points in the general part of the specification to make it apparent that there are links between the two, and that there is more information than that stated in the specific purchase description.

When material is specified as being completely in accordance with some national standard, such as commodity pipe, fittings, and flanges, the actual manufacturer is often not very important as long as the manufacturers to be used are known to be acceptable, say by virtue of having a known track record or being on the acceptable bidders list. In other words, a tee is a tee is a tee. This is somewhat less true for valves. Valves, in the commercial markets at least, are known primarily by manufacturer and figure number ("figure number" is an old term that dates to nineteenth-century graphic art, where the "figures" were woodcut illustrations that were added to catalogs and patent documents) rather than by strict adherence to national standards. For valves, also, the possibilities for variation are much greater than for more commodity material. It is less true, then, that one manufacturer's valve is *precisely* equivalent to another's. Subtle variations in design or construction can make one valve subtly superior to, or vastly superior to, another in service suitability.

For this reason, it is common practice to pick a particular figure number as a sample or a reference, and either list all the other valves that are considered equivalent to it or indicate to the bidders that they are to offer an item that is equivalent. This is done with a phrase such as "or equal" or "or approved alternate." Sometimes manufacturers help this process along by publishing tables of equivalence between their products and those of other companies. The pitfall here is that, if the selection of figure numbers is poorly done, or if the data is unchecked and the figure number does not precisely agree with the text of the description above it, chaos can ensue. Remember, also, that with the practice of referring to figure numbers comes the habit of looking only at the figure numbers and not bothering to read the rest of the description. In a more rigorous specification system, you may prefer to drop this practice, but it is in near-universal use in industrial valve specification systems.

The choice of references is a critical one, for the reason mentioned above that it can be shortcut on the part of buyers and sellers alike. Figure-numbering systems vary widely, from very good and very comprehensive, to so highly comprehensive as to be nearly impenetrable, to being bare bones, to being nearly nonexistent. Some figure-numbering systems still in use are 80 years old, and are literally numbers of figures from a catalog from long ago, with no perceptible rhyme or reason. Most of the ones developed in the last 20 years are fairly sensible, where the individual digits and characters are all meaningful. Often these manufacturers publish a "how to order" page in the catalog that allows the specification person to build a figure number easily and (usually) accurately. It somehow seems to happen that manufacturers who have good, understandable figure number systems are the ones most likely to be referenced.

### *The Relationship of Valves to Piping Specs*

Valve specifications do not exist in a vacuum. A piping specification engineer will typically also be responsible for specifying pipe and other piping components, various specialties such as steam traps, sometimes also small equipment, and possibly also pressure vessels or rotating machinery such as pumps. For a control systems engineer, the responsibility often includes instrumentation items, meters and gauges, control equipment, and sometimes electrical gear. Some organizations have valve engineers who are responsible for all valving, whether manual or actuated. And, of course, there are thousands of locations where everything within sight is your responsibility, and you will have to know enough about everything in the plant to keep it running or replace it.

The same codes, and in many cases the same standards, that apply to one aspect of valve design also apply equally to every other component. The purpose here is not to give you the information required to select a piping material or a flow meter, but to remind you that the processes described with specific reference to valving are also useful in many other areas of plant design.

It may seem that it would be perfectly suitable to spec out valves to match the piping exactly. That is, the valves and the fittings would have the same type of end connections, and the flanges on valves would be exactly the same as the flanges on anything else. This is generally true, but a few refinements will go a long way toward optimizing the system specifications.

To start with a relatively simple case, let's think about utility services in a large plant. Most of this piping is outdoors; it is noncritical in terms of pressure and temperature, and, for that matter, routing and just about any other consideration. There is a great deal of other small size piping in the plant, mostly carbon steel with socket-weld fittings, flanges, and valves. Because socket-welded construction is faster to construct and does not require the clearances or extra unions that threaded construction requires for assembly, and does not leak as much, there are thousands of socket-weld fittings and flanges in the plant. There is not much point in stocking a few threaded fittings just for utility services, when the socket-weld fittings are cheap and are certainly suitable for utility services. But valves are a different matter. A valve may be worth $20 to $30 in steel, and $10 in bronze or brass, as opposed to a dollar or so for an elbow, and you would probably as soon have ball valves in most utility services anyway. So it makes perfect sense to have a mixed bag of socket-weld steel fittings but threaded bronze valves in air and water lines.

A similar situation occurs whenever you need steel valves with special trims, such as Monel or alloy 20. Commercial practice in valve manufacturing today, especially for imports, is to make long production runs of standard products, meaning in the case of steel valves that the trim is normally F6 with hardfaced seats, since that is by far the best-selling trim. If you want anything else, especially in the case of cast-steel valves, the predominant practice is to take existing valves off the shelf and modify them to the desired trim by making new stems and overlaying the seats and disc with the desired trim material after gouging out or machining back the existing trim material. It may turn out, especially in small-sized flanged valves, that it is cheaper to buy an all alloy valve than to go through the modification process, besides taking a week or two for the modification. Thus the system would end up with carbon steel piping, fittings, and flanges, and some carbon steel valves and some alloy valves.

Another good example is lube oil and seal oil piping for large rotating equipment. The oil used is "turbine oil," a considerably higher grade of lubricating oil than would be found in your car. Absolutely no pickup of rust or scale from the piping is permissible. Lube oil reservoirs are usually lined with an impervious epoxy or internally sandblasted prior to service. Neither is feasible with the interconnecting piping. One approach is to use carbon steel piping, with an acid wash circulated through the system to remove every possible bit of debris. Another is to use stainless steel piping, since it will never corrode and needs only a flush to remove dirt and debris prior to service. The total installed cost of the stainless steel is generally lower than that of the carbon steel because the expensive chemical cleaning and purge has been eliminated. But the valves do not have to be stainless steel. The valves, first of all, are clean enough as they come from the factory that a simple internal inspection and flush will suffice for them also, and the solid stainless construction is not needed. The combination of stainless piping and carbon steel valves appears often in lube oil piping.

Yet another case where this might occur is in low-temperature services between −20°F and −50°F (−29°C to −45°C). In this range, carbon steel is still acceptable by code, provided it has proved to be resistant to brittle fracture. These steels are normally special grades, but they must have been impact tested (Charpy tested) to demonstrate their notch toughness. Impact-tested steels look and act exactly like nontested steels at room temperature, so a steel component that is not capable of brittle fracture resistance can very easily be installed. Only the markings identify the low temperature steels, and if the markings are lost, the piece of material becomes useless because there is no way of demonstrating its suitability without a destructive test. Stainless steel components that have no welds, like forged valve bodies, are acceptable by code without testing, and are easily identified visually. Since the impact testing can cost as much as $75 per set of tests, the stainless steel is often cheaper in initial cost, let alone the cost of uncertainty during installation. So here, also, it may be better and more cost-effective to have a mixture of carbon and stainless steels in the same system. Sometimes conformity is more expensive and less suitable than purpose-built specification.

The use of butt-weld valves warrants some thought also. If butt-weld valves are used in a system, it is generally because it is necessary to eliminate as many flanged joints as possible—for instance, in high-pressure steam (Fig. 1-8), where leaks can easily occur due to thermal movement and are impossible to stop except by bringing the system down for repairs or by doing a very hazardous boxing-in procedure on the leak. Control valves are no exception, because even though it may be desirable to pull the valve every now and then for calibration or because the instrument shop is not equipped to do on-site work, the leak problem generally is felt to take greater precedence. But occasionally there is no way around having to install a flanged component in an all butt-welded system, and usually it turns out that the flange location has to be next to a valve. It is possible to build valves with one flanged end and one butt-weld end, and in transmission line work that is fairly common. Elsewhere, though, it is not very common, and many casting patterns do not lend themselves to that configuration. When a flange is required immediately adjacent to a valve, it is generally better to specify a valve with butt-weld ends and simply provide a weld-neck flange to be welded to the valve.

### *The Influence of Size and Rating*

Other factors to consider in specifying valves are the size of the valve and the material required. These two factors will significantly limit the available valves to choose from for your intended application. In fact, these two factors are perhaps the first ones the specifier should look at after deciding what the valve is supposed to do.

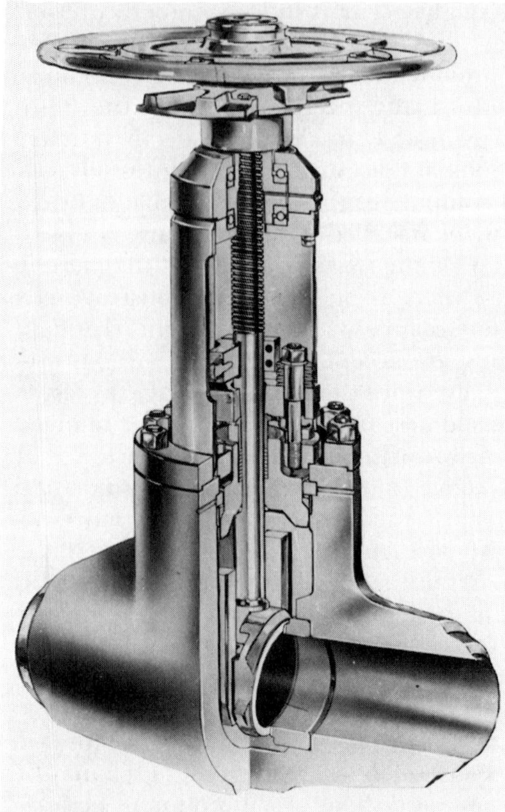

**Fig. 1-8.** Butt-weld-end pressure-seal bonnet gate valve for high-pressure steam service. (*Courtesy of Velan Valve Corporation.*)

Sometimes the type of valve and the material are related. That is, low-cost materials and low-cost designs often are found together, as are high-cost materials and special designs, since the selling price of the valve is determined by the cost of the material and the complexity of the manufacturing process. There are many examples of crude or ordinary valve designs in exotic material, often not the best combination to have chosen, but arrived at because of thoughtlessness or unwillingness to "go out on a limb" on the part of the valve specifier or some other user involved in the selection process. On rare occasions, you may encounter sophisticated valve designs made in a low-grade material, which is an even more pointless exercise since all the features of the design are useless if the material does not hold up. But these two extremes are more the exception than the rule.

Let's assume you want a simple block valve. As the examples in the previous section serve to illustrate, you will look at the piping specification for the selection of material and end connections for the size you need. Then, knowing the material and the line size, you can determine more easily what you can get. Let's say that this is plastic piping, specifically PVC, and that the line size is 1 in. or 1½ in. (DN 25 or DN 40). It will not take very much research on your part to determine that the most widely available type of block valve is a ball valve with union ends (Fig. 1-9). Although it is possible to get small plastic gate valves and even butterfly valves, the ball valves are the most widely available and the cheapest. In this case, you will want to ask yourself if there is any service condition or external problem that might make the ball valve a poor choice. If you do not come up with any, then you know that the ball valve is the way to go.

Now let's consider another example, a 24 in. (DN 600) stainless steel line carrying hot acid vapors or some other equally severe service. If you need a block valve, say in a location about 50 feet (15 m) above grade on a vessel nozzle, you could very well choose a gate valve. But, if you consider a metal-seated high-performance butterfly valve, first it will be apparent that you could save thousands of dollars over the gate valve price. Yet there might be other factors to consider. Will your plant maintenance people trust such a valve? Can you be sure, if you have not put in many high-performance valves before, that this valve will really close tight? Do not forget that the butterfly valve will fit in places that the gate valve will not, and that it will be easier to put a chain operator on the butterfly if you cannot have it accessible. Also, do not forget the weight savings, which is important not just for the number of pounds of stainless steel that you will be buying, but is important every time a crew has to install one and every time you design a support (remember, this is probably light wall piping and a 1500- to 2000-lb [700- to 900-kg] gate valve weight is something to be concerned about, while a 200- to 300-lb [100- to 150-kg] butterfly valve is a lot easier to support). If you look at all these factors, you might easily conclude that you are happy with the gate valve, but it is more likely that the butterfly valve will be the choice.

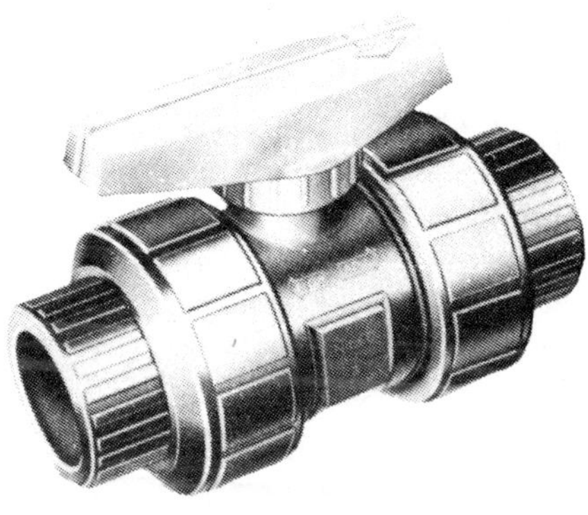

**Fig. 1-9.** All-plastic ball valve with union ends. (*Courtesy of Nibco Inc.*)

Let's suppose now that you need a flow-control valve. A globe type body with a diaphragm actuator is the classical choice. But suppose this is a severe service. You might need a high alloy body. Suppose this is also a fairly high-pressure service. You could easily end up with a valve that is worth as much as your house, if you simply pick a standard valve that is capable of class IV shutoff in addition to its throttling ability. But do you need a shutoff ability? Is this valve going to be continuously modulating, or is it going to be a fairly steady service? You might find that, as in the last example, there is a butterfly valve that would be a better choice (Fig. 1-10). You could still put a diaphragm actuator on it, and again you would save a great deal in space and weight. If you do not like butterfly

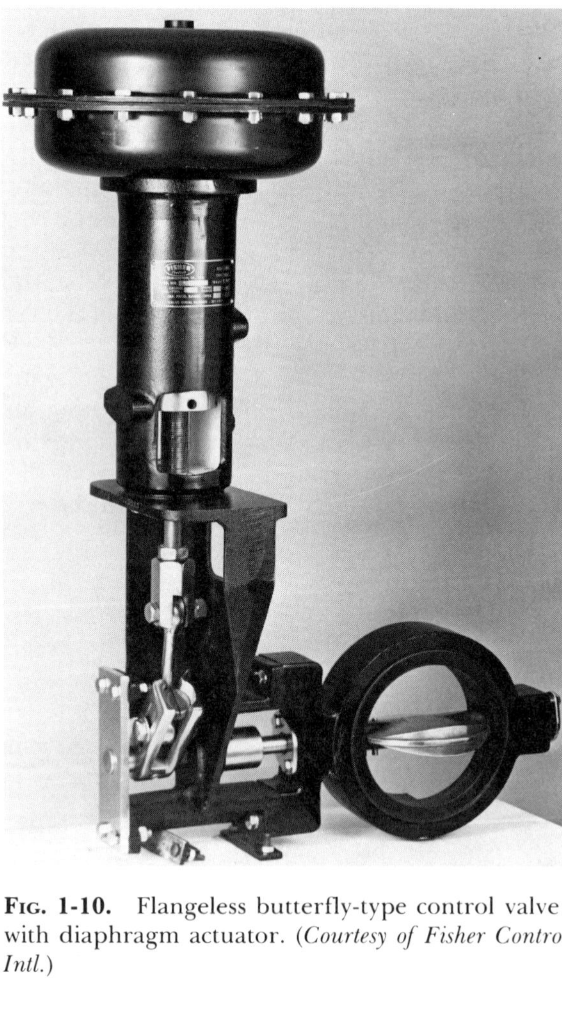

**Fig. 1-10.** Flangeless butterfly-type control valve with diaphragm actuator. (*Courtesy of Fisher Controls Intl.*)

valves, or if you have debris or slurry in the line that would make a butterfly valve uncloseable, you could use a characterized flangeless ball valve (Fig. 1-11) and get the same savings over a globe-type valve. In fact, if it turns out that you have solids in the line from somewhere, you would want to choose the ball valve even if it was more expensive than the globe-type valve. But, on the other hand, if this is a clean service with no solids, still high pressure and still requiring a high alloy, but with a very small flow rate such that a 1 in. (DN 25) valve body is plenty large enough, then the globe type would be an appropriate choice because the

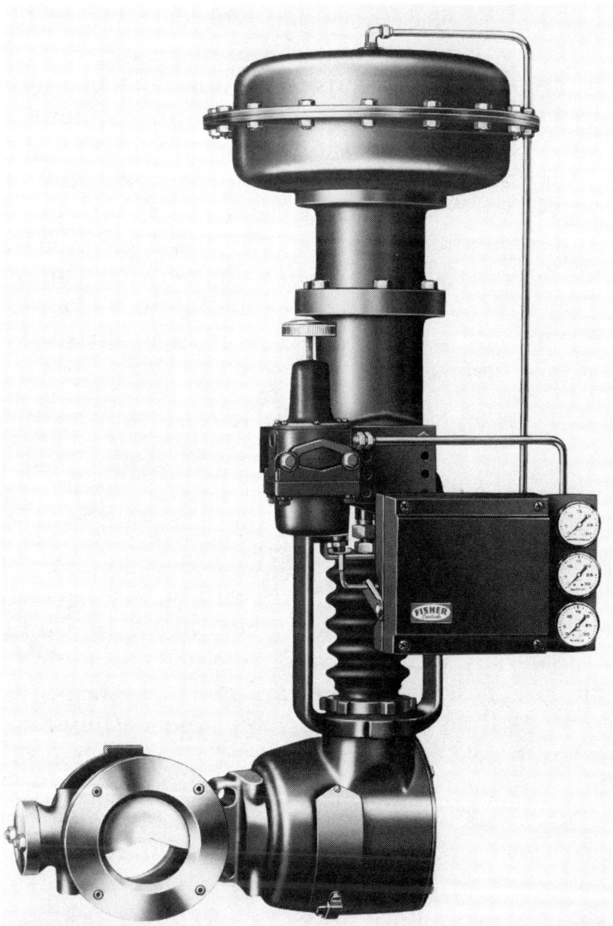

**Fig. 1-11.** Flangeless ball-type control valve with diaphragm actuator and pneumatic positioner. (*Courtesy of Fisher Controls Intl.*)

butterfly valve is not made that small. Although the characterized ball is available in a 1-in. (DN 25) size, if you suspect that the valve may still be oversized, the globe can be easily retrofitted with smaller trim or a fluted trim cage. If the cost of the valve is still a concern, you could consider one of the designs that have separable carbon steel flanges in place of the all alloy bodies.

The point is that the size and the pressure rating of the valve have a great impact on your possible choices. The range of available sizes and pressure ratings of the different valve types will be discussed more thoroughly in Chapter 2, in the sections that describe each valve and its operation. As you set about to choose valves for your service, the help of valve specialists can be invaluable in that these people are more likely to know the market and what is available in practical terms. Many valve manufacturers show off their special designs, in part out of pride of accomplishment, but it can be hard to tell from reading catalogs which items are specials and which are standard production. Special designs and unusual sizes are often a necessity, but if you are on a tight budget or schedule, you would be better off to avoid them. If at all possible under these circumstances, be flexible and be aware that sometimes great savings in time and expenditure can be made with a very simple substitution. Salespeople are occasionally able to volunteer alternates to your design, but are often reluctant to do so since they are not aware of all the parameters. The people who are in a position to do so are generally within your own organization.

## Construction Details

This section will address the various design options available to the specification engineer, the "building blocks" that can be arranged in various ways to produce different valve styles. The same valve can be built in many different materials, with many different end connections, body styles, and operators. These differences are quite significant, in many cases more significant than what basic valve type is chosen. A knowledge of available and feasible construction details is important for successfully specifying valves for purchase and use. It is true that you can insist on, and get, just about anything you want. But some choices are more feasible than others, and in some cases the differences are more significant than others. Your knowledge can mean the difference between a custom design with a long lead time order and a stock item purchase. There are some valve types that are not available under any feasible circumstance with certain types of end connections, for example, and the reasons why may not be apparent until you study the valve design. The same valve design will be available with different end connections, depending on size, because of piping design conventions that relate to ease of fabrication and do not have much to do with the internal configuration of the valve.

The conventions used for end connections, bonnet details, choice of body materials, and the like vary from industry to industry because of the differences in conditions, just as is true for the types of valve that are preferred. Within any industry, individual preferences flavor these choices also. But unless you are pioneering in a field, you will be best off following the common logic of design in choosing these construction details. This is because of the types of product that manufacturers are prepared to make, as much as anything else. You will, if you are working within the confines of an existing system, prefer to follow the conventions that have already been adopted unless you have reason to believe that the configurations that already exist are not good enough, perhaps because they leak too much or because they might be overkill, or you now know you can use a much cheaper valve that will do the job just as well.

In these days of computer-aided design, you can build a library of details that can be endlessly combined to form complete drawings. This is precisely how some valve manufacturers now do their engineering. For many years this could be done only in your imagination. Where you used to have to imagine this valve design with socket-weld ends instead of threaded ends, now in about 5 minutes you can have a complete drawing ready for issue. It is still true, though, that just because something can be drawn does not mean that it is practical to build. It may be that after you commit a design to paper, it is more obvious that some combination of design features will not work than it would be turning the design over in your head.

Let's think about end connections first, since end connections obviously have to be the same as, or at least compatible with, the rest of the piping system. The first types of end connections were all mechanically joined, and the first of these was the bolted joint. There is a famous picture of Louis XIV's waterworks project to bring water to the gardens of Versailles, completed in 1664 as a cast-iron piping system with flanged bolted joints. The use of threaded joints evolved gradually with the ability to cut reliable threads, until these two end connections became standardized near the end of the nineteenth century. There have been recorded installations of very large threaded connections, and the use of small flanges is common to this day, but in general the threaded connections are common in smaller sizes and flanged connections in larger sizes. Flanged joints are easier to make up in place, since the pipe and fittings do not have to rotate, while threaded joints make up faster in very small sizes than do bolted joints and are a great deal cheaper. Threaded joints larger than 2 in. (DN 50) can be difficult to make up and very often are difficult to make up tightly.

In the refining industry, the break between threaded and bolted joints has moved back and forth over the years. In current practice the most common arrangement is to use threaded joints up to 1½-in. pipe size (DN 40), with bolted joints starting at 2-in. size (DN 50). Formerly, the break often was above 2-in. size (DN 50), with 2½-in. size (DN 65; not often used) being the smallest size where most connections were flanged. In industries where leakage is not as consequential, and where lower cost is more important, such as waterworks, threaded joints are common up to 3-in. or 4-in. size (DN 75 or DN 100) (Fig. 1-6). Even in refinery work, utility air and water connections are usually threaded up to 2-in. size (DN 50) or even 3-in. or 4-in. size (DN 75 or DN 100), especially with bronze valves (Fig. 1-12) and galvanized steel pipe, neither of which can be welded.

The use of welding as a pipe joining process began in the 1910s on an experimental basis and was routine by the beginning of World War II. There were a lot of crude welding processes used, which eventually gave way to more reliable processes that were recognized by the piping codes with higher allowable stresses. Today, the codes recognize some welding processes as being less strong than the adjoining base metal by derating the allowables, while recognizing some as being equivalent in strength to the base metal provided that certain examination techniques are carried out to ensure a lack of various defects in the weld area. The design adaptations necessary to carry out these welding examinations influence valve design somewhat.

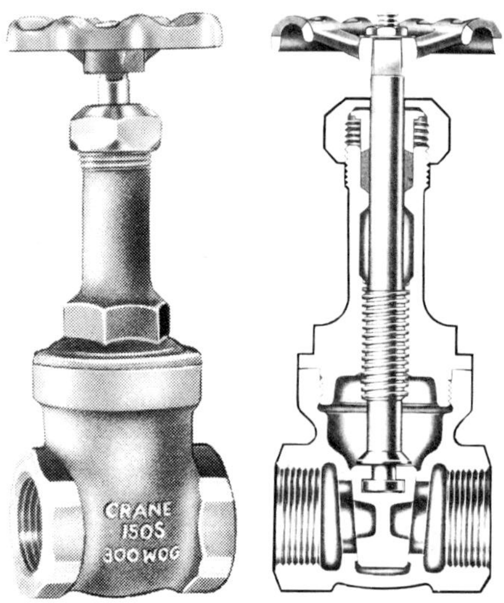


**FIG. 1-12.** Threaded-end threaded-bonnet union gland solid wedge gate valve. (*Courtesy of Crane Co.*)

The first use of welding was for butt-welding (groove welding in terms of the weld type) the ends of pipe and pipe fittings together. Valve designs were also developed, adapted from flanged-end valves, to allow butt-welding of valves. Today butt-weld-end valves occupy the same size range as flanged valves, with 2-in. (DN 50) being the smallest normally available size. Not long after the introduction of butt-welding, the socket-weld end (fillet weld in terms of weld type) was developed from the threaded end. Socket-weld dimensions are close to those of threaded ends, although the socket-weld requires slightly less wall thickness or outside diameter than threaded ends since the stress raisers associated with the thread roots are not present (Fig. 1-13). This closeness of dimensions means that it is possible, and in fact very common, to manufacture valves with threaded ends and, when socket-weld valves are ordered, to machine out the internal threads to produce the socket-weld end dimensions while leaving the external dimensions slightly larger than might be absolutely required. This is an interchangeability that is not practical with flanged and butt-weld ends.

A consequence of the increasing use of socket-welding ends is that, since the socket-weld is faster to prepare and weld than a butt-weld, the break between socket-weld and butt-weld or flanged ends has moved back up to 2-in. pipe size (DN 50) in many plants. Sometimes a composite construction is used in 2-in. size (DN 50) where valves and some other components are socket-weld, but pipe fittings and flanges are butt-welded. A socket-weld valve of any type is cheaper than the corresponding flanged valve. This is true in steel valves but even more so in alloy and stainless valves, and not just because of the flanges, but because the socket-weld valve is likely to be a forged compact pattern while the

**SOCKET WELDING END DIMENSIONS (INCHES)**

**ANSI B16.11**

| Nominal Pipe Size | Minimum Bore Diameter (M) | Minimum Bore Depth (P) |
|---|---|---|
| 1/4 | .555 | .38 |
| 3/8 | .690 | .38 |
| 1/2 | .855 | .38 |
| 3/4 | 1.065 | .50 |
| 1 | 1.330 | .50 |
| 1 1/4 | 1.675 | .50 |
| 1 1/2 | 1.915 | .50 |
| 2 | 2.406 | .62 |
| 2 1/2 | 2.906 | .62 |
| 3 | 3.535 | .62 |

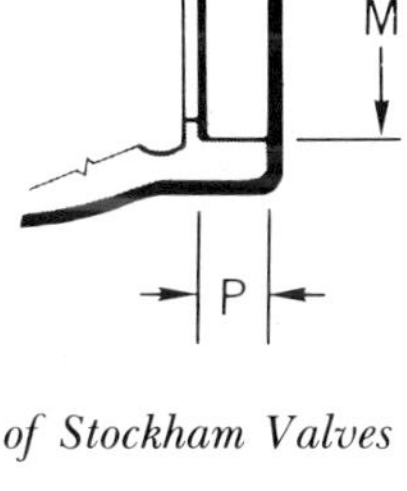


**Fig. 1-13.** Dimensions of socket-weld end bores. (*Courtesy of Stockham Valves and Fittings.*)

flanged valve is cast, so the socket-weld valve may weigh one-fourth of what the flanged valve weighs.

Many other types of end connections are common, usually intended for specific types of pipe and specific services. Often, the valves are the only components made of cast iron or steel, since these materials are standard for the valves while the remainder of the piping system may be plastic or aluminum or almost anything else. Where cast-iron piping has continued to be used, such as in underground water in utilities and industry, the use of bell and spigot joints or mechanical joints (a special mechanically tightened spigot joint) has continued even with valve end connections. Here, often there are no pipe sizes smaller than 2 in. (DN 50) because of joint design considerations and the available pipe sizes.

Cast-iron valves for water service, built to AWWA standards, are available with end connections flanged on both ends to ANSI B16.1 flat-face dimensions, with spigot ends to ASTM-A74 or ANSI A21.1 (dimensionally identical), with mechanical joint ends to ANSI A21.11, with ends for PVC pipe to AWWA C900, with (almost obsolete) ends for asbestos cement pipe, or with various combinations thereof. Smaller valves are also available with threaded ends, but this is not common in underground service.

Cast-iron and bronze valves in sizes up to 2 in. or 3 in. (DN 50 or DN 75) for other than underground use are almost always provided with threaded ends, National Pipe Thread (taper) to ANSI B1.20.1, formerly designated ANSI B2.1. Valves larger than this have flanged ends, to ANSI B16.1 for iron valves and ANSI B16.24 for bronze valves. Class 125 (PN 10) iron flanges are flat faced, while class 250 (PN 20) flanges have a raised face. Both have drilling templates identical to steel ANSI B16.5 drilling, class 125 being the same as class 150 steel, and class 250 being the same as class 300. (There are no similar correspondences in the ISO standards.) Bronze flanges also have drilling templates the same as steel flanges, but the face of a bronze flange is full face and machined smooth with two grooves to help retain the gasket. Except for the class 250 iron, even though steel flanges can be bolted to the iron or bronze valve, the steel flange must be flat face to avoid breaking or bending the valve flange, which is less strong. In addition to the flanged ends, some iron valves are available with grooved ends for clamp couplings.

Brass and bronze valves are often designed with solder ends (Fig. 1-14) for domestic and utility water service, sometimes for air and steam services, although the melting point of the solder limits their use in steam to very low pressure or domestic steam systems such as steam heat. Solder ends are not usually modified from threaded ends because the female end of the valve needs to be fairly thin in cross section to avoid acting as a heat sink. The areas of the joint must be sufficiently hot during soldering to allow the solder to flow throughout the joint. Sizes of

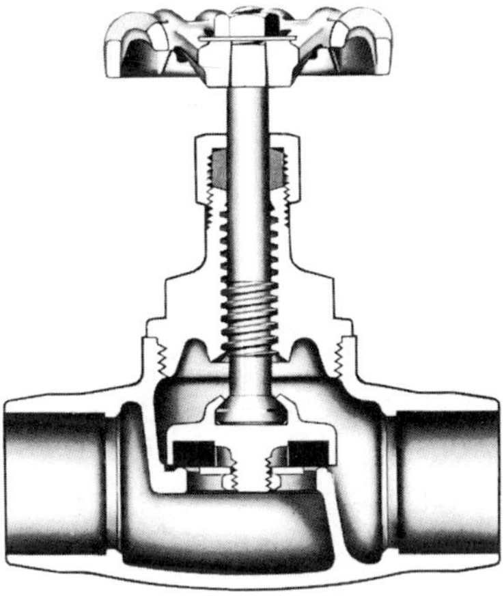

**FIG. 1-14.** Solder-end threaded-bonnet threaded-gland globe valve. (*Courtesy of Crane Co.*)

solder end valves are normally ½ in. to 2 in. (DN 12 to DN 50), but solder end valves are available as large as 4-in. size (DN 100). The old tin-lead solders are being replaced with tin-antimony solders to decrease the health hazards of lead leaching, and the tin-antimony solders are actually stronger and capable of higher temperatures than the tin-lead. Some bronze valves are available with ends prepared for brazing, although this is not common (Fig. 1-15). These valves are usually modified from threaded-end bronze valves much as steel socket-weld valves are modified from threaded-end valves.

The National Pipe Thread (NPT) taper pipe thread is used on all metal threaded ends in standard use, with the exception of fire protection equipment, where the National Hose Thread (NHT) or various nonstandard city hose threads are in use. Another different hose thread is the Garden Hose Thread (GHT), which is a standard size (¾ in.) regardless of hose size. Outdoor faucets, known as sillcocks, have NPT thread on the pipe end and GHT thread on the outlet end.

Nonmetallic body valves use a variety of end connections, but the threaded ends are the most similar to those of metal valves. The thread is basically NPT, but a blunt-ended thread form is used since the threads are usually molded rather than cut. Most valves have the end connection styles that are common to the particular type of material used. PVC and chlorinated polyvinyl chloride (CPVC) valves usually have solvent-weld ends, which look very much like socket-weld ends in a metal valve. The joining process uses a solvent cement, usually preceded by a softening agent, that breaks up the molecular structure in the joining area and re-fuses the molecules together, in effect welding the two surfaces into one structure. Many valves have integral union ends on one or both ends, so that the end can be solvent-welded to the pipe but the valve can

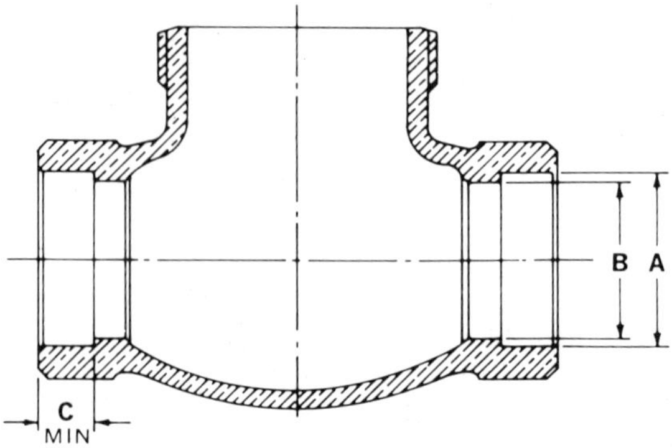


**DIMENSIONS**

| SIZE | A | B | C |
|---|---|---|---|
| 1/4 | .540 | .413 | 17/64 |
| 3/8 | .675 | .548 | 5/16 |
| 1/2 | .840 | .714 | 3/8 |
| 3/4 | 1.050 | .923 | 13/32 |
| 1 | 1.315 | 1.191 | 7/16 |
| 1 1/4 | 1.660 | 1.534 | 1/2 |
| 1 1/2 | 1.900 | 1.778 | 5/8 |
| 2 | 2.375 | 2.247 | 21/32 |
| 2 1/2 | 2.875 | 2.745 | 25/32 |
| 3 | 3.500 | 3.341 | 53/64 |

**Fig. 1-15.** Dimensions of socket-brazed ends. (*Courtesy of Stockham Valves and Fittings.*)

still be removed by undoing the union. Polypropylene and polyethylene cannot be cemented, so a heat fusion process is used. Polypropylene valves are usually similar to PVC and CPVC valves in design, but have ends equipped for socket fusion in which external heat or a resistance coil embedded in the joint is used to melt the plastic and fuse the joint into one solid structure. Polyethylene valves are uncommon except in gas distribution, but they are normally installed either by socket fusion or by butt fusion, in sizes 2 in. (DN 50) and larger, in which the ends are heated and forced together, very much like a steel resistance weld.

Depending on the standard to which they are made, steel and alloy valves are available threaded or socket-weld in sizes up to 2 in. (DN 50), and flanged end in all sizes. Cast-steel valves are normally not made with threaded or socket-weld ends, while forged-steel valves are produced with flanged ends, either forged in one piece or with flanges welded on. The welded-on flanges (Fig. 1-16) sometimes are special designs that fit into a standard socket-weld end and are electric fusion welded, and sometimes are friction welded on without the addition of filler metal.

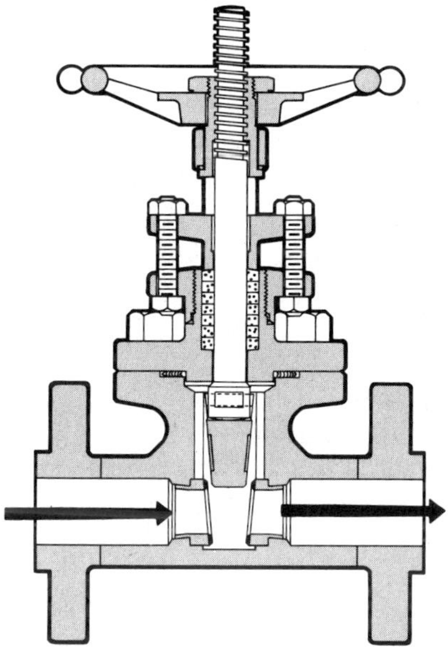

**FIG. 1-16.** Forged body compact pattern (API 602) gate valve with welded-on flanges. (*Courtesy of Henry Vogt Machine Co.*)

Cast-steel and alloy valves are available with butt-weld ends, normally no smaller than 2 in. (DN 50), but made from different patterns than the flanged end valves since the end contours of the butt-weld and flanged valves do not completely overlap for either type. Other types of proprietary end connection, such as the high-pressure hub and clamp type, are also available. Valves with pressure-seal bonnets are available usually as butt-weld ends. Some industries, such as cross-country pipelines, commonly buy steel valves with one end flanged and one end butt-weld, or some other combination, but in most markets such valves are not available.

It is very common for vent and drain valves to be specified with different ends, such as one socket-weld end for attachment to the piping and the opposite end threaded for a plug. This arrangement is convenient, but an even stronger vent and drain valve is the design that has an integral male end designed to fit a threaded or socket-weld piping connection (Fig. 1-17a, b) or a reinforced integral male end designed to be welded directly to a header (Fig. 1-18).

Flange standards are usually ANSI B16.5, but can also be API 6A or other specialized standards. The flange facing is usually raised face, with $\frac{1}{16}$ in. (1.6 mm) height in class 150 and class 300 (PN 16 and PN 25) and $\frac{1}{4}$ in. (6.3 mm) height in class 400 and higher, but a number of other facings are recognized by B16.5. These include the ring joint (Fig. 1-19), large tongue and groove, small tongue and groove, large male and female, and small male and female. These are listed in the approximate

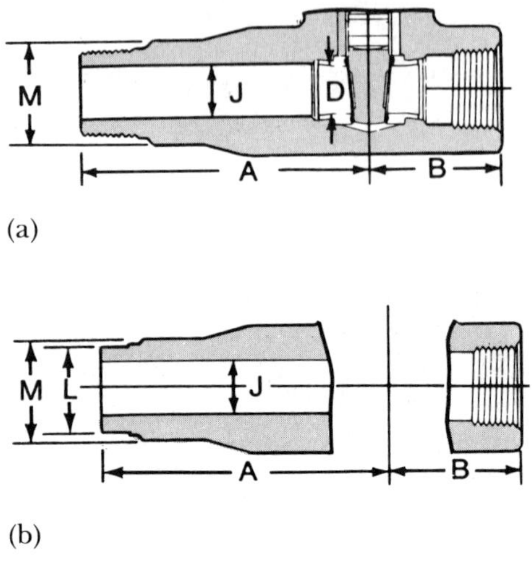


**Fig. 1-17.** Extended body (male threaded end and male end suitable for either socket-welding or butt-welding) valves for installation in a branch fitting. (*Courtesy of Henry Vogt Machine Co.*)

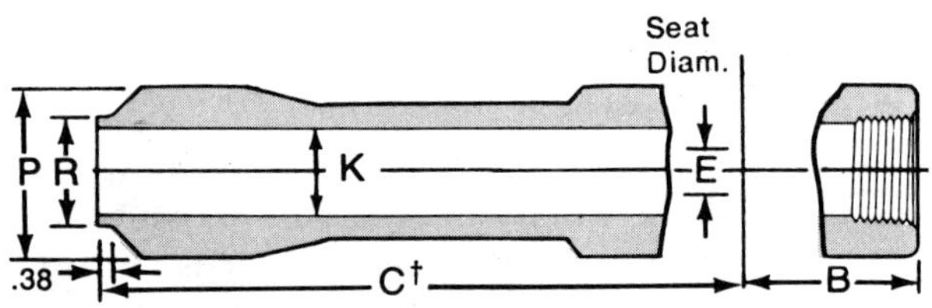


**Fig. 1-18.** Extended body valve with integral reinforcement for direct welding to header pipe. (*Courtesy of Henry Vogt Machine Co.*)

order of frequency of use, although all the tongue and groove and male and female joints are used so rarely that many users will never encounter them in a career. The flange surface finish can vary also, ranging from the serrated, fairly rough finishes appropriate for sheet gaskets to smoother finishes appropriate for spiral-wound gaskets. Very fine surface finishes are too smooth to effectively grip any type of gasket.

Before the use of standard flanges was complete, it was sometimes possible to buy valves with undrilled flanges to make your own. Some valve standards allow for ANSI standard drilling pattern, but the flange on the valve may be thinner than ANSI standard. Large flanged valves over 24 in. size (DN 600) may have any one of several standards since there are a number of flange dimensions to choose from. However, most valves of over 24-in. size are available only with flanges drilled to ANSI B16.1 / MSS-SP-44 / AWWA-C207 class E (all the same drilling template,

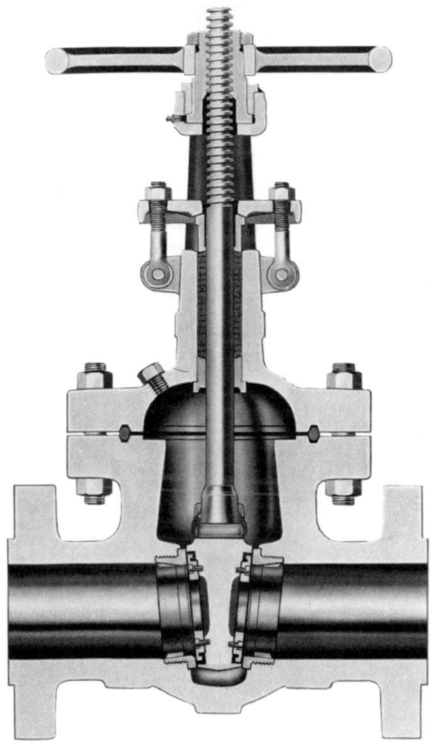

**Fig. 1-19.** Cross section of class 600 bolted-bonnet raised face flanged-end cast-steel gate valve with pressure-energized seats, ring joint bonnet gasket and bonnet tap. (*Courtesy of Pacific Valves, a unit of Mark Controls.*)

but different ratings and flange thicknesses). Other flange standards, such as API 605, are usually not possible to produce on valves since the bolt circle is sometimes too tight and would drill through the body wall or other components.

Where higher process temperatures must be maintained to avoid solidification, the piping may be encased in another pipe filled with steam or a heat-transfer liquid. This jacketed pipe usually has flanges with inner diameter the same size as the core pipe and an outer diameter appropriate for the jacket pipe. The valves that are used in these systems must either have oversize flanges (Fig. 1-20) for full jacketing on the valve or, if oversize flanges are not used, can only be partially jacketed (Fig. 1-21). Standard valves can also be fitted with clamp-on jackets (Fig. 1-22) instead of integrally welded jackets.

For special services, such as sanitary or high-purity gas work, valves are available with very small butt-weld ends, either to pipe sizes or to tubing sizes, which are smaller than pipe sizes and have closer tolerances. For sanitary services, these valves are also available with various approved sanitary service clamp joints (Fig. 1-23), which permit quick

**Fig. 1-20.** Sleeved plug valve with full steam jacket and oversize end flanges (to accommodate jacketing and to mate with jacketed pipe). (*Courtesy of Xomox Corporation.*)

**Fig. 1-21.** Sleeved plug valve with partial steam jacket. (*Courtesy of Xomox Corporation.*)

**Fig. 1-22.** Ball valve with bolted-on steam jacket consisting of aluminum heat-transfer castings enclosing a steel pressure-retaining steam chamber on each side of the valve. (*Courtesy of Neles-Jamesbury, Inc.*)

disconnection and complete cleaning. Even the actuators of control valves are polished stainless steel to allow for complete sterilization.

Many more ordinary valves in small sizes (⅛ in. to 1 in.) (DN 2 to DN 25) are available with SAE tubing ends or the flareless proprietary tubing ends (see Fig. 2-181), for instrument piping, small process piping, board-mounted small process or control panel configurations with bulkhead nuts to allow the piping to be behind the board with only the valve handles projecting above the board (Fig. 1-24), special vacuum services with O-ring joint tubing ends, or other specialized joints. Some very high pressure services use threaded gland ends that positively seal the tubing ends against the valve or fitting body (Fig. 1-25).

The butt-weld joints in common use are the ones shown in ANSI B16.25, either the 37½ degree taper for wall thicknesses up to ⅞ in. (22 mm) nominal, or the compound 37½ degree and 10 degree bevel for thicknesses over ⅞ in. (22 mm). The J-bevel or U-bevel designs, which are harder to machine but easier to inspect by radiography, are also used. Some industries use only the taper bevels, while some prefer the J-bevels for specialized use. The power industry and a few others use backing rings, which requires a different internal contour, while other industries prohibit backing rings. The point is that the exact end configuration must be specified, although generally you will get taper bevels unless otherwise stated.

The other major joint to be concerned about in a valve is the bonnet-joint or the body joint, whichever way the valve is disassembled to get at

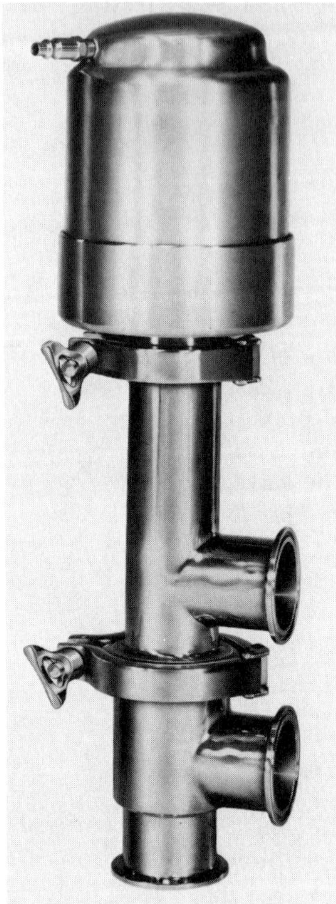

**Fig. 1-23.** Sanitary service three-way control valve with sanitary actuator. (*Courtesy of Tri-Clover, Inc.*)

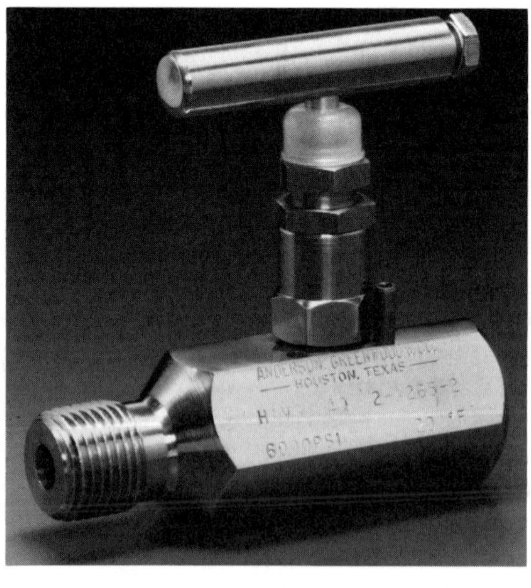


**Fig. 1-24.** Barstock needle valve with bulkhead nut (above bonnet and below packing nut) for board mounting. (*Courtesy of Swagelok Co.*)

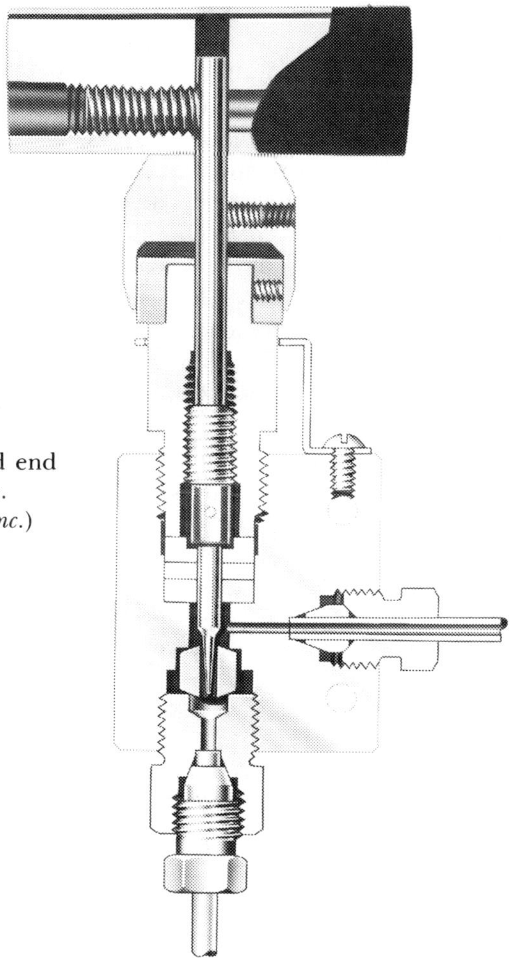

**FIG. 1-25.** High-pressure gland end on throttling angle needle valve. (*Courtesy of Autoclave Engineers Inc.*)

the insides. Usually, gate and globe valves have bonnets; check valves do also, but they are usually called caps; and some other types of valve, such as plugs, also have the same arrangement, but it is called a cover plate, cap, or other term. Ball valves that do not have a bonnet are usually split body, with a joint that is larger than the end flanges, or they have an end piece that allows entry through one end. Many butterfly valves have nothing, or a retainer plate which again is accessed from one end. These joint designs are even more varied than the end connections, since they do not have to conform to any standard and do not have to mate to anything except the matching parts of the valve. The opening into the body, by whatever name it goes, is similar in character to the end connections. That is, it is almost never less tight than the end connections but not usually much better.

Lower cost bronze valves usually have threaded bonnets, while better ones have union bonnets. Only fairly large bronze valves have bolted bonnets. The union nut on union bonnet designs comes off with the bonnet—that is, the threads are on the body. Union bonnet designs are better because the entire bonnet does not have to rotate while seating (Fig. 1-26). Small iron valves also have threaded or union bonnets, but some small iron valves have U-bolt bonnets in which the bonnet is retained by two bolt holes, one on each side, with a nut on top of the bonnet, each of the nuts being on one end of a U-bolt that passes around the body (Fig. 1-27). Almost all flanged-end iron valves and some of the larger threaded-end iron valves have bolted bonnets. For gate valves, these are normally oval or somewhat elliptical bonnet flanges, flat faced like the end flanges and equipped with a full-face gasket and low-strength steel bolts or sometimes bronze bolting. Globe valves and some check valves have circular bonnet flanges, or at least ones that are more rounded than the gate valves are.

Steel and alloy steel valves have a wide variety of bonnet designs. Small screwed and socket-weld valves have bolted bonnet flanges, usual-

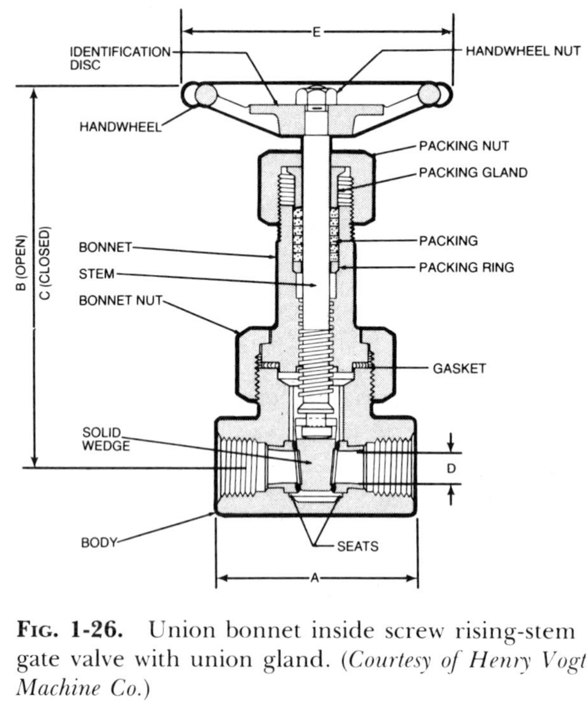


**Fig. 1-26.** Union bonnet inside screw rising-stem gate valve with union gland. (*Courtesy of Henry Vogt Machine Co.*)

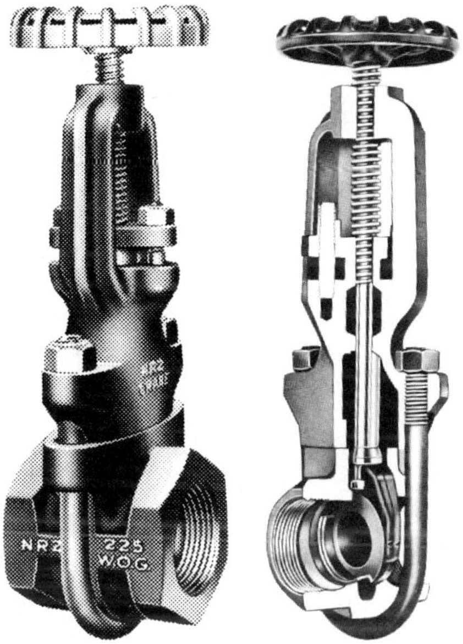


**Fig. 1-27.** U-bolt bonnet gate valve. (*Courtesy of Crane Co.*)

ly, with four bolts in a rectangular or square pattern although the gasket surfaces are usually round. Some of the bonnet exterior surfaces are rectangular or square, while others are round. Welded bonnets are common on steel valves (Fig. 1-28), and generally consist of a threaded body-bonnet joint with a seal weld around the outside. There are some joints where the weld is a strength weld also, but this is less common. The welded bonnet styles weigh less than the bolted bonnets, and certainly leak less. With the cost of labor being so much more than the cost of materials, small steel valves are not normally economical to repair, so the bolted bonnet does not serve any particular advantage.

Another adaptation of valve designs for specific services is the extended bonnet for low-temperature service. Whenever the process stream and the piping are significantly below 32°F (0°C), atmospheric moisture will begin to freeze on the piping surface. The loss of "cold" from cold lines is also much more significant in terms of the cost of producing the "cold" in the first place than the conservation of heat is. The common method of dealing with these problems is to use a heavy insulation with a moisture-tight vapor barrier. In order to insulate over as much of the surface as possible and to keep the packing gland from freezing over, valves in low-temperature services are often built with extended bonnets and stems (Fig. 1-29 and Fig. 1-30). These extended stems can be applied to any type of valve. The bonnets must be long enough to put the packing gland completely outside the insulation, but if

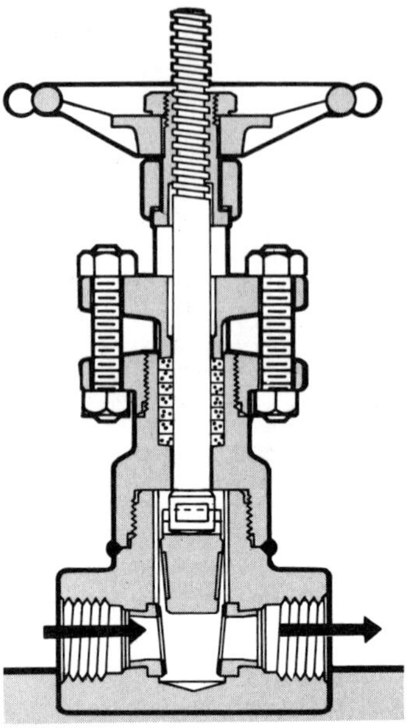

**Fig. 1-28.** Seal-welded bonnet, rising-stem, outside screw gate valve with bolted gland. (*Courtesy of Henry Vogt Machine Co.*)

they are too long they can become a tripping hazard, as well as mechanically unstable (Fig. 1-31).

Larger valves usually have flanged bonnets. In class 150 (PN 16), the body-bonnet joint is usually flat faced, but with a gasket that is retained inside the bolt area. Higher ratings have a captured or retained gasket design in which the body gasket surface is recessed somewhat while the bonnet seating surface is raised, so that the gasket is entirely below the plane of the body-bonnet split (Fig. 1-32). Gate valves in class 150 (PN 16) have oval or sometimes almost rectangular bonnet flanges, while globe and check bonnet flanges are usually round. Even gate valves in higher classes have round bonnet flanges.

One way to tell the rating of a steel gate valve at a glance is that a class 150 (PN 16) valve has a fairly short face-to-face dimension, with an oval bonnet and oval body, while a class 300 (PN 25) gate has a round bonnet flange but still a somewhat oval or flattened body cross section, and a class 600 (PN 40) or higher valve almost always has both a round bonnet flange and a round or cylindrical body cross section.

All sizes of steel valves can have bolted bonnets, while the seal-welded bonnets are normally found in sizes 4 in. (DN 100) and smaller. For higher pressures, where bolted bonnet joints get very large, the pres-

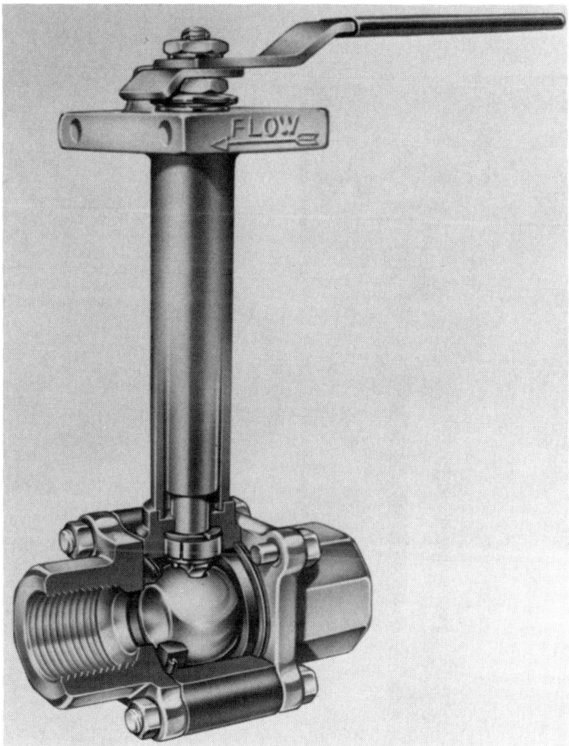


**Fig. 1-29.** Extended bonnet (stem and top-works) for cold insulated lines to allow the packing to be above the insulation and remain at near-ambient temperatures. (*Courtesy of Worcester Controls Corp.*)

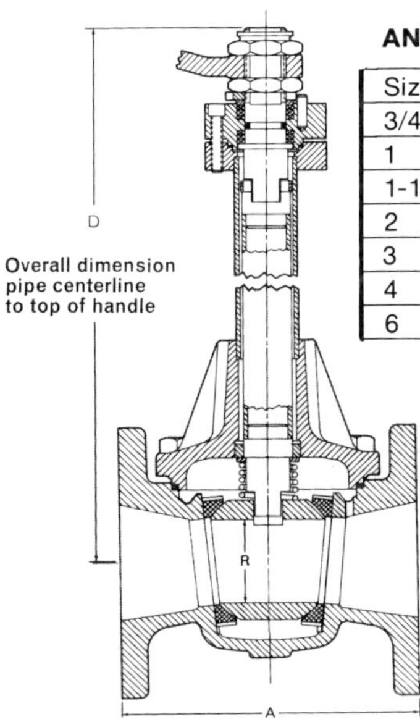


**ANSI CLASS 150**

| Size | A | D | R |
|---|---|---|---|
| 3/4 | 4-5/8 | 11-3/4 | 13/16 |
| 1 | 5 | 11-3/4 | 13/16 |
| 1-1/2 | 6-1/2 | 12-13/16 | 1-3/16 |
| 2 | 7 | 14-3/16 | 1-1/2 |
| 3 | 8 | 17-3/4 | 2-1/4 |
| 4 | 9 | 18-1/16 | 3 |
| 6 | 15-1/2 | 20-15/16 | 4-1/2 |

**ANSI CLASS 300**

| Size | A | D | R |
|---|---|---|---|
| 3/4 | 6 | 11-13/16 | 13/16 |
| 1 | 6-1/2 | 11-13/16 | 13/16 |
| 1-1/2 | 7-1/2 | 12-7/8 | 1-3/16 |
| 2 | 8-1/2 | 14-1/4 | 1-1/2 |
| 3 | 11-1/8 | 17-3/4 | 2-1/4 |
| 4 | 12 | 18-1/16 | 3 |
| 6 | 15-7/8 | 20-15/16 | 4-1/2 |

**Fig. 1-30.** Cross section of extended bonnet on top-entry ball valve. (*Courtesy of McCanna, Inc.*)

**Fig. 1-31.** Extended bonnet gate valves with different stem extension lengths. (*Photograph by R. C. Merrick.*)

sure-seal bonnet is often used (Fig. 1-33). Pressure-seal bonnets are used in class 600 (PN 40) but are more common in class 900 (PN 65) and up. The smallest size normally available is 2½ in. (DN 65). There are a number of variations of the pressure-seal bonnet, including some in which the bonnet has intermittent buttress threads like the breech of a cannon. Occasionally the pressure-seal bonnet also is seal-welded, but this really should not be necessary. Since the pressure seal is used most often in high-temperature, high-pressure steam, the most common materials are carbon steel and low chromes, but some stainless steel pressure-seal valves have been built for high-pressure process work.

The principle of operation of the pressure-seal bonnet is that the sealing surfaces are under compression while pressure forces are push-

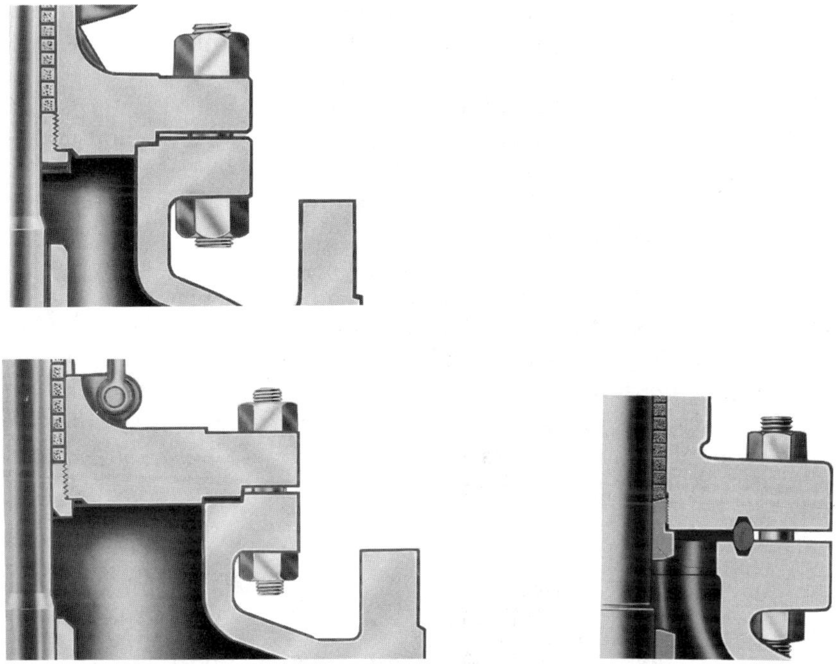

**Fig. 1-32.** Typical configurations of bolted-bonnet gasket designs. (*Courtesy of Pacific Valves, a unit of Mark Controls.*)

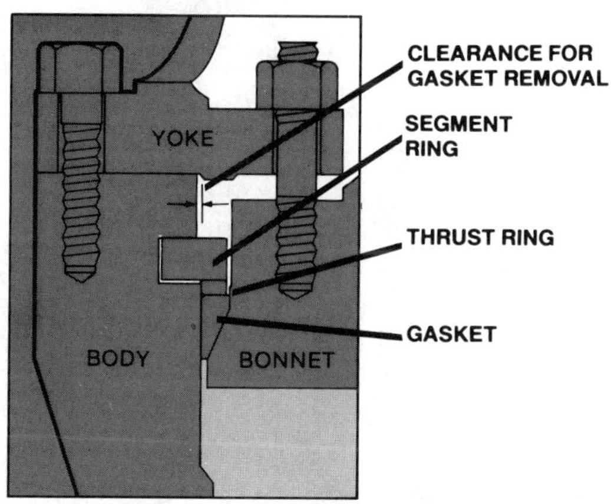


**Fig. 1-33.** Pressure-seal bonnet design. (*Courtesy of Pacific Valves, a unit of Mark Controls.*)

ing up on the bonnet, rather than under tension, as in a bolted bonnet joint. The gasket is solid metal, triangular in cross section, straight sided on the outside and tapered on the inner side, while the bonnet has a similar taper with a slightly greater angle. The bonnet is lowered into the body with the gasket resting on it; then a separate ring, in several sections, is fitted above the gasket. This ring fits up against an overhang at the top of the body to provide a stop, then the bonnet is drawn up tightly or preloaded by means of a few small bolts. This begins to compress the gasket, with the seating taking place at the bottom edge of the gasket first. Any amount of internal pressure that is applied serves only to increase the seating force and the tightness of the gasket. The bonnet can be removed later, after depressuring, just by knocking out the pieces of the segmental ring.

Welded bonnets can be used in valves larger than the socket-weld-screwed size range, but they are unusual. In 3 in.- and 4-in. sizes (DN 75 and 100), there are several designs of forged body valve, often Y-pattern globe, built for high pressure with seal-welded bonnets (Fig. 1-34). There are some types of valves suitable for buried service with modified pressure-seal closures that are then welded up (Fig. 1-35). All-welded or strength-welded closures are rare, but they are available. Some very heavy body styles, in almost any type of valve, even including ball and

**Fig. 1-34.** Seal-welded bonnet (at right) on Y-pattern globe valve. (*Courtesy of Edward Valves, Inc., formerly Rockwell Edward.*)

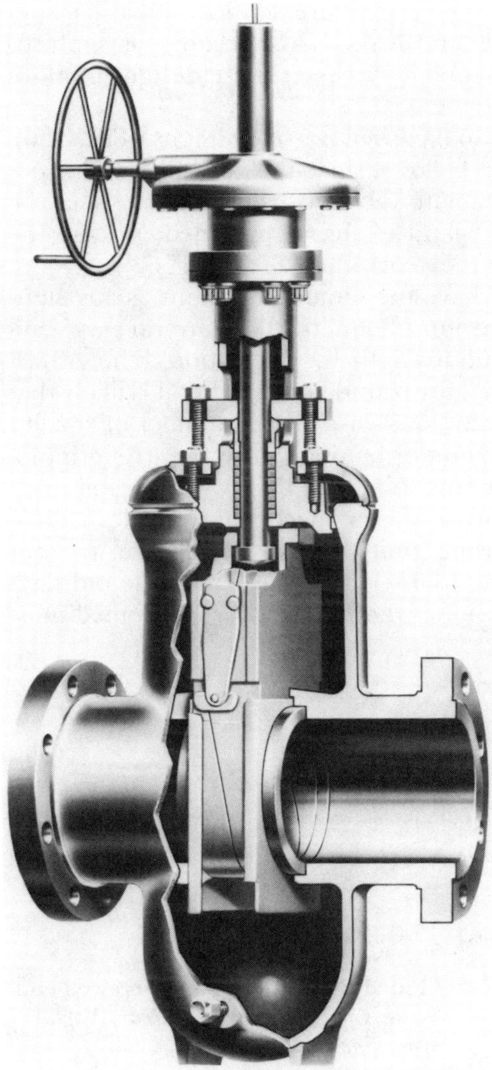

**Fig. 1-35.** Pressure-seal bonnet design through-conduit gate valve. Bonnet can also be seal-welded. (*Courtesy of Flow Control Division, Cooper Industries, Inc.*)

butterfly valves, have been built with welded closures. The difficulty with these designs lies in the very heavy strength welds required, which require a lot of heat input with the consequential body distortion. Any design that does not depend on very close metal seating and that can tolerate the heat generated during welding can be built as a totally welded-up design, but these cases are rare.

Ball and plug valves, since they have rotating and not rising stems, are easier to seal. Some have a single O-ring for the stem seal. Many others have a more complicated packing design, but relatively little adjustment ability. Still others, for very severe services, have packing chambers with adjustable glands and leak-off connections as sophisticated as any packing design. Chapter 2 discusses these designs in more detail.

Some globe and gate valves also have bellows seal designs. (Fig. 1-36). The globe requires a fairly short bellows, since the vertical travel is not that great, but to accommodate a gate valve's travel the bonnet may be quite long. A separate packing gland is always provided, in case the bellows is overstressed, damaged, or corroded.

Bonnet or body joints rarely have the same dimensions as any standard flange. Especially with nonround bonnets, there are no easy ways to demonstrate the flanged joint strength by calculation. The ASME rules for flange design can be used for round bonnets, but it is likely that bonnets could have six bolts or ten bolts, or a larger number of smaller bolts, or any convenient combination of bolting. Since the strength of a bolted bonnet, or any type of bonnet joint, can be proved during pressure testing, that is not normally a matter of concern. Modern valve designs are often developed using finite element analysis, or other sophisticated computer modeling tools (Fig. 1-37), but in the old days valve designers simply added metal to the places where it seemed most

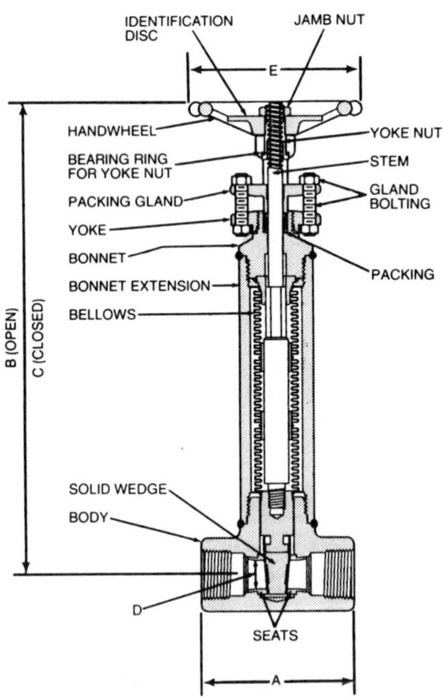


**Fig. 1-36.** Bellows seal screwed-end gate valve. (*Courtesy of Henry Vogt Machine Co.*)

**Fig. 1-37.** Weld overlay on gate valve body, with bolted bonnet and ring joint ends. (*Courtesy of Vetco Gray Inc.*)

needed and tried it out. The complex shapes of a gate valve, for instance, were not easy to model, and the effects of corrosion and allowable stresses were harder to predict than they are now, so many gate valve designs and gate valve standards have very thick walls compared to the way they might be designed now.

Other features that could be applied to almost any valve include body and bonnet taps. The taps are usually threaded, but socket-weld and other configurations are possible.

# 2

# Valve Construction

## Block Valves

### *Gate Valves*

In most industries, the first choice for block valves is the gate valve. In modern times, there have probably always been more gate valves than any other single type of valve, and perhaps more than all other types put together. This will not likely be true in the future, but some form of gate valve will always hold a large share of the market.

There are a number of different types of gate valve, but very few industries have occasion to use more than a few of these types. This is at least in part due to the different requirements for flow control that are customary in different industries.

The gate valve is probably the oldest valve type in existence. It gets its name from its similarity to the gates in city and castle walls, which were less vulnerable to being broken down (that is, they had better shutoff ability) than hinged doors were, because the gate could slide up and down in a set of guides that retained it and kept it from being carried away. The first type of gate valve was what we now know as a slide gate, used in irrigation ditches. For centuries it consisted of a wooden board thrust by hand into the mud at the bottom of the ditch, and it throttled quite well. This form of flow control is still widespread today.

The development of wooden and iron pipe led to a corresponding development in valving, although rather slowly and haphazardly. The first pipes were intended as nothing more than enclosed ditches, with gravity providing the force for movement of liquid. The use of steam for propulsion and pumping provided the first real need for pressure in piping systems and for leak tightness and shutoff ability under pressure.

There are two major types of gate valves. The first type seats by wedging the gate between two inclined seats, and is commonly known as the wedge gate. The second type seats by sliding a gate across parallel seats, and includes virtually all other gate designs. Among these are the parallel-seat gate, the slide gate, the knife gate, the sluice gate, the through-conduit gate, and several other variations. The wedge gate will be discussed first, because of its more universal application. Even though the slide gate came first, the parallel-seat gate styles are of secondary importance now in most industries.

Gate valves have been made in virtually all materials ever used for valves (Fig. 2-1). Almost every size and pressure class ever devised has been well populated with gate valves, although as pressures and sizes get higher the proportion of gate valves generally declines. The vast major-

**Fig. 2-1.** Flanged-end, bolted-bonnet, rising-stem gear-operated stainless steel gate valve. (*Courtesy of Newmans Inc.*)

ity of gate valves are handwheel or manual gear operated, but it is quite common to add power actuators. The usual practice is to avoid throttling with gate valves, since they have very poor control characteristics and do not stand up well to throttling conditions, but half-open gate valves can be found in virtually every industrial facility.

There are certain definite shortcomings of the gate valve. Besides its poor throttling ability, it generally does not have extremely tight shutoff ability. It does not open or close well against high differential pressure, it does not tolerate entrained solids well, it has large entrapped volume spaces, and if the valve comes apart internally it might fail closed without an outward visual indication. In addition, the gate valve falls toward the high end of the weight range, it can take a long time to open or close, it cannot tolerate a wide range of orientation, and it usually occupies a substantial space extending perpendicular to the pipe.

There are, however, reasons for the popularity of the gate valve. It is cheap and widely available. Valve mechanics generally understand gate valves better than they do some other types. All-metal designs are inherently fire-safe, which has a high value in flammable services. Many types can provide the full cross-sectional dimension and/or the full cross-sectional area (two different requirements) of the pipe line. They generally shut off well enough, and many types are designed so that a longer wrench or a bigger operator can shut them off tighter. With some internal modifications, such as bypasses, vents, or guides, almost all of the inherent drawbacks of gate valves can be overcome.

The gate valve that is of most interest to valve and specification engineers and users is the wedge gate (Fig. 2-2). This valve is the dominant type of block valve in many fluid-processing industries, including the hydrocarbon and chemical markets. It is also found in general industrial work, municipal water systems, home plumbing, and many other applications. Wedge gate valves have been made of virtually every useful material, although some materials, notably plastics, lend themselves better to other types of valving. They have been designed for pressure ratings ranging from feet of water to thousands of pounds per square inch, although there are other valve designs that take high pressure (specifically high *differential* pressure) much better. They have been built in every size, although very small valves are more likely to be other types such as needle valves, and very large valves are more likely to be other types such as slide gates or butterflies. Methods of body construction also vary widely depending on the application.

The most common type of wedge gate is, as the name implies, wedge shaped, or wider at the top than the bottom, with seats inclined at the same nominal angle such that when the stem pushes the gate down into the seating area, the gate becomes wedged between the seats. This arrangement allows a great deal of seating force to be applied because of mechanical advantage, makes adjusting for wear fairly easy, and does

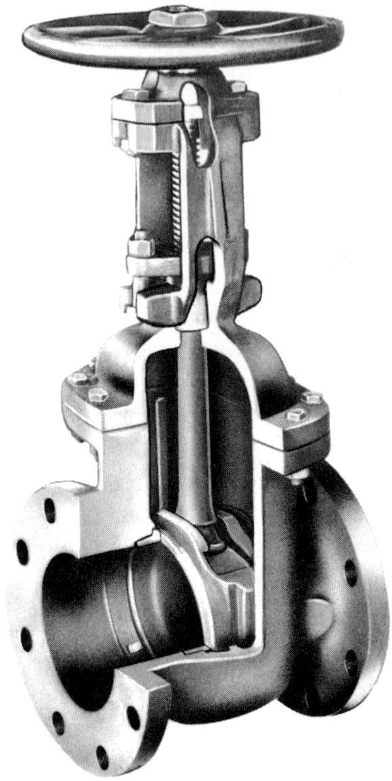

**Fig. 2-2.** Cutaway of flanged-end, bolted-bonnet, rising stem, outside screw and yoke gate valve with threaded-in seat rings. (*Courtesy of Crane Co.*)

not take up a great deal of space. The wedge shape makes the gate come loose easier when thermal expansion tries to "lock" the gate in place, and only a slight vertical movement of the gate will separate the seating surfaces from each other. In many common gate valve designs, the wedge angle is 5 degrees from vertical (the included angle of the wedge is 10 degrees).

Some types of wedge gate valves do not look like wedge gates—that is, they appear to be parallel-seat gates, but their seating ability does derive from wedging action. Examples of these are the municipal water service valves that have two interlocking gate halves with a stem attachment device in between, with parallel seats (see Fig. 1-3). Between the halves of the gate is a very small wedge that bottoms out when the gate touches the inside bottom of the valve and thus forces the two halves of the gate outward and into the seats as the stem continues to push down until it reaches the bottom of its travel.

The wedge itself comes in many forms. The simplest is a solid casting (Fig. 2-3a) or, as is more usual in small steel valves, a forging with flat wedge faces, flat sides, and a stem attachment at the top. The method of attaching wedges to stems must permit the wedge to float slightly with respect to the stem, so the normal configuration is a stem with a T-shaped head that fits into a slot in the top of the wedge (Fig. 2-4). Larger

(a) (b) (c)

**Fig. 2-3.** Solid wedge, flexible wedge, and two-piece split wedge. (*Courtesy of Pacific Valves, a unit of Mark Controls.*)

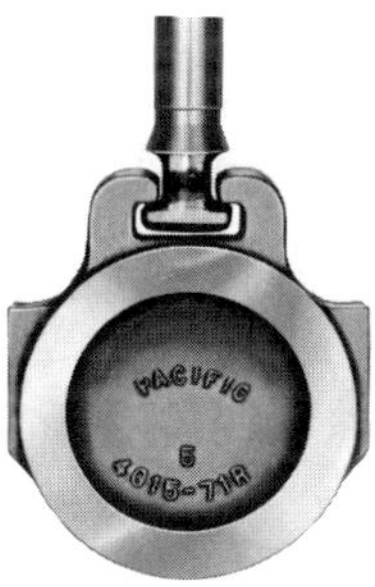


**Fig. 2-4.** Typical T-head stem to wedge attachment. (*Courtesy of Pacific Valves, a unit of Mark Controls.*)

wedges are usually cast, with seating surfaces raised above the surface of the wedge and with guide rails on the sides to engage wedge guides cast into the sides of the body cavity.

The solid wedge has no flexibility. When a hot valve is closed, the wedge may be slightly cooler, because it is up in the bonnet out of the flow stream when the valve is fully open. Since when the valve is closed there is no longer a flow of hot fluid through it, the downstream side starts to cool off. The body will start to contract, and if the valve was closed tightly the wedge can get very stuck. This means that when someone attempts to open the valve, it may not open at all without the operator exerting a great deal of force on it, and the high load imposed on the seat and wedge interface can result in galling. This is very damaging, since the rough surfaces created by galling will then scratch and gouge other areas of the seating surface, leading to fluid leakage across the seats, which causes seat erosion and finally gross leakage across the valve.

One way to combat this problem is by the use of the flexible wedge (Fig. 2-3b). The flexible wedge is one solid piece, but there is a groove or gap in the edges all the way around the circumference. The flexible wedge is often made by casting in special patterns, but it can also be made by machining a groove about an eighth or a quarter of an inch (3 to 6 mm) wide in a solid wedge. Approximately the center third of the wedge is solid metal from one side to the other.

This allows considerable elastic deformation of the wedge as it is seated, which not only permits it to "give" a little as the seats move under thermal or pipeline stresses, but also allows the wedge to conform slightly to seats that are out of plane. An additional benefit of the flexible wedge is that it has a slight springing action as the wedge is pushed into the seats, and thus contacts the seats slightly more tightly. In certain cases, this can mean the difference between leaking and not leaking.

The thickness of the wedge increases as the pressure rating of the valve increases, since it is subjected to bending forces caused by the differential pressure of fluid acting on one side. In valves designed for 3000 to 6000 psi (200 to 400 bar), which can be built as fairly standard wedge gate valves, the thickness of the wedge equals or exceeds half its diameter. Wedges this thick are very rarely built as flexible types, because the thickness means they would not be very flexible anyway. In terms of ANSI pressure classes, the class 600 (PN 40) valve (1480 psi [104.2 bar] maximum design pressure) is generally the highest pressure class in which flexible wedges are commonly used.

High-pressure valves inherently have to withstand a high differential pressure. The forces associated with opening and closing a gate valve against high differential pressure mean that the stem must be larger in diameter and the operator torque must be higher. Higher operator torque means either larger handwheel diameters or greater gear ratios in manual gear operators, and occasionally manual gear operators may need double reduction gears. Higher loads are imposed on the stem threads, so the stem thread areas of high-pressure valves usually need to be surface hardened to withstand the higher unit stresses.

One technique that is occasionally used to defeat the high differential forces that occur when a valve is opened (which is generally a more severe problem than when the same valve is closed) is to supply a bypass around the valve. This can be installed as part of the piping from above one end of the valve to below the opposite end, but it is also commonly supplied as part of the valve. This is most commonly done in utility steam services, to solve two problems at once. One is the differential force caused by steam pressure on one side and no pressure on the other, and the other problem is the fact that the piping downstream of the valve may be hundreds of degrees cooler than that upstream. Hitting this piping with hot high-pressure steam is a tremendous shock, so the bypass around the gate valve is opened first. The bypass is generally a

¾-in. to 1½-in. (DN 18 to DN 40) line with a gate or globe valve in it. Flowing this small quantity of steam into the downstream piping starts to warm it and to relieve the differential pressure. The main valve is then opened after the pressure equalizes.

There is a standard, MSS-SP-45, for both the size of bypasses and the location of bypasses and various other taps and drains that are commonly installed in gate valves (Fig. 2-5). This standard applies to any common type of flanged valves, not just gate valves, but the great majority of bypasses and a sizable percentage of drain taps are applied to gate valves, regardless of material.

By the nature of their on-off service, aerodynamic or fluid dynamic forces are rarely a significant consideration in gate valve design. There are a couple of instances to consider, though. One is where the gate is partly open. The gate will "flutter" in the flow stream and bang around on the seats, especially if the flow rate is high. Excessive flow rates can even flutter a gate that is in the fully open position, especially when the gate is in the stem-vertical position. A gate valve with a cylinder actuator, where the piston force is pushing or pulling directly on the stem, can be subject to uneven loading when the valve is opening. When the valve first starts to open, if there is significant differential pressure or thermal differential, the gate will tend to stick and then "pop" off the seats. This takes up the slack in the wedge-to-stem connection and then in the stem-to-actuator shaft connection, and if there is high pressure in the valve, the differential force of the line pressure will tend to shove the stem upward very fast. This imposes very severe loads on whatever device is being used to stop the stem travel, as well as high column loads on the stem. Mounting brackets and stops on cylinder-actuated high-pressure gate valves need to be designed for severe impact loading.

The orientation of gate valves is important. The standard gate valve (and most other valves, as well) is designed to be installed with the stem vertical and above the pipe. This is the most desirable installation since the gate is hanging freely from the stem when not seated. If and when the stem packing leaks, when it is located above the pipe there is less liquid head acting on the leak and the packing gland is generally in the best position for maintenance. The second choice for gate valve orientation is in vertical lines with stem horizontal. In this orientation the gate slides across the lower seat while it is being opened or closed. In horizontal lines, if the valve cannot be positioned with stem vertical, an angled stem is adequate, but the closer to vertical it is the better. A horizontal line with the valve stem horizontal is adequate in small valves, but the relatively heavy wedges found in large or high-pressure valves require support in the form of special guide rails or even rollers under the wedge. The worst possible orientation is with stem downward, since the force of gravity is opposing the closing force and aiding the path of packing leaks while making maintenance more awkward. A well-de-

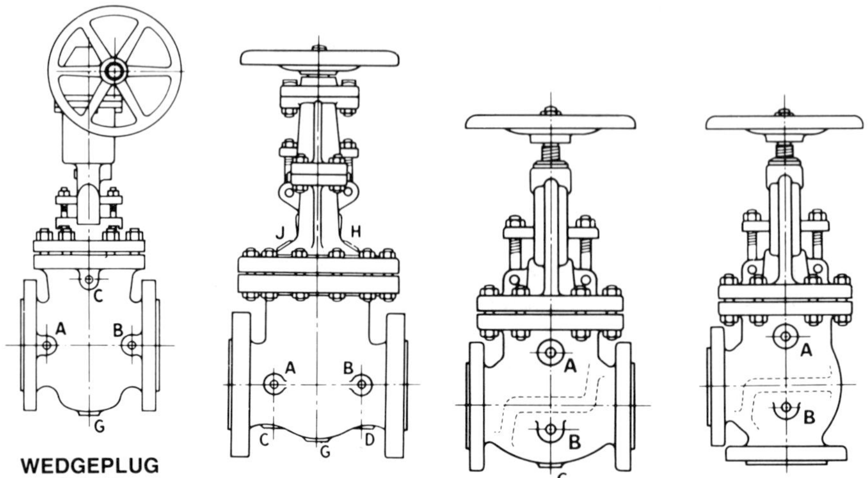


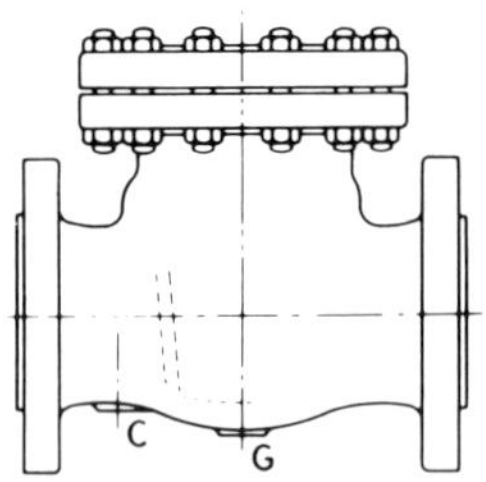


**MAXIMUM SIZE OF TAPPED HOLE (INCHES)**

| Size of Valve in Inches | Gates | Globes and Angles | Swing Checks | Wedge-plug |
|---|---|---|---|---|
| 1 | | | | ½ |
| 1½ | | | | ½ |
| 2 | ½ | ½ | ½ | ½ |
| 2½ | ½ | ½ | ½ | ½ |
| 3 | ½ | ½ | ½ | ½ |
| 4 | ½ | ½ | ½ | ½ |
| 6 | ¾ | ¾ | ¾ | ¾ |
| 8 | ¾ | ¾ | ¾ | ¾ |
| 10 | 1 | | 1 | 1 |
| 12 | 1 | | 1 | 1 |
| 14 | 1 | | | 1 |
| 16 | 1 | | | 1 |
| 18 | 1 | | | 1 |
| 20 | 1 | | | 1 |
| 24 | 1 | | | |
| 30 | 1 | | | |

**Fig. 2-5.** Standard locations and designations of bosses for taps and drains. (*Courtesy of Stockham Valves and Fittings.*)

signed packing chamber is deep enough that pressure from inside the valve is dissipated somewhere in the middle of its depth. When a valve is upside down, though, any fluid that is forced up that far through the packing is still under the influence of gravity and will tend to come out to the end of the packing gland.

The wedge gate has been built with virtually all combinations of end connections, bonnet and packing gland types, materials, and pressure ratings. Generally, as with other types of valves, bronze bodies are available only with threaded ends or solder joint ends. Cast-iron body gate valves can have threaded, flanged, mechanical joint, or other special end connections to mate with iron pipe. Steel and alloy valves can have threaded, socket-weld, flanged, butt-weld, grooved, high-pressure clamp, or special ends. Plastic body gate valves are virtually nonexistent. Bronze and iron gate valves commonly have threaded or union bonnets, occasionally bolted bonnets, with threaded or bolted packing followers. Steel and alloy valves occasionally have threaded bonnets and threaded gland followers but generally have bolted bonnets and bolted packing gland flanges. High-pressure valves often have special bonnet designs that make use of wedge-shaped pressure-seal gaskets, or welded bonnets. Pressure-seal valves are almost always available with butt-weld ends only, and are normally used for steam service.

Both low-pressure brass and bronze valves and corrosion-resistant stainless steel and nickel-alloy valves are often built with integral seats. Here, the seat area is cast integral with the rest of the body, and the seating surface is merely faced off and lapped. This is adequate where either no corrosion is expected or the entire body is resistant to corrosion. In iron, steel, and chrome-steel valves, the seat area is made of a more corrosion- and wear-resistant material than is the rest of the body. Examples are iron bodies with bronze trim and steel bodies with 13 percent chrome steel or stainless steel trim. For these valves, the area surrounding the seat is machined back and large-diameter threads are cut in; then the seat (Fig. 2-6), which is made of machined bar or ring forging, is threaded into the body. There are notches or lugs in the seat to permit this. The threads also allow the seats to be removed and refinished during maintenance. With the current economics of labor and material, though, not as many valves are being repaired as before. In addition, the thread area is a source of corrosive attack and a leak path. Many steel gate valves now are built with seat rings unthreaded, simply set into place and fillet-welded at the back edge (waterway edge).

**Fig. 2-6.** Threaded seat ring for gate valve. (*Courtesy of Pacific Valves, a unit of Mark Controls.*)

When corrosive attack is severe, sometimes the edge inside the body cavity is seal-welded also.

In steel valves, the seat material normally used is a 13 percent chrome steel, known as 410 stainless, or F6 (the forged grade) or CA6NM (the cast grade). For improved wear resistance, the seats are often overlaid with a weld deposit of hardfacing material. This is especially important if the valve is designed for service at temperatures of 400°F (200°C) or more. The most common hardfacing alloy is a cobalt-chrome-vanadium alloy. Other common seat materials have included nickel-chrome alloys or 300 series stainless steels. A differential hardness is desirable to avoid galling, and the usual, but not universal, custom is to make the seat the harder side since the wedge is more easily removable for remachining. Differential hardnesses are a requirement of some gate valve standards, and they can be created by material selection, hardfacing, nitriding or other surface treatment, or heat treatment.

On the wedge side, small wedges are generally cast solid. They may be a block shape or cast with a raised area for the seat. In steel and alloy valves, forged wedges are common. Larger wedges are generally cast, almost always with the center area depressed somewhat from the seating plane. This reduces weight while making it easier to machine and lap the seating surfaces. In larger wedges it is common to use cast steel to match the body material, and to use a welded overlay for the seating surfaces in the form of the 13 percent chrome steel or other chosen seat material.

The stem is machined from barstock, and in steel and alloy valves the bottom design for the stem attachment, known as the T-head, is generally forged. Iron body valves have bronze stems, although it is possible to get an all-iron valve (without any copper-bearing components) with a steel stem. There are two different stem designs: the rising stem that is used in the majority of valves and virtually all steel and alloy valves, and the nonrising stem, still in use today only in utility service valves. Whereas the rising stem is pulled upward by the rotation of the handwheel around it, while the stem itself does not rotate, the nonrising stem rotates by means of an attached handwheel and the valve disk travels up and down when the stem rotates (Fig. 2-7). The nonrising stem has a thrust collar to keep it from moving vertically. The position of the rising stem is easily deduced from the amount of stem visible above the handwheel, while no such visible indication is possible with the nonrising stem. Some nonrising-stem valves can be fitted with position indicators (Fig. 2-8).

When valves are installed underground or behind walls, a different type of position indicator (Fig. 2-9) can be installed. The indicator and the handwheel are often the only visible parts of the valve.

In outside diameter the stem threads are the same as or slightly smaller than the portion of the stem that goes through the packing chamber. Some stem threads are cut, and some are rolled. The normal

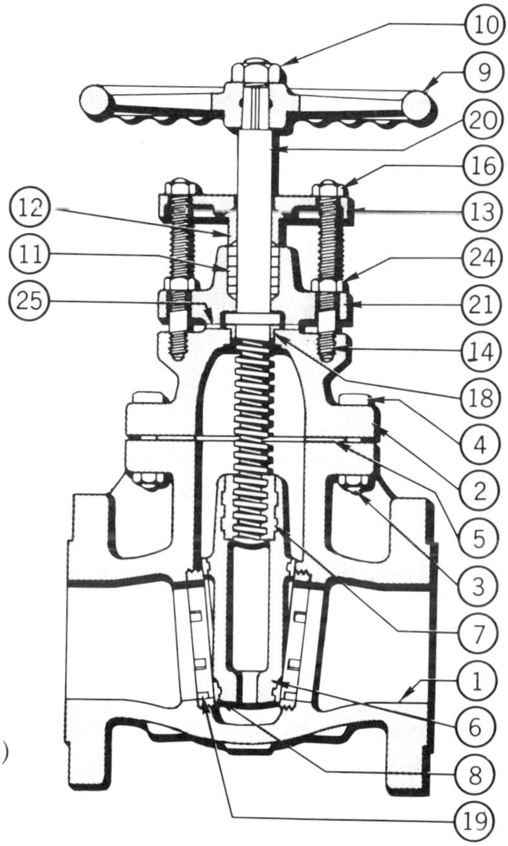


(a)

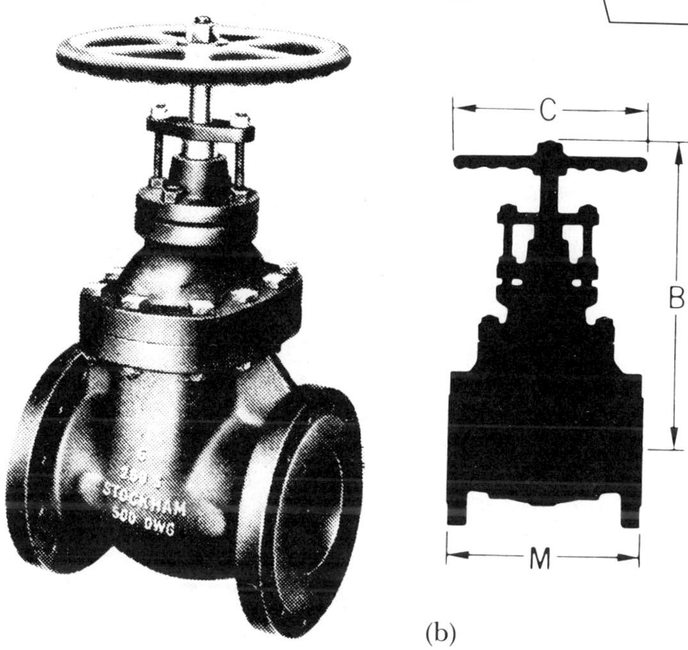


(b)

**Fig. 2-7.** Nonrising stem iron-body gate valve. (*Courtesy of Stockham Valves and Fittings.*)

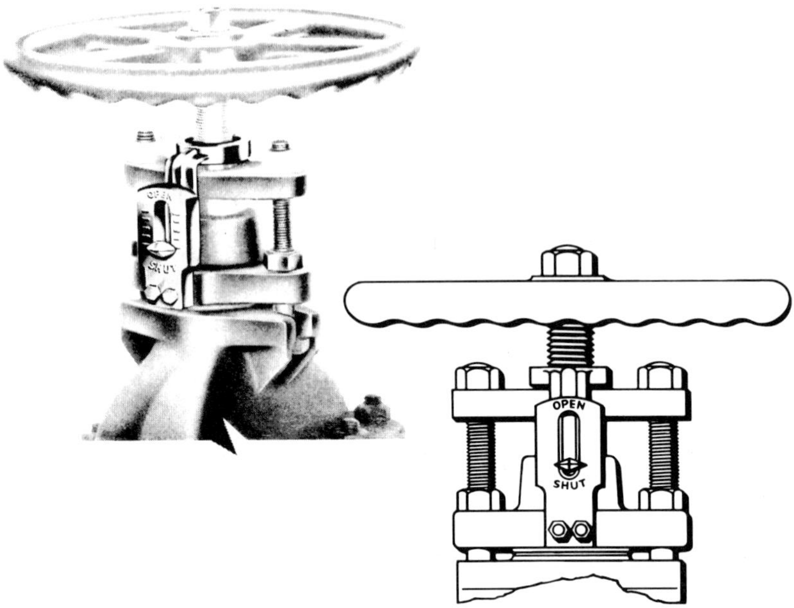


**Fig. 2-8.** Indicator fitted to nonrising-stem gate valve to show the valve position. (*Courtesy of Stockham Valves and Fittings.*)

thread pattern is the Acme thread, usually double threaded. The portion of the stem that contacts the packing is polished to avoid grabbing or cutting the packing as the stem moves up and down. Formerly, a surface finish of 32 RMS (1.2 $\mu$m) or so was considered adequate, but graphite ribbon packings require about 16 RMS (0.6 $\mu$m) finish for successful sealing. Stems should be designed such that, if they fail in tension, they will fail at a point just at the bottom of the thread area and not in the disc-to-stem attachment area. Such a failure could easily escape notice and would almost always allow the disc to fall closed, irretrievably. This design feature is a requirement of some gate valve specifications.

The threads in the stem convert rotary motion from the handwheel into sliding motion of the stem (rising-stem valves). The tool that does the converting is the stem nut or yoke nut. This nut has internal threads to match the stem, with either a lubricating point or a set of ball bearings, depending on the size and sophistication of the valve. The outside of the handwheel nut interfaces with the handwheel or other actuating device. Most handwheels are attached directly to the operating nut, while most special actuators contact the nut by means of a pair of lugs that engage a pair of lugs on the yoke nut, which allows for a hammer blow action.

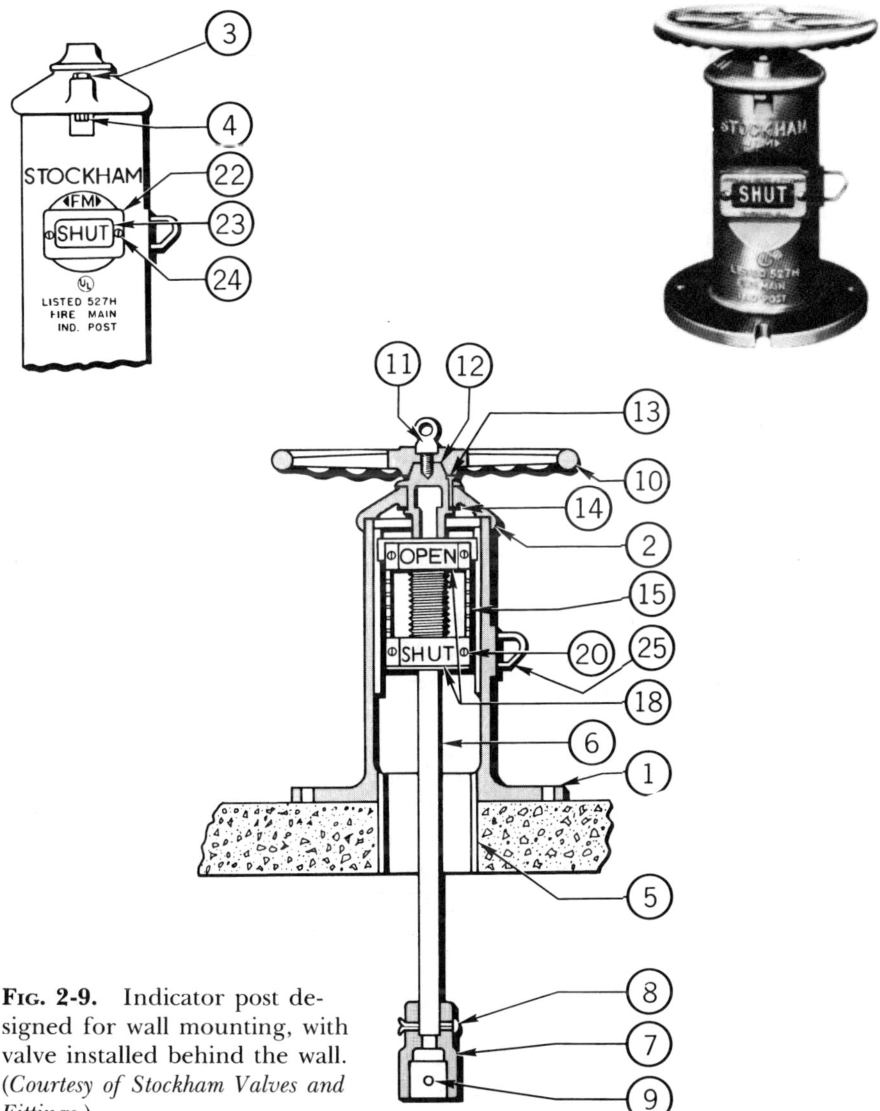


**Fig. 2-9.** Indicator post designed for wall mounting, with valve installed behind the wall. (*Courtesy of Stockham Valves and Fittings.*)

Many standards require the yoke nut to be significantly less hard than the stem, for wear resistance, but not a soft alloy that will melt quickly under exposure to fire conditions.

The packing chamber or gland itself is simply an opening, larger than the stem diameter, that allows the insertion of packing to seal the stem-body joint while allowing stem movement. Common packings in the old days, in simpler services, included woven fibers, soft metal foils, and even rags. For many years the standard packing in most hydrocarbon and steam services was a square braided packing made of asbestos fiber, often lubricated with powdered graphite and interwoven with a corrosion-resistant wire for support, sometimes also dusted with a sacrificial metal powder. Although not much asbestos woven into valve packings escapes into the air, its manufacture certainly is hazardous and today no asbestos packing is manufactured in the United States. Much of it is available in the overseas market and is still imported, including that installed in some valves made overseas. Today the packings of choice in many services are either a braided aramid or other synthetic fiber, still with powdered graphite impregnation and wire reinforcement, or a graphite foil and/or braided pure graphite packing. In more corrosive or contaminant-sensitive services, a fluoropolymer packing is preferred, either braided into packing or machined into solid V-shaped rings that are inserted into the packing gland in such a way that the escaping fluid makes them tend to seal tighter.

The packing is retained by the gland follower, which sits on top of the packing and is just slightly smaller in diameter than the inside of the packing box, and the gland follower flange, which has two bolts, one on each side of the stem, to hold it down and to tighten the packing when necessary. By convention, the gland follower is parallel to the line orientation, and the yoke, which comes up on either side of the packing area to support the top of the stem and the operating mechanism, is perpendicular to the line orientation.

Under the packing gland, at the point where the stem emerges into the interior of the valve, is a bushing and backseat that both guides the stem and provides a conical metal-to-metal seating surface when the stem is fully raised. On the stem, this surface is the top of the T-head area. This is supposed to allow packing to be removed and replaced while the valve is under pressure. The seating ability of the backseat is one of the items normally required to be tested during valve inspection, but in the real world this is definitely recommended only as a last resort in emergencies. Often, the backseat bushing forms the bottom of the packing chamber. This makes for easier machining. The material of the backseat is generally the same as the rest of the trim.

A number of specialty gate valve variations exist. One is the double wedge (Fig. 2-3c), commonly used in stainless steel valves in place of the flexible wedge. Here, the disc is in two halves, which interlock and hold between them a carrier that is rigidly attached to the stem. This design is much more flexible, more suited to the lighter wall construction used in many stainless steel valves, and suited to the extreme thermal differentials found in low-temperature services, down to around −325°F (−200°C) or below. Despite the flexibility of the double wedge, it is still advisable to drill a small hole through one of the wedge halves, to prevent cryogenic liquids from boiling off and causing a severe overpressure of the body cavity because of the lack of a means to leak off.

The hole must be in the upstream side, or toward the side with higher pressure, since the downstream side uses the differential pressure across the disc to help provide the seating force.

Also used in low-temperature services, and in valves intended for tighter shutoff, is the soft seated gate. Many soft-seated gates modified from standard steel gate valves are unsuccessful, because the resilient seating materials tend to get quickly worn down or pinched as the gate closes. Many of these are simply O-rings in a dovetailed groove machined into either the disc face or the seat. In Europe these designs are popular, but they are probably more successful there because the maintenance climate is somewhat different from ordinary American practice. A more successful resilient-seated valve (Fig. 2-10) uses a pressure-energized seat installed in the wedge, much like some butterfly and ball seat designs in concept, in which the line pressure gets behind the

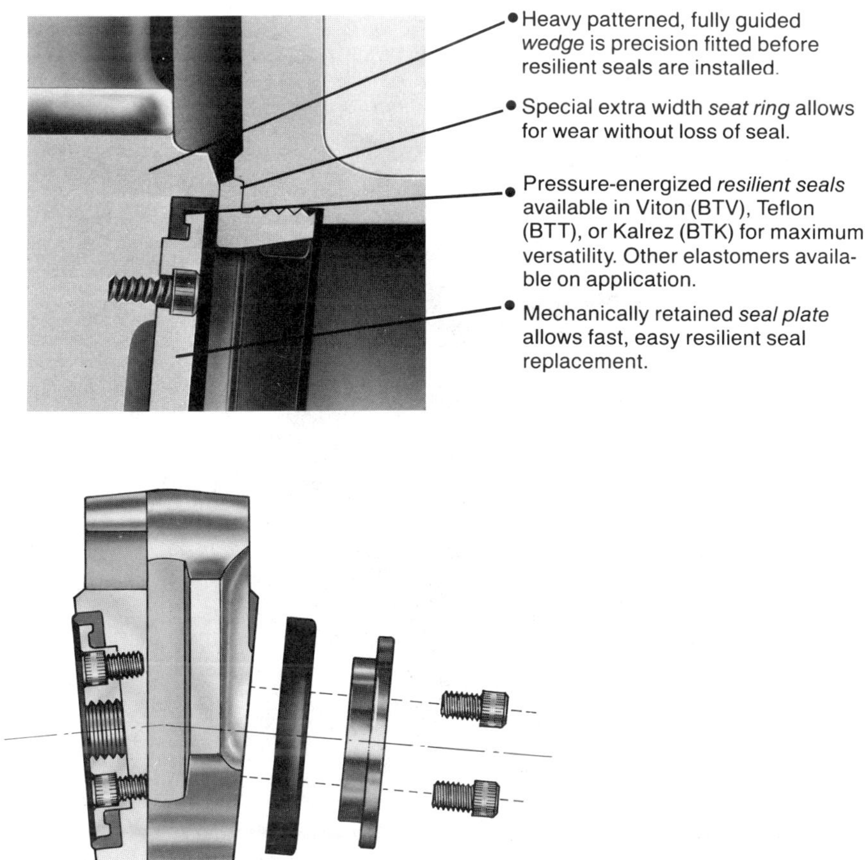


**Fig. 2-10.** Pressure-energized soft-seated wedge design. (*Courtesy of Pacific Valves, a unit of Mark Controls.*)

resilient seat and pushes it into the hard seat with enough unit pressure to seal tightly.

There are two other types of resilient-seated gate valve that have appeared on the market in recent years. One is for the plumbing market, where there is a soft-seated valve that acts like a cross between a gate valve and a globe valve. It has a rubber plunger-type gate that seals across the inside surface of the waterway. A second type, also based on European designs, is for better shutoff in wastewater and municipal water services (Fig. 2-11). An iron body gate valve is designed with a resilient seat that seals against the bottom of the waterway and along the top, with a slightly wider internal cross section. Municipal water service

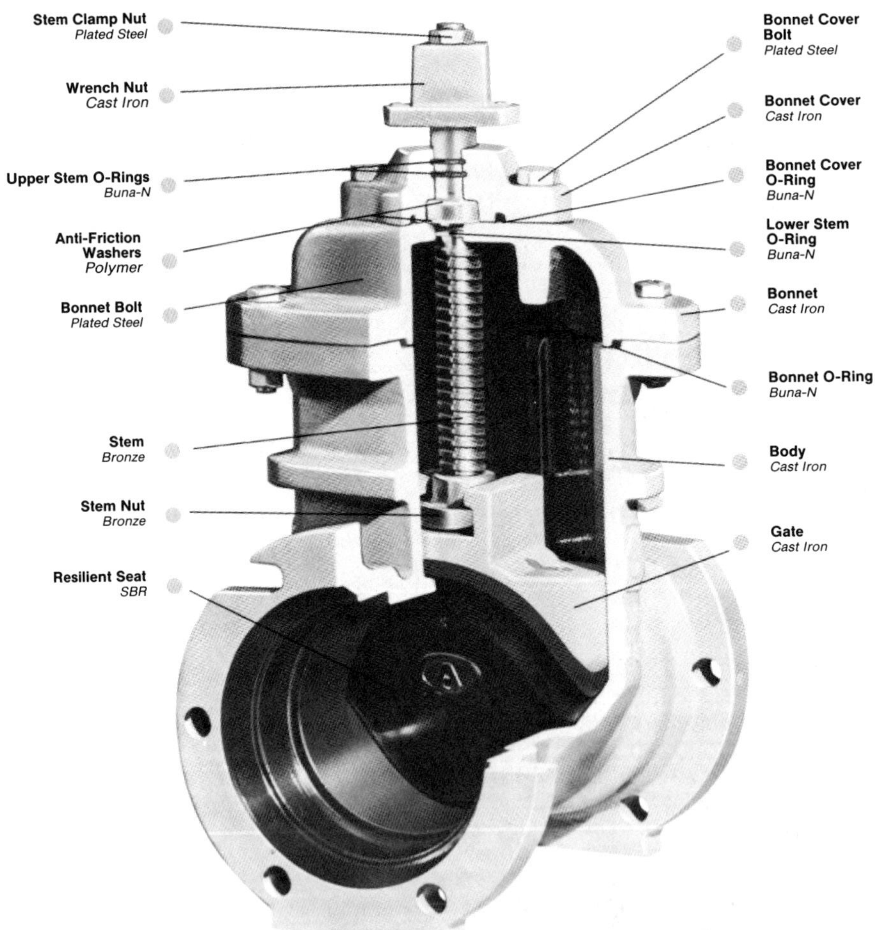


**Fig. 2-11.** Resilient-seated gate valve for wastewater or municipal water service. (*Courtesy of American-Darling Valve, division of American Cast Iron Pipe Company.*)

valves are also sometimes built with bolted openings in the side (Fig. 2-12a,b) for cleaning out the sediments that accumulate under the gate, in the seating area. Some small gate valves have been built with a small hole completely through the disc, to act as a restriction orifice when the valve is closed and to permit full flow by opening the valve.

For toxic or other services where atmospheric leakage is intolerable, there are several steps that can be taken. The first is to use a welded or seal-welded bonnet, which eliminates any possibility of gasket leaks. Eliminating packing leaks is more difficult. The use of graphite packing helps, as does the use of a lantern ring in the packing with a sealant injection plug in the side of the bonnet opposite the lantern. The best way to eliminate bonnet leaks is with a bellows seal design (see Fig. 1-36).

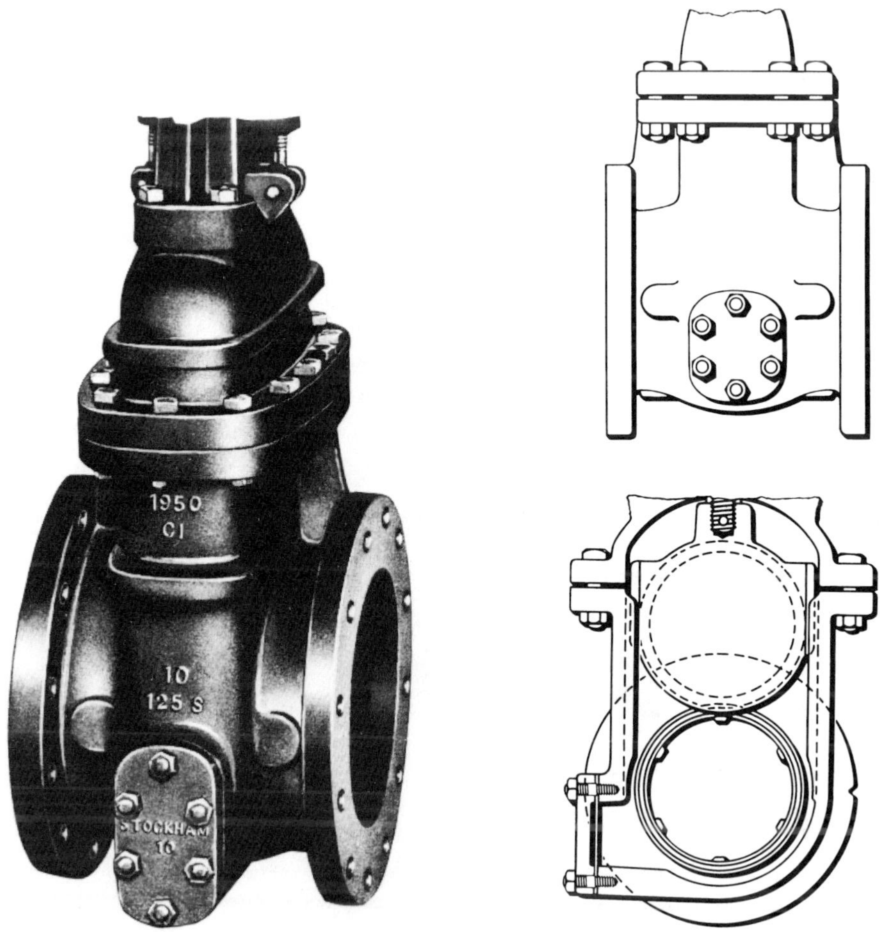


(a) (b)

**Fig. 2-12.** Exterior view and cross section of cleanout. (*Courtesy of Stockham Valves and Fittings.*)

In a bellows seal, there is an expandable bellows that surrounds the stem. The bellows is welded to the stem at the bottom, just above the T-head. At the top, the bellows is welded to the underside of the bonnet. The bonnet then is seal-welded to the top of the body, theoretically eliminating all leak paths. Since the packing chamber still is present, the valve is normally still furnished with packing to help stabilize the stem and to act as a backup in case the bellows springs a leak. Most bellows seal gate valves have quite a long bonnet extension, to accommodate the rather large bellows expansion required for gate valves.

The other major class of gate valves is the parallel-seat type. These range from the very simple to some very special custom designs. The simplest type of commercially available gate valve is the slide gate. Many slide gates, known as sluice gates, are used to this day in the same service for which the first "gate valves" were used—namely, irrigation and drainage water (Fig. 2-13). Such applications generally are designed for differential pressures equal only to the amount of head resulting from water depth—in open channels no additional pressure is possible. In larger valves the net differential force acting on the gate can be quite large. Most gate valves for open-channel installation are designed with the gate being mounted in a frame that is bolted to a concrete wall or

**Fig. 2-13.** Sluice gate designed for attachment to concrete waterway. (*Courtesy of Waterman Industries, Inc.*)

section of concrete waterway, which actually takes the thrust imposed on the gate by the differential pressure of water. Most larger valves have heavy ribbing on the nonpressure side of the gate to resist the bending forces.

Sluice gate valves designed for open-channel mounting are normally rectangular or square, but some have a circular opening, designed to be bolted into corrugated iron culvert pipe. Even these designs are not capable of withstanding substantial pressure. Of course, very rarely would there be any requirement for these valves to withstand a great deal of pressure since they are normally selected for systems, such as cross-country water conveying, where the only pressure available is that of the water head itself. The gate material is either welded low-carbon plate steel or an iron casting, often galvanized. The frames of both the round and the rectangular designs are usually either cast gray iron or fabricated from structural shapes. The stems are usually steel rod, and most have manual handwheels or ratchet levers for opening. The ratchet lever types close by gravity. Some sophisticated designs have rollers or bronze slide bearings on the gate for ease of opening.

The largest valves in the world, apart from the types that are hinged like a door such as canal locks, are undoubtedly slide gates, which are used in dams or water diversion projects, and usually are made of welded steel. Some slide gates that operate on very similar principles to those just described are used for conveying or solids handling, where typically a powdered or grainy solid is moved inside pipe by means of high-velocity air or gas. In these cases, the flowing fluid is a vapor-solid mixture at (normally) very low pressures, typically a few ounces of water pressure. However, these streams must be conveyed in fully pressure-tight systems. The valves are often stainless steel or aluminum, sometimes for corrosion resistance but more often for preventing contamination of the product. Some simple slide gates are operated by sliding the gate open and closed by hand, but normally these valves have traditional operators. The handwheel is often used, but valves in conveying systems are more likely to have power actuators for remote operation. Generally these valves have round openings for installation in round pipe, but rectangular ducting is occasionally used. Internal pressure ratings and differential pressure capability are often low, although many of these valves are designed for the sake of convenience with standard class 125 (PN 10) or class 150 (PN 16) flanged ends for use with standard pipe.

A further variation on this type is the knife gate valve, which is simply a slide gate with a knife edge at the bottom of the gate (Fig. 2-14). The knife gate has the ability to shut off through a layer of solids or pulp at the bottom of the valve, or to wedge its way through hard or abrasive grainy substances such as sand or catalysts. Often the knife edge is beveled on the upstream side and flat on the downstream side, which bears against the seat. With the addition of guides or wedges positioned on the upstream side at the seating location, the knife gate can be made

**FIG. 2-14.** TAPPI standard knife gate valve with gear operator. (*Courtesy of Valen Valve Corporation.*)

to wedge tightly into the seat to provide tighter shutoff. Knife gate valves are almost always unidirectional because of this feature.

A variation on this design uses V-shaped inserts in the body next to the seat and an inverted V in the gate, such that as the valve starts to close the shape of the flow opening changes to diamond shaped. As the valve approaches fully closed, the diamond-shaped flow opening gets smaller but does not significantly change in shape or in proportion of length to width, which is unlike a conventional gate with rounded bottom in which the flow opening becomes more and more crescent shaped as the gate approaches the seat. This difference enables a valve with V-shaped seats to effectively throttle, and perhaps to more effectively cut through the solids in the way of the seating surface, much more than a standard knife gate is capable of doing. In control valve terminology, this would be known as a characterized seat.

The knife gate valve is available in several combinations of material. These valves are commonly made with stainless body and gate and fluoroplastic or elastomeric seat, and less commonly with aluminum body and stainless or aluminum gate and elastomeric seat. Also, many are made with carbon steel or low-chrome body and stainless gate with metal seat and hardfacing on the knife edge or over the entire gate surface, for hot catalyst or other high-temperature service. Many catalysts used in hydrocarbon processing are hard, often alumina based, and handling oil-catalyst slurries or pneumatically conveyed hard powdery solids is a very severe valve service. Many of these valves are actuated, both because of the power sometimes required to close a valve of this type and because of automation requirements, since catalyst handling still is often a batch operation.

Because the knife gate and slide gate valves are usually used in transfer of solids, which tend to fall out of air-solids flow streams extremely readily, most conveying piping tends to be gravity flow and as nearly vertical downflow as possible. It is also true that most of these valves function better when the gate is horizontal, such that the force of gravity is not so much trying to move solids into the path of the gate closing against the seat. When slide gates are used in horizonal lines, or when the solids are sticky, as in oily catalyst services, it is common to provide a purge line or lines into the area right below the seat so that air, inert gas, or steam can be fed into the seating area to help push solids out of the way. As the gate approaches the seat, the area of open space between the seat and the gate decreases, such that the velocity of the purge increases past the gate, which helps to sweep debris out of the gate's path.

All of the valves described above, although quite different in appearance, seat in a similar manner. A slightly different type of seating mechanism is the through-conduit gate, also known as the slab gate (Fig. 2-15). This valve is most common in oil production and cross-country pipe-line applications, but it is occasionally used in refining and other industries. Its gate consists of a flat plate with a round opening equal to the pipe size cut in it. The opening is normally located near the bottom of the gate. When the valve is in the open position, the opening is concentric with the waterway, which provides an almost smooth continuous passage through the valve. This arrangement gives a very small pressure drop, not much more than that of an equivalent length of straight pipe, but its primary value lies in the fact that this valve will easily pass scrapers or pigs, as well as sand and other entrained matter from oil wells, because of its lack of internal obstructions. Because of its positive tight shutoff ability, it is also used in places where high degrees of separation are required, such as in tank farms.

The seating mechanism of the through-conduit valve is the opposite of that of most other gate valve types, in that the seats move in relation to the gate. The seats are usually spring-loaded, occasionally with an active or pressure-assisted seat mechanism, and the smooth gate surface permits the use of elastomer or fluoroplastic seats (Fig. 2-16). Even so, most through-conduit valves are also designed so that the gate is somewhat free to move downstream under pressure, which means that it tends to seat somewhat better on the downstream side than the upstream side.

This feature of the through-conduit valve is shared with many ball valves, as is the sales pitch that if the downstream seat fails or becomes damaged, "you can always take the valve out, turn it around and put it back in, and have a whole new seating surface." In theory, this is true,

**Fig. 2-15.** Through-conduit valve with expanding gate for tight shut-off. (*Courtesy of Flow Control Division, Cooper Industries, Inc.*.)

but the number of instances of that practice actually occurring is presumed to be rather low.

Some more sophisticated through-conduit valves have a two-piece gate that is held together by brackets at the sides of the gate. As the gate is being lifted or lowered, one side of the gate is supported by the other side, and the first side drops slightly so that the seating faces back off just slightly from the guides. As the sealing position is reached, the suspended side stops and the other side continues slightly, which wedges the two halves outward and seats the gate firmly, helping to provide a

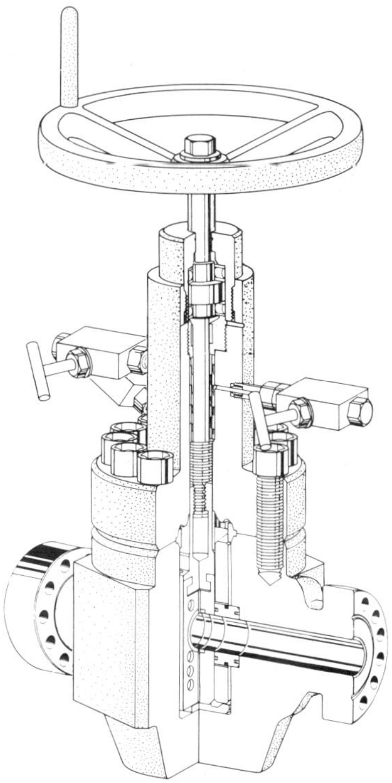

**Fig. 2-16.** Through-conduit valve with nonrising stem and packing purge connections for high-pressure oilfield service. (*Courtesy of Vetco Gray, Inc.*)

tight seal. Because the valve is designed to be equipped with soft seals, it is also necessary to qualify it as fire-safe for many hydrocarbon services. The fire-safe designs have a metal backup sealing system that the gate can seat against when the soft seats are burned or melted away.

The through-conduit gate valve is distinctive in appearance, having a very long top-to-bottom dimension with the end connections almost centered vertically in the body. This is necessary because the total length of the gate, including the area surrounding the opening, must be accommodated inside the body when the valve is in the closed position and in the open position. Most through-conduits are built in carbon steel, often with a cast body. This valve is also available in very large sizes, which are generally fabricated from steel plate. Many of the fabricated-steel and larger cast-steel valves are heavily reinforced with ribs to resist the internal pressure. In higher pressure ratings reinforcements are no longer sufficient, so the body cross section is changed to circular. In higher pressures, it is possible to use a type of pressure-seal bonnet that can be seal-welded closed. Accessories available with this valve include body grease injectors to provide positive hydraulic pressure between the body and the gate for the purpose of keeping sediment or other foreign matter in the flow stream from getting behind the gate. With these injectors, it is often possible to completely bury the valve underground with only the actuator projecting above grade. This is very useful for block valves in buried pipelines. Available sizes are from 2-in. to 48-in. (DN 50 to DN 1200) or larger pipe sizes, and from class 150 (PN 16) up to class 900 (PN 60) ratings, with the larger valves not available in the higher ratings.

These valves can have handwheel or manual gear actuators driving a normal threaded stem, or can be direct actuated by piston actuators. A standard spring-loaded actuator can be used for fail-open or fail-closed configuration, as with almost any other rising-stem valve, but the through-conduit valve possesses an additional degree of freedom in that the gate can be built with the opening near the top, instead of near the bottom, such that moving the stem down or into the valve moves the gate to the open position instead of to the closed position as in any "normal" gate valve. This arrangement is most useful when it is more convenient to have a fail-open design or with a push-down-to-open actuator. As with a number of other special application valves, they can be equipped with special direct-coupled actuators for emergency shutdown service (Fig. 2-17). This valve with a direct-mounted fail-closed actuator is the one installed every few miles on cross-country pipelines and often on wellheads.

There is one other type of parallel-seat gate valve that is occasionally used in the power generation industry for high-pressure steam lines (Fig. 2-18). Its body looks very much like that of the wedge gate described previously, but the disc is in two halves separated by springs. The seats are parallel. As this valve closes, the gate slides across the seats until it bottoms out, so it does not seat tighter in the full closed position. The advantage to this design is that thermal expansion in the valve does not force the gate tighter into the seats. Sealing action comes only from line pressure. Such a valve is good for shutting off flow in a high-pressure, high flow-rate line such as a main steam line. It would not necessarily shut off well at low differential pressure. This valve is designed to be capable of opening and closing with full differential pressure across the seats, so the disc guides or tongues are often hardfaced in addition to the seats and disc surfaces, although in normal operation for main steam lines, there are small bypasses around the block valves to maintain a small flow of steam to keep all portions of the piping hot. For main steam lines that may operate at up to 3000 psi (200 bar), a great deal of force is required to open and close a valve against full differential pressure, and the velocity induced by a small opening like a slightly open gate is very great and very destructive to the valve seats. The bypass solves that problem, and permits the use of a valve whose only job is to shut off mass flow of steam.

### *Plug Valves*

The various types of modern-day plug valves are all descended from the cock valve that has been in use for hundreds of years. The cock, in its simplest form, is just a section of the waterway that can be turned at right

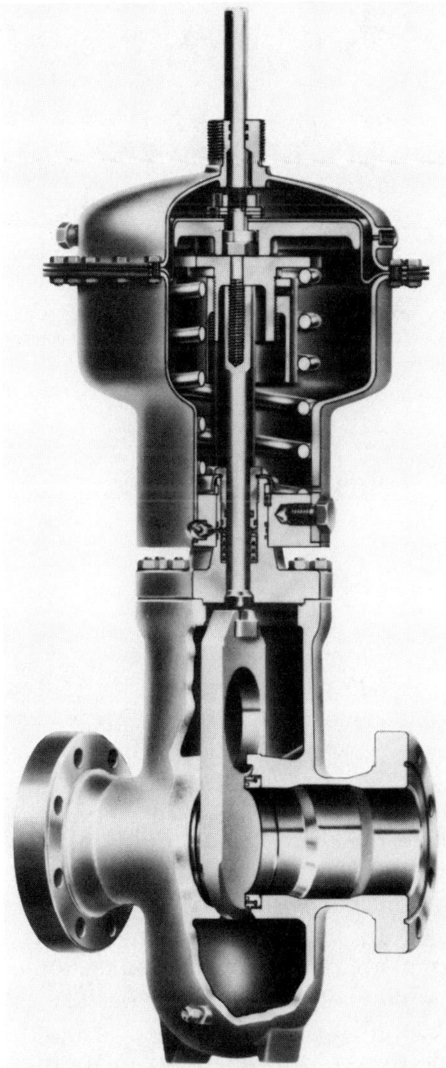

**Fig. 2-17.** Through-conduit valve with direct-mounted diaphragm actuator, air-to-open (fail-safe) spring return and position indicator for pipeline isolation. (*Courtesy of Flow Control Division, Cooper Industries, Inc.*)

angles. Perhaps all other so-called quarter-turn valve designs are derived from this principle also, but the plug is the oldest and the simplest in concept.

Today there are several distinct types of plug valves. All work in the same general fashion—that is, a section of the waterway turns at right angles to the flow and exposes a solid surface that blocks flow passage. Ball valves work on exactly the same principle. The sealing mechanisms employed are considerably different for the plug and the ball valve, and each has quite different applications. There are a number of services for which both would be considered suitable.

Because of the geometry of the plug itself, when the plug is smaller in

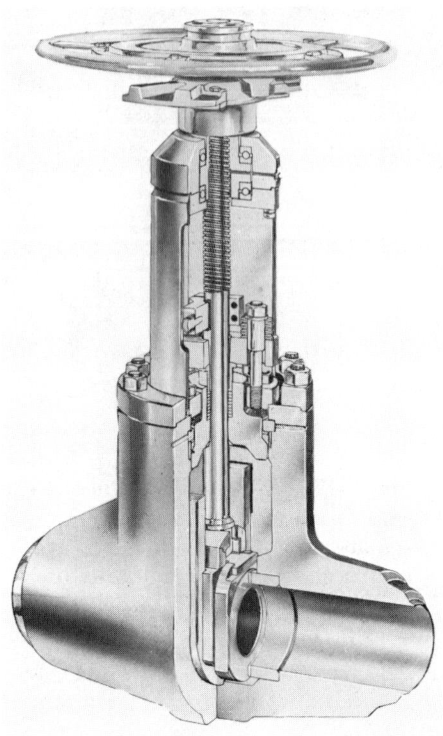

**Fig. 2-18.** Parallel-seated pressure-seal gate valve with live-loaded packing gland. (*Courtesy of Velan Valve Corporation.*)

diameter the valve is smaller and easier to operate. Since the flow path must be somewhat narrower even than the diameter of the plug, that reduces the port opening considerably. Most plug valves thus have a considerably reduced port, although the narrowness required by the plug diameter can be compensated for by making the opening taller. The change in shape and cross-sectional area naturally introduces some pressure drop, and can affect the ability of the valve to pass large particles. Most plug valves have a rectangular or trapezoidal port opening. The width is often one-half or one-third of the height, in an attempt to reduce the overall diameter of the plug as much as possible. As a result, the change in shape and the overall decrease in port opening compared to the area of the pipe combine to produce a greater pressure drop and less flow capacity than would be found in some other types of valve.

It is possible to build a full-port plug valve. The full-port designs have a round opening in the plug, as opposed to the rectangular openings in regular-port plugs, so the plug size must be considerably larger than in the regular-port valve. Full-port plugs have a body the equivalent of two or three sizes larger than the regular-port version, compared to one size larger as would be the case with ball valves. As would be expected, then, full-port plugs are relatively unusual and considerably more expensive than the regular-port type. Typical uses are extremely abrasive or erosive slurry or catalyst-bearing services or locations where the ability to pass a solid object such as a scraper or pig would be required. It is also possible to build special plugs for throttling and control services, although there are not a lot of them around.

The simplest form of plug valve is the cock, which today is found mainly in household and laboratory gas and water lines. The cock has the plug and handle made in one piece, with a simple low-pressure packing device. The sealing surface is either cylindrical or tapered (cone shaped) and generally lubricated, although the lubrication only helps to make the valve gas-tight. Normally one-hand operation is expected from this valve, and it is almost never made in pipe sizes over ½ in. (DN 12). The one design that is available in sizes up to 2 in. (DN 50) or so is an all-plastic valve, generally in polyethylene (Fig. 2-19) designed for natural gas utility distribution lines. This valve uses a cylindrical plug and depends on the natural low friction of its body and plug components.

An extension of this concept, and the oldest type of large plug valve, is the lubricated plug (Fig. 2-20). The plug in these valves is always conical, since the sealing adjustment is obtained by moving the plug up or down to permit easier opening force or a tighter seal, depending on which way the plug is moved. The plug in these valves rides on a film of grease or similar lubricant (Fig. 2-21). There is a grease fitting or some other means of injecting sealant into the valve, since fluid flow and the opening and closing action of the valve tend to dissipate the sealant. Smaller low-pressure plug valves use only sealant to lift the plug and adjust its fit, while plug valves for severe services have a stop screw on the bottom that can lift the plug away from the body sealing surfaces slightly. Most of these valves, especially in high-pressure services, must be greased every time they are operated, before the plug is moved, but

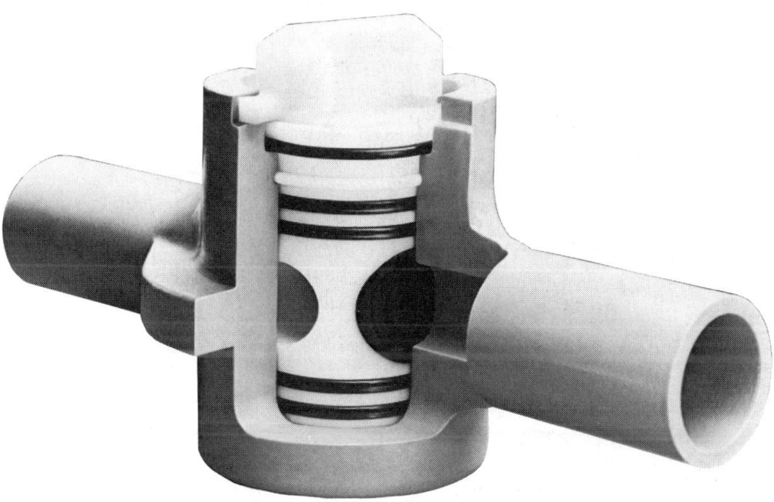

**FIG. 2-19.** Polyethylene-body plug valve for gas distribution service. (*Courtesy of Nordstrom Valves, Inc.*)

**Fig. 2-20.** Lubricated plug valve. (*Courtesy of Nordstrom Valves, Inc.*)

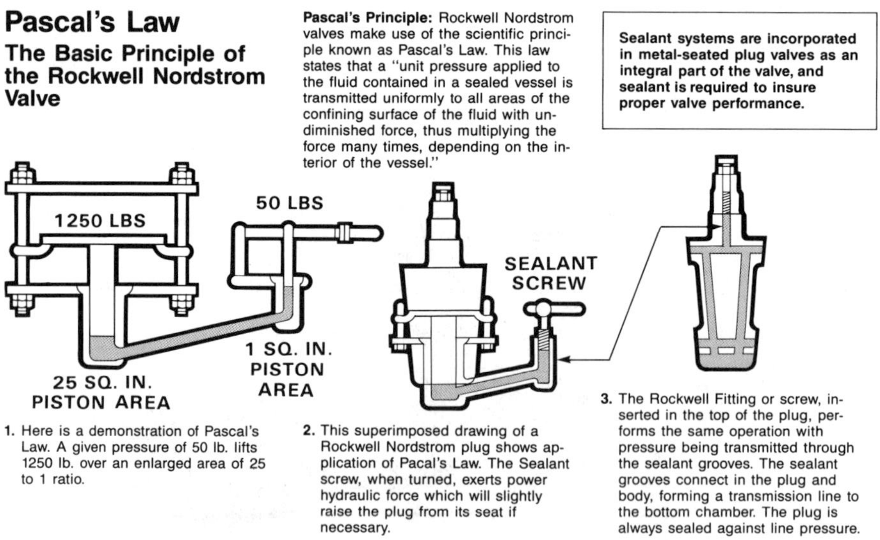


**Fig. 2-21.** Principle of operation of lubricated plug valve. (*Courtesy of Nordstrom Valves, Inc.*)

this is not universally true (Fig. 2-22). Provided the sealant used is adequate, the valves can be operated at very high temperatures because the clearances are able to be adjusted with the stop screw. Because the sealant is injected into the body-plug clearances, the lubricated plug is ideal for services that carry abrasive or erosive particles. This is because the sealant is actually the primary barrier between the high- and low-pressure sides of the valve in the closed position, and because the injection of sealant flushes out any particles that may have worked their way behind the plug, thus minimizing the amount of scratching and scoring that occurs on the plug surface.

For highly abrasive services, the interior surface of the transition tapers and the plug can be hardfaced. This adds to the usefulness of the plug design in such services. In services that are both very hot and highly abrasive, there are basically only three valve choices: a lubricated plug valve, a special metal-seated ball valve, and a knife gate valve.

The lubricated plug valve may be built with the plug tapered either toward the top or toward the bottom. In either case, the cover is on the large end of the plug. Most designs have the cover on the top like a gate valve bonnet, with the adjusting screw in the bottom (Fig. 2-23). Some

**Fig. 2-22.** Sealant injection gun. (*Courtesy of Nordstrom Valves, Inc.*)

fairly sophisticated types have the cover on the bottom, but the adjusting screw is still in the bottom since the screw pushes against both the weight of the plug and the hydraulic force of the grease behind the plug.

The lubricated plug valve is suitable for direct burial, if equipped with an enclosed extension stem, and the injection fittings can be piped down to the valve. The sealant makes the valve capable of bubble-tight shutoff even under difficult conditions, including dirty services and high differential pressures. These features make the lubricated plug one of the most common block valves for gas transmission. However, there is a relatively large mass associated with the plug, so relatively large plug valves become very heavy. Plug valves in sizes over 24 in. (DN 600) are practically nonexistent except in low-pressure water utilities.

Three-way and four-way lubricated plug valves are common (Fig. 2-24). In a lubricated plug, the ports are arranged in a plane that cuts transversely through the valve, perpendicular to the plug's axis. In other words, with the stem vertical, all the ports are connected to horizontal lines. Three-way plugs can be used for diverting or mixing, and can be

**Fig. 2-23.** Bolted cover lubricated plug valve. *(Courtesy of Nordstrom Valves, Inc.)*

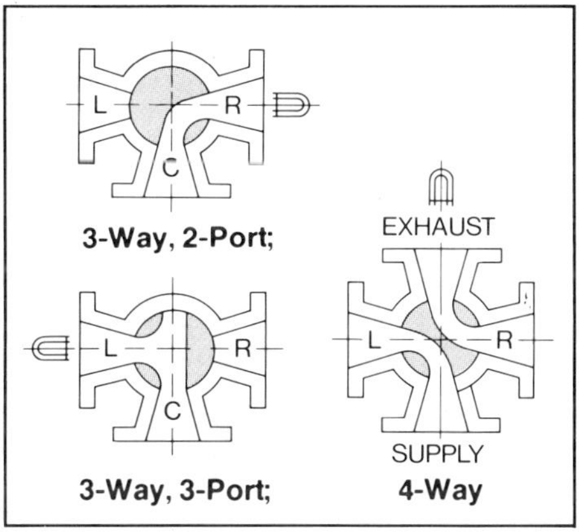


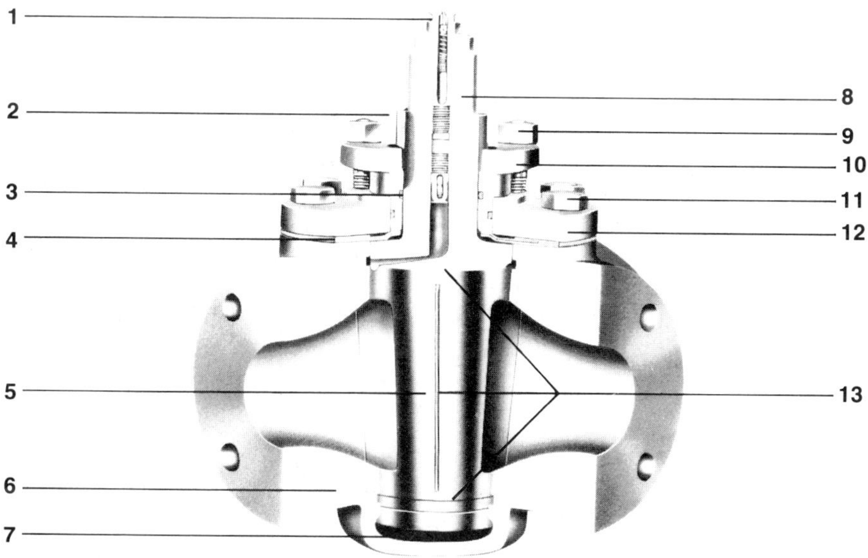


**Bolted Gland Type**

1. **Rockwell Fitting** (Combination Sealant and Gun Fitting)
2. **Stem**
3. **O Rings**
4. **Flexible Metal Sealing Diaphragm and Gasket**
5. **Tapered Plug**
6. **Heavy Wall Body**
7. **Sealant Chamber** (Provides Plug "Jacking" Force)
8. **Wrench Square**
9. **Gland Cap Screw** (Valves smaller than size 3 (DN 80) have screwed instead of bolted gland)
10. **Bolted Fixed Adjustment Screw**
11. **Cover Cap Screw**
12. **Cover**
13. **Sealant Grooves** (Provide "Sealdport" Sealant System)

**Fig. 2-24.** Multiport lubricated plug valve. (*Courtesy of Nordstrom Valves, Inc.*)

built in transflow or nontransflow patterns. In transflow patterns one port begins to open before the previously opened one completely shuts off, so that there is some mixing of streams or flow through both ports for a short time. In some cases this is desirable or required; in other cases it is undesirable. Nontransflow valves must be larger to provide the physical separation required and are not as common. Four-way plugs are used for flow switching and reversing the direction of flow through some piece of equipment, for backwashing for instance. Three- and four-way valves are prone to not sealing very well in directions where the pressure force is trying to assist the hydraulic forces pushing the plug away from the seats—that is, in situations where the pressure is directed toward the plug when there is no equivalent pressure force in the opposite direction on the other side of the plug (Fig. 2-25).

Another common type of plug valve and the one that now has the largest percentage of the plug valve market, is the sleeved plug (Fig. 2-26). This type is a relative newcomer to the valve market, dating from 1950. It looks like the lubricated plug from the outside, except that there is no lubricant fitting and no adjustment screw. The plug rides in a sleeve of plastic material, normally a semirigid fluoropolymer such as PTFE. The sleeve is locked into the body by means of ribs and grooves, and is smooth on the inside. The plug rotates in the sleeve by virtue of the low coefficient of friction between the plug and the sleeve. (The lowest coefficient of friction is stainless on plastic, not plastic on plastic.) To assist in this low friction, the plug is often made of stainless steel or plated with hard chrome even if the service warrants only a less corrosion resistant material like carbon steel. The top of the plug is slightly tapered also, and it bears against a fluoropolymer diaphragm that meets

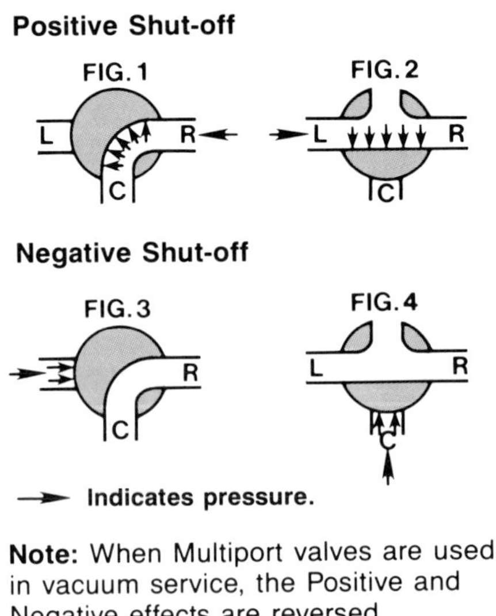


**Fig. 2-25.** Pressure forces on multiport lubricated plug valves. (*Courtesy of Nordstrom Valves, Inc.*)

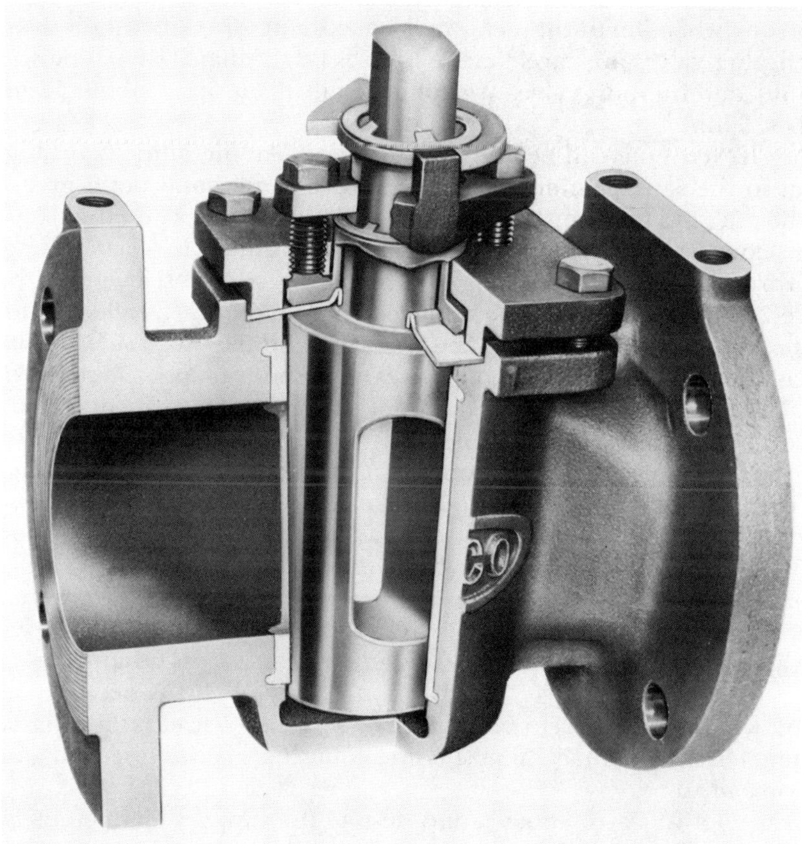

**Fig. 2-26.** Sleeved or nonlubricated plug valve. (*Courtesy of Duriron Company, Inc., Cookeville Valve Division.*)

the sleeve at its outer edge. The diaphragm both separates the inside of the bonnet from the line fluid and acts as a thrust bearing. It also acts as the stem seal in many cases. The diaphragm can be backed up by graphite seals to make the plug valve fire-safe with respect to leakage up the stem and out through the bonnet joint, in other words, leakage to atmosphere. Leakage around the plug and down the line is harder to be eliminated by fire-safe designs, because of the sleeve configuration.

Bodies are often still made of carbon steel or ductile iron, if that is all that is required. The body mass of a sleeved plug valve is about the same as that of a ball valve and definitely smaller than that of a gate valve. Even in moderate sizes the plug, just like the lubricated plug type, reaches a very large mass. This large moveable mass (which is larger than the corresponding gate valve, and which is moved a shorter distance in a rotary motion against greater friction than the wedge of a gate valve) somewhat limits the achievable maximum size of plug valves. Sleeved plug valves are rarely made in sizes larger than 18-in. pipe size (DN 450), and full-port valves are not normally made above about 12-in. size (DN 300).

The sleeved plug can be adjusted for wear like the lubricated plug, but not in the same manner. Since the sleeve is solid and not a grease film, the plug has to be pushed down from the top farther into the sleeve as the sleeve wears. Eventually the limit of adjustment is reached, normally through the sleeve wearing too thin, and then the sleeve must be replaced. The old sleeve can be removed and a new one installed, but it must be hot-pressed into place, and deformed on the outer side so that the plastic flows into the ribs and grooves of the body to lock the sleeve in place. Generally, the plug itself does not see much wear except in very erosive or corrosive services. The body may corrode out behind the sleeve, since the sleeve may be somewhat permeable to acids and other very corrosive fluids and since fluids can easily get behind the liner from the edges. This may cause the sleeve to collapse and fail to seal completely, even before the body develops a leak.

For services that are very corrosive, any type of plug valve is a better choice than most other types of valves, because a somewhat corroded plug valve is more likely to shut off tightly than are most other types of valves with the same degree of corrosion damage. This is because the location where sealing occurs is an area of significant size, rather than a very small area of contact, almost a line contact, as is the case in a gate valve or ball valve.

Because of this area contact, and despite the slippery characteristics of the grease film or the plastic sleeve, plug valves tend to have a higher breakaway torque and operating torque than do most other valves. It is true that higher pressure rated valves require more torque to open than lower rated ones, but the magnitude of the breakaway and operating torques due to the sleeve friction is such that a gear operator is required on almost any plug valve larger than 4-in. size (DN 100). Some non-gear-operated plug valves are offered with longer wrenches to overcome the torque, but there is a limit to the practical size of wrenches. Some are available as long as 3 ft (1 m), but they are generally practical only in valves that are seldom opened and in locations where the clearances do not matter.

Sleeved plug valves are available in three-way designs (Fig. 2-27), but to allow the same size plug to be used for two-way and three-way valves, the third port is generally located in the bottom of the valve. This makes piping up to a three-way valve somewhat more complex, since it is generally better to position the valve with stem upright, and the flow arrangement that is easiest to do is where the flow is into the bottom and out the sides or vice versa. Flows that go from one side port to the other are a little more difficult to do.

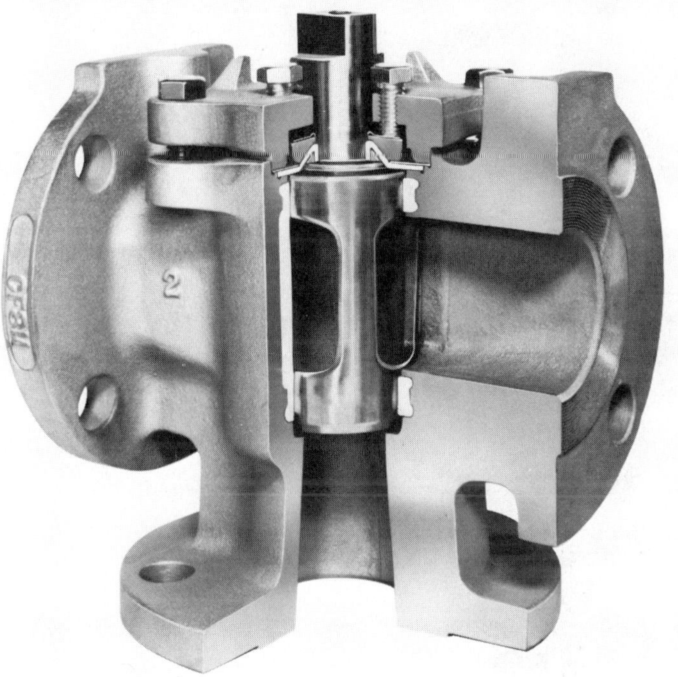

**FIG. 2-27.** Three-way sleeved plug valve with third port on bottom of valve. (*Courtesy of Xomox Corporation.*)

It is also possible to build three-way full-port plug valves, but these tend to be special designs (Fig. 2-28). Sometimes these valves can be configured as 120-degree angle valves for flow diversion in gravity flow or true Y-pattern piping. These are more common in Europe than in the United States.

Most sleeved plug valves have no separate packing adjustment device, because the usual location for packing adjustment bolts is taken up with the plug adjustment screws. Common plug valves seal adequately with the top diaphragm anyway, and in many cases the graphite packing used for backup is not separately adjustable. But since the plug is so well suited to tight shutoff of corrosive or nasty fluids, in recent years some severe service plug valves have been developed (Fig. 2-29) where the primary refinement is better top seals including a full packing gland. Generally this requires raising the top envelope dimensions of the valve, since the only way to provide a packing gland is to build upward from the diaphragm area. Since this new packing gland is so easily accessible, it is also possible to add purge ports and other refinements to make it even lower in atmospheric emissions.

Plug valves used for throttling service have a somewhat different

**FIG. 2-28.** Three-way full-port plug valve with third port on side of valve. (*Courtesy of Xomox Corporation.*)

plug design (Fig. 2-30). The body of the plug and its downstream port opening are of conventional design. The upstream port has a characterized opening (V-shaped) like the one in a control valve cage. This presents only a very small opening when the valve first begins to open, and gives an equal-percentage flow curve. In addition, some plugs can be built with a low noise or anticavitation trim. The same V-port plug is used, but the side opposite the point of the V is machined flat and drilled with the large number of small holes characteristic of anticavitation trims. The entire flow does not pass through this drilled area. Until the valve is almost completely closed, some flow can bypass the holes, which permits larger particles to bypass the hole area and go down the line without plugging the valve.

Still another type of plug valve is the fully lined type (Fig. 2-31), which has a plastic liner completely covering the waterway in addition to

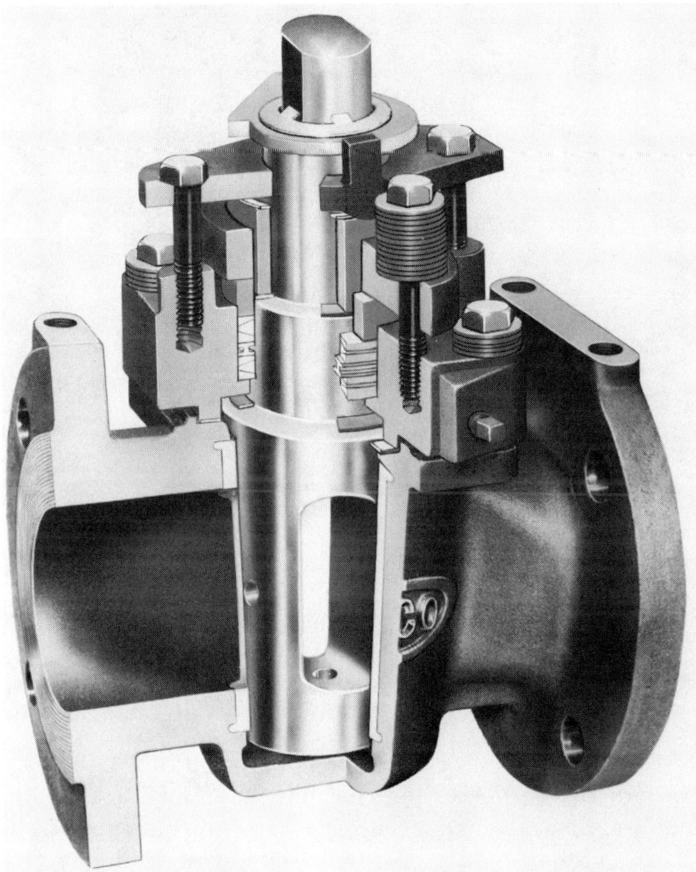

**Fig. 2-29.** Hazardous service plug valve with live-loaded packing gland and packing leakoff connection (plugged). (*Courtesy of Duriron Company, Inc., Cookeville Valve Division.*)

the area behind the plug, and a liner that completely encases the plug except for the top of the stem above the packing rings. This valve operates just like the sleeved plug, except the sliding surfaces are plastic on plastic rather than metal on plastic. This type is used primarily in extremely corrosive services where the plastic liner is an alternative to very expensive corrosion-resistant metals such as nickel-base alloys. An all-alloy valve may be ten times the price of a steel or iron valve with plastic liner, and may not offer as good a performance, since in the presence of many corrosive agents, the plastic liner is almost completely inert and often relatively impermeable. Several different polymers can be used as liners, including PTFE (Teflon), PVDF (Kynar), PVDC (Saran), polyethylene, and polypropylene.

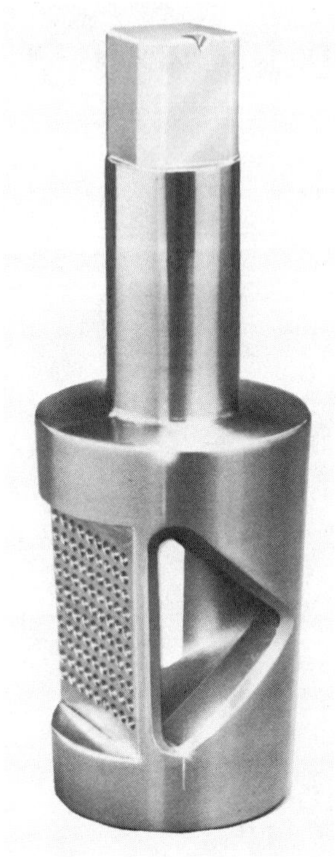

**Fig. 2-30.** V-port plug for throttling, also equipped with anticavitation drilled flow passageways. (*Courtesy of Duriron Company, Inc., Cookeville Valve Division.*)

There is a variant of the lined plug that basically looks like a plug from the outside, but operates like a butterfly valve in that the plug is carved out on both sides instead of being bored through. When this valve is open, flow goes around a central element on both sides. This valve is quite rare.

Still another type of plug valve is the lift plug or wedge plug (Fig. 2-32). This design operates in a totally different manner than other plugs. In fact it more closely resembles a gate valve, both in operation and in appearance. The lift plug has a rising stem and a true bonnet line a gate valve, with a handwheel to lift the stem. The plug is tapered with the large end at the top and looks like the plug in a sleeved or lubricated plug valve, but has an elastomer seal inlaid into the plug surrounding each of the waterway openings. This seals against the metal body cavity. When the valve is opened, the stem rises and lifts the whole plug upward slightly, enough to separate the plug and the inside of the body. Then, because of a curved slot machined into the stem, as the stem rises further it is cammed 90 degrees around. This rotates the plug along with the stem. In the open position the plug sits slightly higher than when the valve is closed.

The lift plug valve provides good shutoff, even more so than most plug valves. It is one of the best designs for tight shutoff because of the wedging action like a gate combined with elastomeric seats (Fig. 2-33).

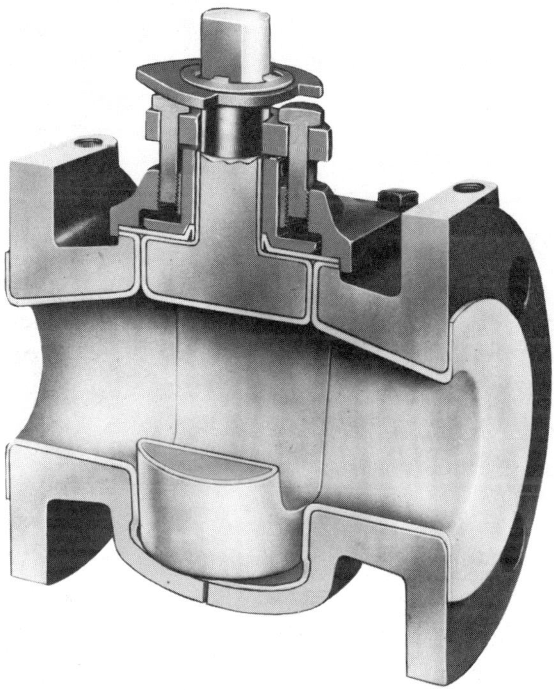

**Fig. 2-31.** Fully lined plug valve. (*Courtesy of Duriron Company, Inc., Cookeville Valve Division.*)

Many valve designs that use elastomeric seats have problems with wear, since by necessity the soft seat must come into intimate contact with a hard mating surface. This problem is difficult to overcome in a gate valve, but in the lift plug it is accomplished by providing a separate plate or slip that the soft seats are set into, which in its turn rides on the plug. This permits the plug to lift independently of the seat surfaces for a short distance and then pull the seals perpendicularly off the seats. In the seating mode the reverse happens: The plug descends only so far before the seals stop moving downward, and with further downward travel the seals are pushed more forcefully into the seats.

Because of its excellent shutoff ability, the lift plug is used for special-purpose shutoff such as isolating two incompatible product streams in a tankage area or as block valves in services with high permeability or low vapor pressure such as butane or hydrogen. For still greater assurance that shutoff has been obtained, this valve can be designed as a double block and bleed. All that this requires is a bleed from the body cavity to the atmosphere. In normal block and bleed service, this bleed is left open when the main valve is closed. This means that the body cavity is at atmospheric pressure, and if a leak occurs across one of the seats, the leaking fluid will drain out of the valve where

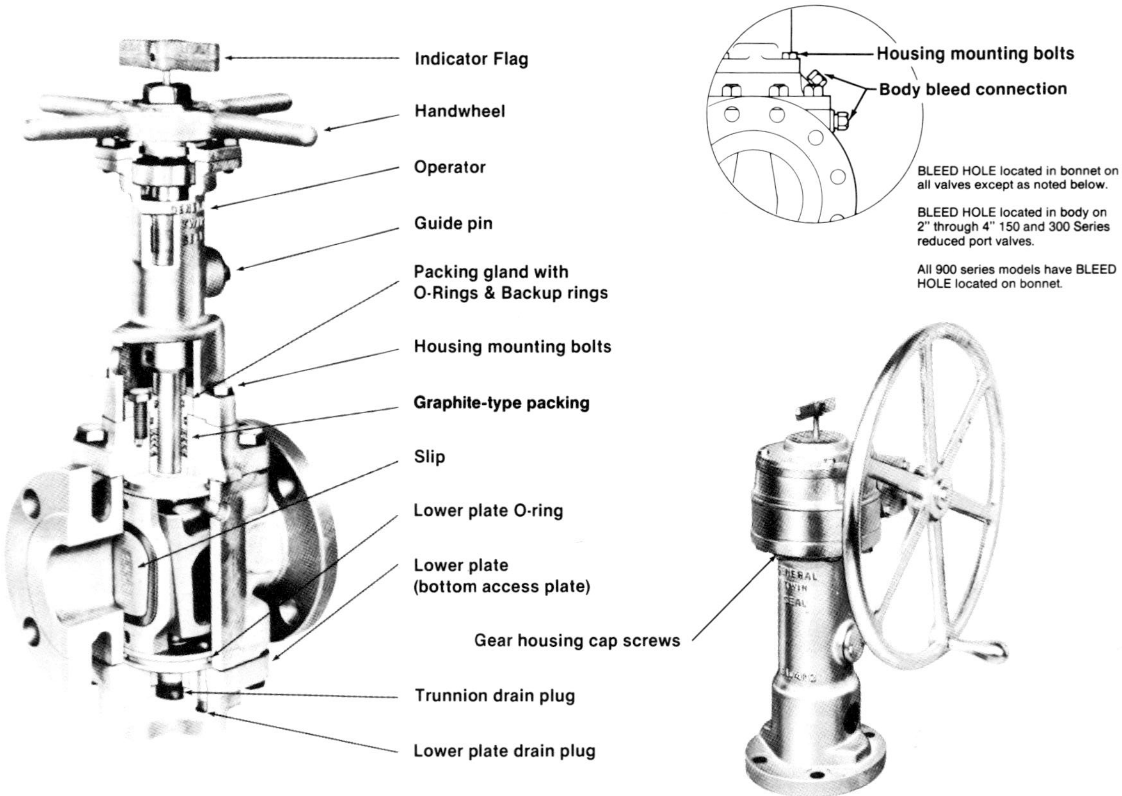


**Fig. 2-32.** Wedge plug valve in the closed position. (*Courtesy of General Valve Company, division of Johnston Pump/General Valve, Inc.*)

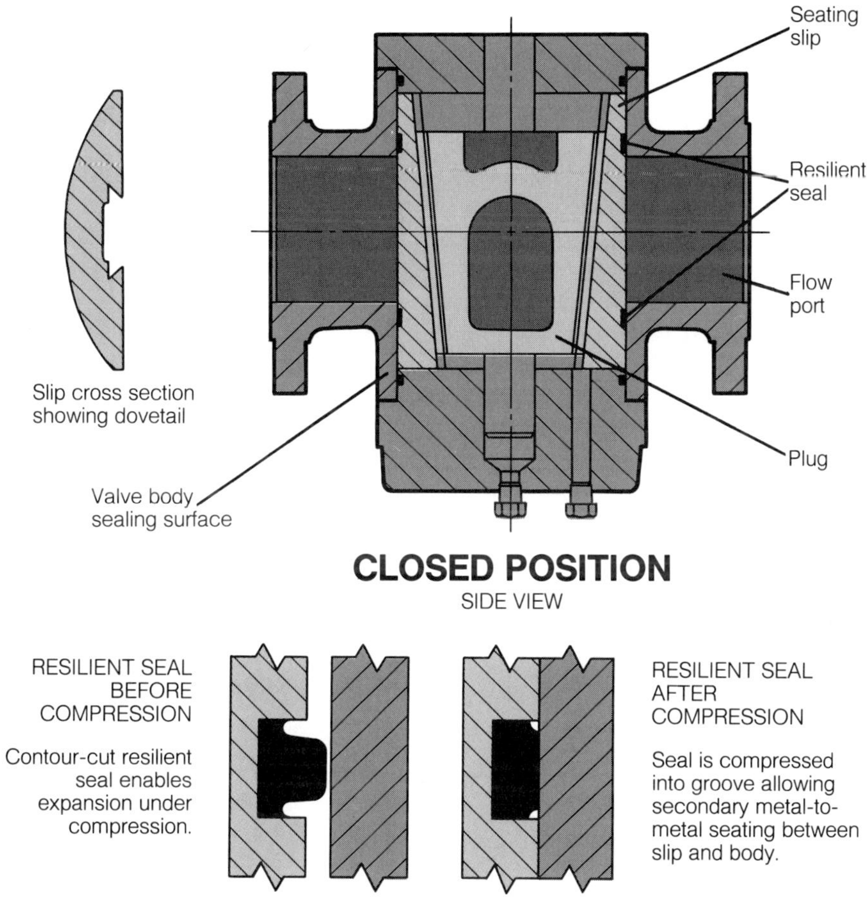


**Fig. 2-33.** Wedge plug seating diagram. (*Courtesy of General Valve Company, division of Johnston Pump/General Valve, Inc.*)

it can be detected and there will be no possibility of it leaking through the other seat into the other portion of the piping system. Because this valve shuts off tight on both seats, a thermal vent is normally required if the bleed is not used, to keep trapped liquid inside the body cavity from overpressuring the body if it expands.

The wedge plug is also available in a full-port model and a four-way design up to 24 in. (DN 600) (Fig. 2-34) that can be used for flow reversal in a section of line. This ability is useful for, among other things, allowing automatic backwash of equipment or sequencing the flow of fluids through a pair of exchangers or cycling a piece of equipment that regenerates a filter bed.

The wedge plug is one of a class of valves that are really an order of magnitude different from standard valves that they appear to resemble.

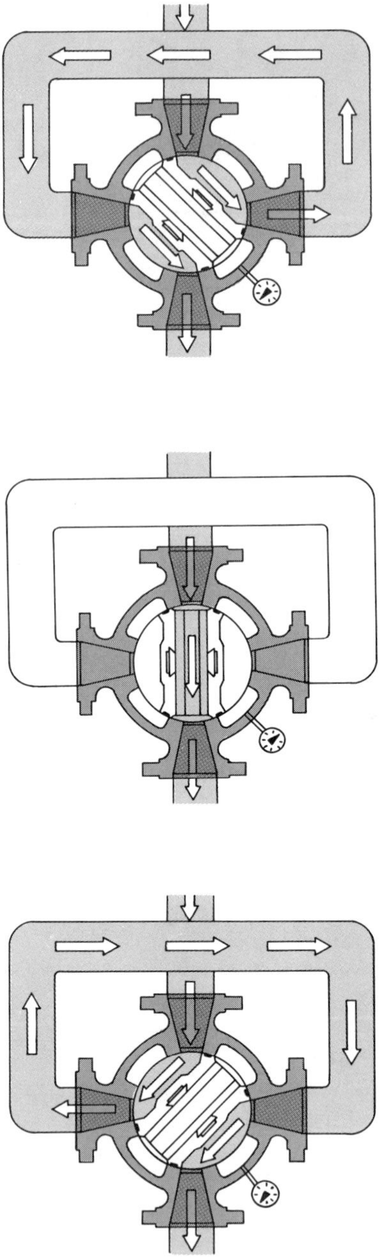

**Fig. 2-34.** Four-way wedge plug operating sequence diagram showing flow isolation and flow reversal. *(Courtesy of General Valve Company, division of Johnston Pump/General Valve, Inc.)*

This class consists of three basic types that, although they are three different body types, really function alike. The wedge plug is one, the pressure-energized soft-seated gate is another, and the single-seated rising-stem ball valve is the third. What they have in common is their method of pressing their soft seats perpendicularly into a mating hard seat without a wiping action, the fact that the stem force helps them seat tighter up to the point that the backup metal seat comes into play, and their ability to seal bubble-tight.

### *Ball Valves*

The ball valve occupies several niches along the valve spectrum, and includes inexpensive types as well as very sophisticated high-performance designs. They all work in much the same way, just like a plug valve, in that there is a section of waterway that rotates 90 degrees when the valve is actuated, and that blocks off fluid flow. The differences in ball valves, as in plug valves, generally relate more to how the valve is put together than how it operates. But as with plug valves, there are some very profound differences between types of ball valves in that some of them function in opposite ways from others. Like the plug valve, a standard ball valve is capable of limited throttling or control, but with some modification it can serve very well as a throttling or modulating valve. It is classified here primarily as an on-off valve because a greater share of its use is as an on-off than as a modulating valve.

The design differences in ball valves relate to how the ball comes out—through the end, through the top, or not at all. These differences are profound from the maintenance and safety standpoints. Another major classification of ball valves is whether the ball is floating or nonfloating, that is, whether the ball moves into the seat when it shuts off or the seat moves into the ball. The floating-ball type is adequate from most services and is the most common. The type where the ball does not float is commonly known as a trunnion type.

Most ball valves are end-entry types, where either one entire end of the body unscrews to remove the internals or a tailpiece comes out from the end, leaving part of that end intact. In smaller sizes (Fig. 2-35) the end or the tailpiece is threaded into the body. Sometimes each end is attached to a center section by bolting or threading (Fig. 2-36). In larger sizes the tailpiece threads into the body until it is flush with the rest of the end flange, or the tailpiece is retained by cap screws. The other common type is the split-body, where the entire body is split slightly off-center of the stem with a bolted joint similar to the joint between body and bonnet in a gate valve. Here end-entry and split-body types will be treated together, since most of their other design features are much the same, in contrast to those of top-entry designs.

The main caution that applies to true end-entry flanged valves that does not apply to split-body types is that, as in the high-performance butterfly, the tailpiece often forms part of the gasket sealing surface. The radial split between the tailpiece and the main part of the body is determined by the outside diameter of the ball, but it can fall near the

**Fig. 2-35.** Two-piece threaded-body, reduced or regular port, threaded-end ball valve showing integral travel stops. (*Courtesy of Watts Regulator Company.*)

**Fig. 2-36.** Flanged-end three-piece threaded-body ball valve. (*Courtesy of Flow Control Division, Cooper Industries, Inc.*)

middle of the gasket sealing surface. This is not as severe a problem as it is in the butterfly, but the tailpiece must end up exactly flush with the remainder of the end surface for effective gasket sealing. With spiral-wound gaskets, which seal more toward the inside diameter of the sealing surface, the problem is actually no more severe than with flat gaskets, but it is a potential problem nevertheless. Many tailpiece designs now are intended to bottom out against a shoulder in the body, which both ensures correct seat loading without a lot of trial and error, and also places the gasket seating surface correctly. Some designs have an enlarged end on the tailpiece that comprises the entire raised face. This is more common on small ball valves.

A standard floating-ball valve, whether end-entry or split-body, is assembled by installing one of the seats into the larger half of the body, then slipping the stem from the inside into the packing gland, then setting the ball onto the first seat and the stem, followed by the second seat and the tailpiece or the smaller body half (Fig. 2-37). During installation the stem and ball are turned as they would be in the closed position, so the ball can be lined up properly with the stem. The two seats are parallel to each other, and perpendicular to and concentric with the flow stream. The packing is installed from the outside, held

**FIG. 2-37.** End-entry reduced-port ball valve, shown here with spring-return air actuator. (*Courtesy of Worcester Controls Corp.*)

down with a regular packing gland. Many ball valves have elastomer O-rings for a stem seal, which are usually installed from the inside along with the stem.

The trunnion-mounted ball types are assembled similarly, except that the trunnions provide rigid support for the ball so the body joint has to be located fairly close to the center line. In a true trunnion design, the trunnions are integral to the ball such that the trunnion supports must be split or accessible from one end. Often the bearings are assembled around the trunnions before the assembly is installed. The other difference is that, since the trunnion-mounted ball does not move, the seats must be spring-loaded with disc or wave springs or sets of tiny coil springs. These assemblies must be put together, usually piece by piece, as the valve is assembled. Some trunnion valves have lower stems that are inserted from the bottom, with a bolted cover that retains the trunnion. These are not the same as true blowout-proof designs, but since the lower trunnion is totally enclosed, it has the same effect.

The second type of floating-ball design is the top-entry valve. (Fig. 2-38). This design has a body in one piece with a cap, like a bonnet, that holds the stem and the packing. The seats in a top-entry valve are normally at an angle to one another, like the seats in a gate valve. This valve is assembled by dropping the ball and seats simultaneously into the body, then placing the cap and stem onto the body (Fig. 2-39). The angle serves to aid in assembly, and to some extent allows wedging of the ball tighter into the seats.

The top-entry design has several advantages over the end-entry designs, one being that there are fewer leak paths to atmosphere, another being that since the body is one piece, the valve is more rigid than some split-body types. The primary advantages of the type are that the

**Fig. 2-38.** Top-entry ball valve with spring loading from the top. (*Courtesy of McCanna, Inc.*)

**Fig. 2-39.** In-line maintenance accessibility of top-entry ball valve. (*Courtesy of McCanna, Inc.*)

wedging action allows for mechanical loading from the top via a spring pushing on the ball to make it seat tighter, and that the valve is repairable in-line; that is, the entire valve does not have to be physically removed from the pipe in order to get inside it for maintenance.

There are several methods of making a top-entry valve seat tighter than the corresponding end-entry type. One is by spring-loading the ball from the top, with a simple coil spring, to force the ball more firmly into the seats. This force is additive to the pressure force of the fluid flow, such that the valve still seats more tightly on the downstream side. Another method is to machine the ball with tapers around the port openings, such that in the open position the seats are not loaded or only lightly loaded, but in the closed position the ball forces itself tightly between the two seats. This helps to avoid the phenomenon known as cold flow, where the fluoroplastic material commonly used for seat material deforms under constant load and loses its shape.

Still another type of ball valve is the true one-piece body (Fig. 2-40), in which the body is literally built around the ball in one piece, by welding the components together, such that the ball can not be removed

**Fig. 2-40.** All-welded one-piece ball valve. (*Courtesy of Nordstrom Valves, Inc.*)

except by cutting the body in half. This means the valve has better integrity in terms of leak paths and strength, and can reliably be direct buried (Fig. 2-41) or installed undersea or used in remote locations where no maintenance would be performed anyway, but it also means that no seat renewal can be done. The inability to maintain the seats can be designed around, for instance, by providing a mechanism to rotate the seats every time the valve is actuated so that a different portion of the seat takes the greatest seating load every time. These valves are also provided with sealant injection devices, so that if the valve is leaking past the seat into the body cavity, the cavity can be filled with grease or sealant like a plug valve to help obstruct the flow of line fluid around the ball. Pipeline service ball valves designed to API 6A or API 6D have this feature, whether one-piece or top-entry valves. The largest market for these valves is pipeline service, but they have also been extensively used in refinery service.

All of these types of valves can be built as both floating-ball or trunnion-mounted designs. The end-entry is normally built as a floating-ball design, and the one-piece is almost always built as a trunnion-mounted ball design. The top-entry can be either type (Figs. 2-42, 2-43).

The essential difference between the floating-ball and the trunnion-mounted ball designs is that in the former, as is apparent from its name,

**Fig. 2-41.** One-piece ball valve for buried installation. (*Courtesy of Nordstrom Valves, Inc.*)

the sealing action is obtained by the ball moving, or floating, into the seat. The driving force for this is line pressure. When the valve is open, the hydrostatic forces on the ball are minimal. In fact, if the ball valve is full port, the inside of the valve appears to the flow stream to be nothing more than a continuation of the pipe. Regardless, the space between the ball and the body wall is exposed to line fluid almost as if it were a gate valve, since the seats do not contact the ball with any great force when the ball is in the open position, and as the ball rotates, line fluid can flow completely around the ball. As the ball closes, the fluid flow forces begin to push on the ball and, in the closed position, the differential pressure is such that the downstream seat takes the full pressure load exerted on the ball. This is not much different from a gate valve, except that no outside wedging or mechanical loading occurs. Sealing occurs by the deformation of the downstream seat, which in standard ball valves is an elastomer or a semiresilient polymer such as PTFE. The seats limit the pressure-temperature rating of the valve (Fig. 2-44). An all-metal seat is possible, but the springiness required is such that low-pressure metal-seated valves are difficult to build. A metal backup seat, which is geometrically different from an all-metal seat, must be provided in any valve designated as fire-safe.

**Fig. 2-42.** Top-entry API-6D floating-ball valve with wrench operator. (*Courtesy of Flow Control Division, Cooper Industries, Inc.*)

**Fig. 2-43.** Top-entry API-6D trunnion ball valve with ring-joint flanged ends, showing sealant injection ports and spring-loaded seats. (*Courtesy of Flow Control Division, Cooper Industries, Inc.*)

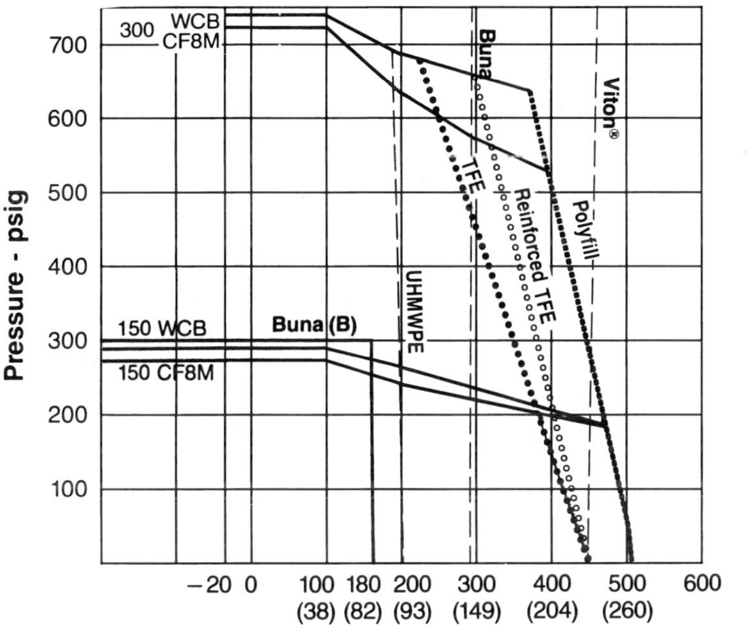


**FIG. 2-44.** Typical pressure-temperature rating chart for ball valves. (*Courtesy of Worcester Controls Corp.*)

The ball ordinarily is machined as a true sphere, which is a relatively easy shape to cut on a machine tool. However, because it is more difficult to lap smooth than a flat surface would be, most balls are machined as smooth as possible (about 16 RMS; 0.6 μm) and then plated with chrome or, preferably, nickel to reduce the surface roughness to about 4 RMS (0.15 μm). When the ball is a corrosion-resistant material like stainless steel, the surface must be polished to this finish. The coating is not totally impermeable anyway, and is normally not permitted to be considered in evaluating the corrosion resistance of the valve. Some very corrosive services will go directly through coatings, and balls for these services must be uncoated. When metal seats are used, the ball and seats must be installed as a set because the ball and seat set must be lapped together to get good shutoff.

The ball in a floating-ball valve has a slot on top to engage the stem, which has a tongue on the bottom. Early ball valves were designed so that the stem could be inserted from the outside, through the packing area, and engage the ball. This design is now prohibited in ball valves for process service, because if the stem was able to go in through the top of the valve it was also able to come *out* through the top of the valve, under pressure and with hydrocarbon or steam coming right behind. Almost all ball valve standards now require a blowout-proof stem, in which a section of the stem below the packing area is larger in diameter than the portion of the stem that extends out to the outside, thus making it impossible to pass the stem completely through the packing gland (Fig. 2-45). This, incidentally, provides a greater area of contact between the ball slot and the stem tongue, which spreads the operating torque over a greater area of the ball. If the ball hangs up or overtorques in some way, it is possible to twist the ball and stem apart, either by twisting the stem in two or by deforming the top of the ball like the screwdriver slot in a household screw.

Many balls, especially in smaller sizes and in valves designed for fairly severe services, are machined out of a solid block or bar. In some valves, the ball is formed out of a tube section, roughly pressed into a spherical shape and then finish machined. The inside of this ball has a roughly spherical shape. Some large trunnion balls are cast with voids in areas where no metal is required, which cuts down the weight and moving mass. In some cases, a large ball may be cast in a hollow spherical shape and a piece of pipe welded inside to form the flow surface. In all cases except the very cheapest, the edges of the ball where they contact the inside flow surface are rounded and polished partway into the rounded edge.

**Fig. 2-45.** Cutaway of three-piece threaded-body ball valve showing blow-out-proof stem and fire-safe integral metal backup seat design. (*Courtesy of Flow Control Division, Cooper Industries, Inc.*)

Some ball valves are built with the ball and stem machined in one piece, and there are a number of trunnion-mounted designs in which the ball has shafts at both top and bottom. Many, but not all, of these "shafts" are really just bearing surfaces or true trunnions, wherein the length is actually less than its diameter. In both these cases the ball can no longer move into the seat at all or can move only in a very restricted manner. This means the seats must move into the ball. This is normally accomplished by placing springs behind each seat. The springs are either a series of small coil springs, or more usually a Belleville or disc spring that pushes on the seat from behind. The reasons for building a trunnion-mounted valve are that the size of the valve makes the ball too heavy to float adequately, or that the pressure is high enough that the seat load caused by line pressure is so high that it would cause crushing of the seat. Placing the ball on trunnions and springing the seats removes the seat loading from any relationship with line pressure and places it under the control of the valve designer. It is often true that trunnion-mounted valves will seal on both upstream and downstream seats, although the downstream seat is still somewhat tighter and the line fluid will still always get behind the ball when the valve is opened and closed. It is possible to obtain very tight sealing with seat springing.

One additional variation on the end-entry design is a valve that has a two-piece tailpiece, one threaded into the body like a small unibody valve, and a second tailpiece that threads or bolts in to provide the flange surface. The inner tailpiece can be used to adjust the seat sealing force by threading it in and out, without affecting the position of the outer tailpiece. This design still requires removal from the line for adjustment.

High-performance ball valves almost always are built with spring-loaded seats and can be expected to deliver bubble-tight shutoff, even under conditions that tend to cause seat damage (Fig. 2-46). This is often accomplished by providing a hard overlay on the seats and a still harder coating on the ball, with a spring force great enough to hold the two in intimate contact. The torque required to operate a valve like this is, naturally, often considerably greater than that required to operate a standard ball valve. Some designs have a spring-loaded seat on only the upstream side, to add to the fluid force on the ball and to provide some sealing in the "wrong" direction, but in the "right" direction the seat is machined directly into the inside of the body so that the ball is pushing against a solid metal seat.

At the opposite end of the spectrum are the low-cost ball valves, which are generally the lowest priced valves of any type because they have the smallest metal mass and the least amount of machining required. A brass valve with elastomer seats will shut off adequately under domestic water pressure and is normally available at any hardware store for $5 or less. Even cheaper valves can be injection molded from PVC or other resins, usually with an acetal plastic ball, sometimes even without

**Fig. 2-46.** Metal-seated ball valve with adjustable seat loading design. (*Courtesy of Marpac, Inc.*)

elastomer seats. Such valves are even suitable for industrial use in water service and will perform adequately, but they should be considered throwaways. Their value makes it uneconomical to expend any labor repairing them.

The lowest cost metal ball valves, which can be available in brass, steel, and even stainless steel, are the two-piece threaded-body types in which the entire valve body is split with a threaded joint (Figs. 2-47 and 2-48). They can also be built as a three-way pattern (Fig. 2-49). The disadvantage of this design is that when someone attempts to apply torque to the piping or directly to the valve—for instance, to remove a plug or other piping attached to one end of the valve—it is possible for the tailpiece to unscrew. When this happens, there are two consequences. One is that, since the valve itself has come apart, it is not possible to shut off the fluid flow there. The second is that, since the ball is no longer retained, it can shoot out under line pressure. Both of these occurrences are undesirable. This inherent problem can be remedied at slightly higher cost by tack-welding the two pieces together (in steel or stainless) or by assembling the valves with a thread adhesive.

An alternative design, which is generally no more expensive, has a one-piece body with a tailpiece that threads into the body through one

**Fig. 2-47.** Cutaway of two-piece threaded-body ball valve showing threaded stem seal on gland. (*Courtesy of Watts Regulator Company.*)

**Fig. 2-48.** Two-piece threaded-body, full-port, threaded-end ball valve. (*Courtesy of Watts Regulator Company.*)

end (Fig. 2-50), such that both end connections are machined into the same part. Often the tailpiece has straight threads that are the same diameter as the bottom of the taper pipe threads. There is a shoulder just above the seat area so that the tailpiece will bottom out. The tailpiece has a slot in its outer end like a screwdriver slot for installation. Pipe threaded into the end of the valve will not affect the tightness of the seat loading, because of the body shoulder. With this design, no amount of wrenching can cause the valve to separate since the tailpiece is inaccessible from the outside. This design is known as the unibody.

**Fig. 2-49.** Three-way ball valve with side outlet. (*Courtesy of Watts Regulator Company.*)

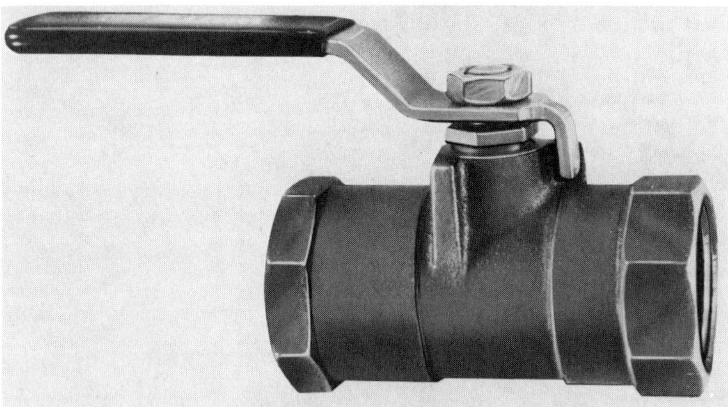

**Fig. 2-50.** One-piece body (unibody) threaded-end ball valve. (*Courtesy of Watts Regulator Company.*)

There are a couple of other slightly more expensive designs in common use. One is the two-piece bolted body (Fig. 2-51), in which the valve is split slightly to one side of the centerline and held together with four cap screws threaded into the larger half of the body. This design is somewhat larger than the one-piece body. Another type is the three-piece bolted body (Fig. 2-52), in which there are two identical end pieces bolted into or around a center piece that contains the ball and seats. This design is even larger than the two-piece bolted body, but it has a couple of advantages. One is that, if the end pieces must be of special metallurgy for welding compatibility with the piping, the center section need not be since it is not being welded to. Another advantage is that, since socket-welding and seal-welding in small ball valves are notorious for melting the seats and destroying the valve before it even goes into

**Fig. 2-51.** Two-piece bolted-body ball valve. (*Courtesy of Neles-Jamesbury, Inc.*)

**Fig. 2-52.** Three-piece bolted-body ball valve with oval handwheel and flareless tube ends. (*Courtesy of Swagelok Co.*)

service, a three-piece ball valve can easily be unbolted and the center section moved out of harm's way (Fig. 2-53) while the end pieces are being welded. Seats and seals are also relatively easy to replace in-line since the complete valve does not have to come out of line. This style is easy to recognize, since it is often almost cube shaped.

There are two different types of three-piece bolted-body designs, depending on whether the center section can be swung out or not while

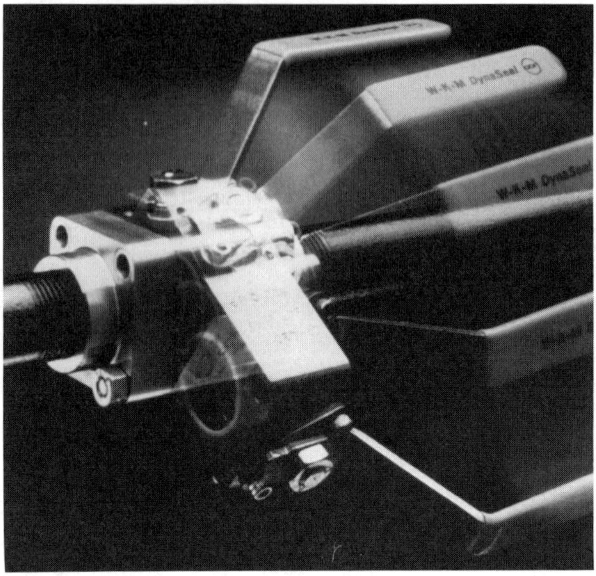


**Fig. 2-53.** Swing-out capability of three-piece bolted-body ball valve. (*Courtesy of Flow Control Division, Cooper Industries. Inc.*)

the end pieces stay held together with a single bolt. The types that can swing out are easier to change seats in or to gain access to the interior of the valve or pipe, but the types that do not swing out, in which the body bolting is completely enclosed by the center section of the body, have more reliable sealing of the end pieces and are good for higher temperatures and pressures. These types always have four bolts for each end, while the swingout types sometimes have three bolts. The three-piece design can also be built as a three-way (Fig. 2-54a,b).

The three-piece ball valve lends itself well to sanitary services, where the valve must be disassembled for cleaning. These valves can be equipped with a cavity-filler seat, designed to leave essentially no cavities around the ball so as not to allow a buildup of contaminated material in the valve (Figs. 2-55 and 2-56). The three-piece ball valve can also easily be used for high purity gas service, where bleeds and purges are installed in the valve (Fig. 2-57). Both services often use welded-end bodies, which can be easily built because of the three-piece design.

Another style used in oilpatch work is the true-union type (Fig. 2-58). This is actually a three-piece design also, but with union nuts on each end piece. The center section is still removable from the line after the end pieces are welded or threaded to the adjacent piping. In addition to the steel variety, this design is used often in plastic body valves where it is probably the most common type. Some plastic body valves are the

(a)

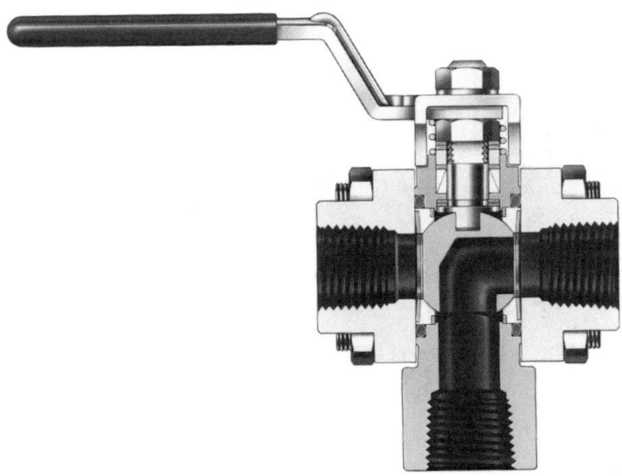

(b)

**FIG. 2-54.** Three-way ball valve with bottom outlet.
*(Courtesy of Swagelok Co.)*

single-union type (Fig. 2-59), in which one end of the body is cemented or solvent-welded solidly to the pipe and the other end has a union for breakout.

Still another design often used in oil production services, but also used in process work, is a three-piece threaded body (Fig. 2-60). This design has two ends that thread into a center section, each including the seat. These valves are generally somewhat sturdier than the two-piece threaded designs discussed above, and are sometimes available with

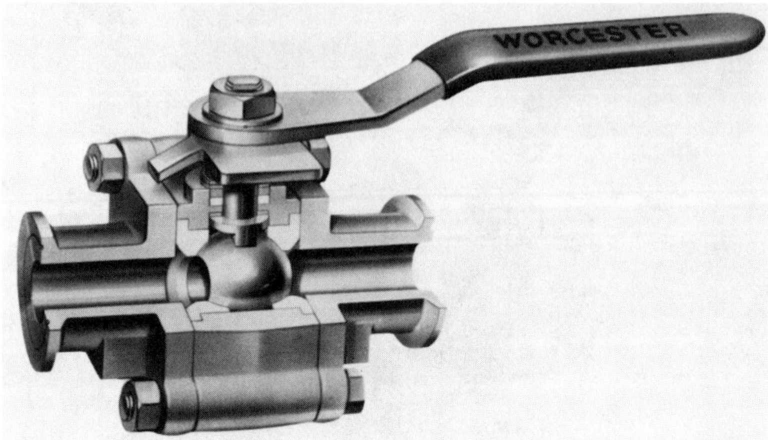


**FIG. 2-55.** Threaded-end three-piece bolted-body ball valve with clamp ends and cavity-filler seat design for sanitary service. (*Courtesy of Worcester Controls Corp.*)

**FIG. 2-56.** Detail of cavity-filler seats. (*Courtesy of Swagelok Co.*)

flanged ends. To help in the prevention of accidental disassembly, the ends are often assembled with a thread adhesive.

All the designs described above are for small valves, from about ¼-in. pipe size (DN 6) to 2-in. or 3-in. pipe size (DN 50 or DN 75). In larger sizes, and in many cases starting from ½-in. pipe size (DN 12), ball valves are generally flanged like any other valves. Even small valves are often built with flanged ends for corrosive or high-pressure services, since flanged joints hold pressure better than do threaded joints and socket-weld end ball valves are so difficult to install reliably. In flanged-end construction, the end-entry is the cheapest. For cross-country transportation piping or for services where leaks are unusually intolerable, welded-end ball valves are common and are generally of the one-piece welded-body type or the top-entry type.

**Fig. 2-57.** Three-piece bolted-body ball valve with butt-weld tube ends and purge connections for high-purity gas service. Valve has air actuator. (*Courtesy of Worcester Controls Corp.*)

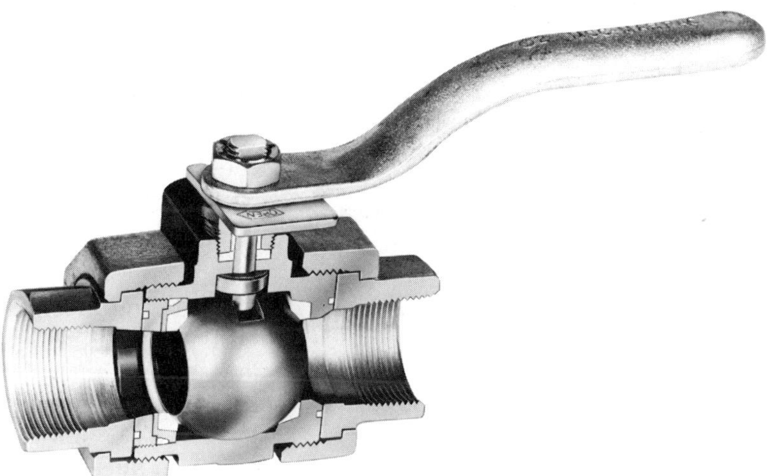

**Fig. 2-58.** Cross section of double-union ball valve. (*Courtesy of Marpac, Inc.*)

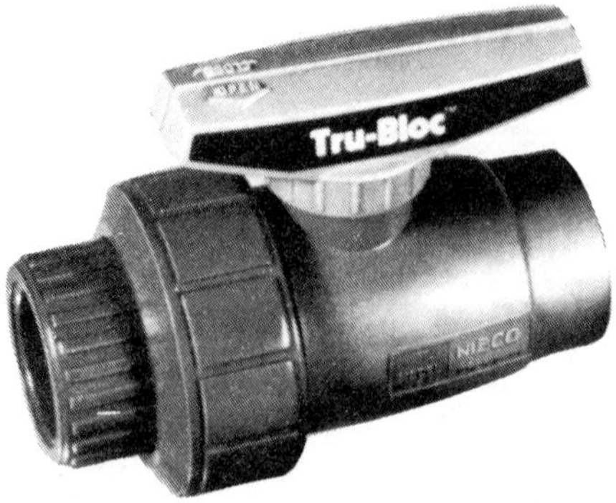


**Fig. 2-59.** Single-union PVC ball valve. (*Courtesy of Nibco, Inc.*)

**Fig. 2-60.** Three-piece threaded-body ball valve with socket-weld ends. Extensions and end-piece contours are designed to dissipate heat from welding. (*Courtesy of Watts Regulator Company.*)

For higher pressures and larger body sizes, the split-body and the one-piece valves have become more common. The split-body design can also be plastic-lined (Fig. 2-61). Split-body valves in larger sizes are often made with two body joints, with two identical end pieces and a center section, somewhat like the smaller three-piece bolted types. This design lends itself to an interesting variation known as the tank-bottom valve (Fig. 2-62), where one of the end pieces is designed to be integrally welded to a tank or vessel.

A quite different type of ball valve is known as the single-seat type (Fig. 2-63a), as well as by proprietary names. This valve has a rising stem, like a gate valve or a lift plug valve. Like a lift plug, it has a slot in the stem that cams the stem 90 degrees around as the handwheel turns and lifts the stem. This valve has a ball, also referred to as a core since it really is not a complete ball, that is loosely attached to the stem with a T-head similar to that of a gate valve, so that the ball can be lifted. There is only

**Fig. 2-61.** Split-body full-port plastic-lined ball valve. (*Courtesy of Atomac, Durco, GmbH.*)

**Fig. 2-62.** Tank-bottom ball valve designed to be welded directly into the bottom of a vessel. (*Courtesy of Worcester Controls Corp.*)

(a)

**Fig. 2-63.** Single-seated rising-stem ball valve. (*Courtesy of Orbit Valve Corp.*)

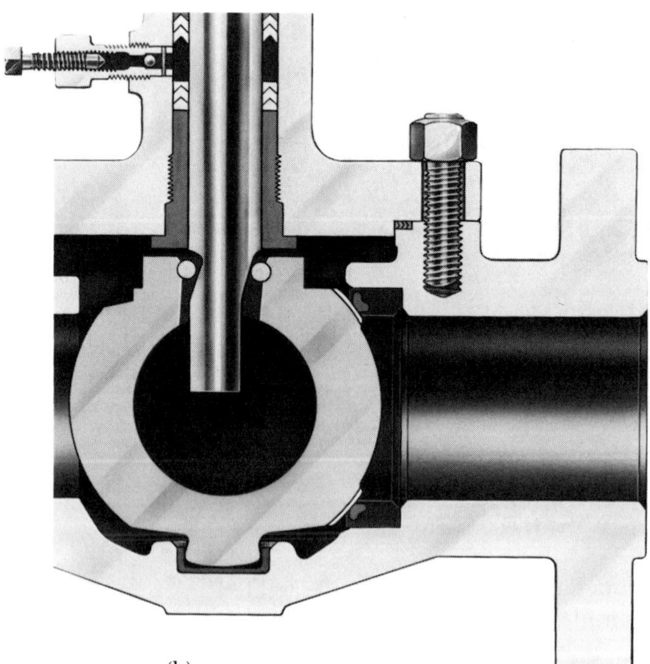

(b)

one seat, on the downstream side (Fig. 2-63b). As with any floating-ball type, in the closed position line pressure holds the ball against this seat. The difference is that the stem can exert force on the ball to seat it more tightly. The single-seat ball valve is capable of very tight shutoff. It will seal in the opposite direction, but generally not as well. Its application, like that of the lift plug, is for absolute separation of two products or for better shutoff of light fluids. The single-seat type can even be machined with grooves and a bleed to atmosphere for a double block and bleed. Because there is only one seat, a special body vent is not required to prevent overpressuring since the body cavity is always open to the piping on one side.

It is theoretically possible to build a standard ball valve in double block and bleed configuration, but it is not practical because of the poor seating ability of the upstream seat. Only a trunnion-mounted sprung-seat design is practical to build as a double block and bleed. Many valves are built with bleeds, but generally this is done when it is desirable to purge the body cavity of some fluid right after the valve is closed. Even though it is true that the seats do not hold pressure that well, it is also true that the leakage rate can be very slight. A valve that has been closed under pressure, then removed from the line, may retain a surprising amount of pressure that can cause serious injury if not properly relieved prior to maintenance activities. The body cavity may also retain quantities of line fluid, which would require thorough purging if the fluid is hazardous. This can be avoided somewhat by means of a ball vent, which is drilled through the ball to equalize pressure in the body cavity with the inside of the flow area. Sometimes this hole is drilled through the top of the ball, under the stem slot, which is the thinnest point in a solid ball. Occasionally it is drilled through the side of the ball. There are services where a vented ball is required, such as in cold services where a liquid trapped behind the ball could vaporize and build up high pressure so quickly as to explode. There are other cases where vents are prohibited, such as in food service where it represents a pathway that would allow buildup of uncleanable contaminants.

There are very small barstock ball valves (Fig. 2-64), but the barstock valve is just as likely to be built as a plug type. Quarter-turn valves in general are unusual in barstock or tubing-type valve usage. The needle or globe type is more common. Ball valves are chosen in these sizes for specific abilities, such as having full-port straight-through flow passages, which provides the capability of being rodable. This feature is being used more commonly now on gage valves and other small instrument takeoffs, or root valves. Externally, these valves do not really look any different from the corresponding plug or globe type. Another specialty adaptation of the ball valve is the hermetically sealed design with a stem containing magnetic inserts, inside the sealed bonnet, with a handle containing magnets outside the stem area (Fig. 2-65). This permits the

**Fig. 2-64.** Small barstock ball valve. (*Courtesy of Swagelok Co.*)

ball to be operated with no physical contact between the handle and the stem.

Very large ball valves, above 24 in. (DN 600), are primarily found in cross-country pipeline applications because they will shut off quickly without prelubrication. Their primary competitor is the through-conduit gate valve. The ball valve is at a disadvantage, though, because of its metal mass, which means increased cost. It also means serious dynamic problems, since the ball valve can be made to operate rather rapidly. There are actuators capable of opening and closing quarter-turn valves very quickly. However, the valve must be capable of withstanding the high forces required to get the ball in motion and those that result when the ball stops at the end of its travel.

The advantage of a ball valve over the through-conduit valve is that the overall envelope of the ball valve is much smaller. Even including the actuator, the ball valve is not much larger than the pipeline, while the through-conduit gate can be four to six times the pipeline size in overall length or height.

The dynamic forces involved in sudden openings and closings also apply to smaller valves as well, but of course size and speed of operation are both factors. The problems become more severe in metal-seated valves, since there is very little shock-absorbing capability. The consequences of high operating forces are heavy wear and dynamic forces that are sometimes transferred into the piping and sometimes even into the supporting structure. If you suspect you may be in a circumstance where this could be a problem, such as in valves over 18-in. size (DN 450) at reasonable pressures or as small as 6 in. (DN 150) in high pressures, or

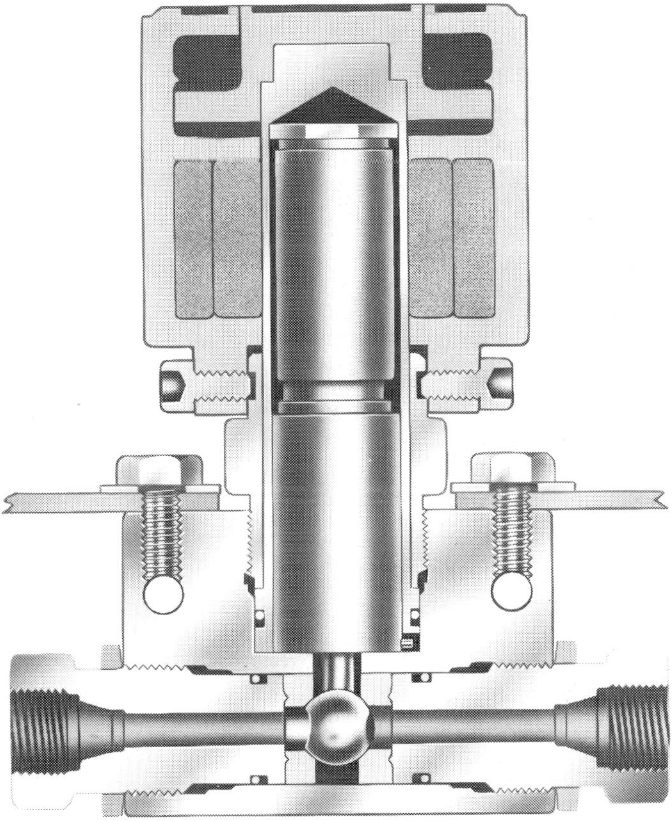

**Fig. 2-65.** Hermetically sealed ball valve with magnetic operator, requiring no stem seal, for high-pressure service. (*Courtesy of Autoclave Engineers, Inc.*)

with closing speeds much less than 2 seconds, it is time to solicit assistance from a more sophisticated valve designer.

Another type of ball valve, which is most commonly used as a control valve, is the eccentric ball valve (Fig. 2-66). It is essentially one-half of a ball with one seat, and as such it is very directional, more so than the rising-stem ball described above. The eccentric ball is a true quarter-turn valve, like other ball valves. It is often referred to as a "rotary" valve or other names to distinguish it from a more standard ball valve. Because of its eccentric construction, the ball is rigidly mounted with a trunnion and sometimes with a top bearing where the stem does not serve to guide the ball at all, only to rotate it.

Almost all eccentric ball valves, and some of the high-performance or special-purpose ball valves, have a distinctive shape to the edge of the ball that enters the flow stream first (Fig. 2-67). These are known as characterized balls. Just like special globe disks, these contours provide some type of special flow characteristic. Some are V-shaped, to provide

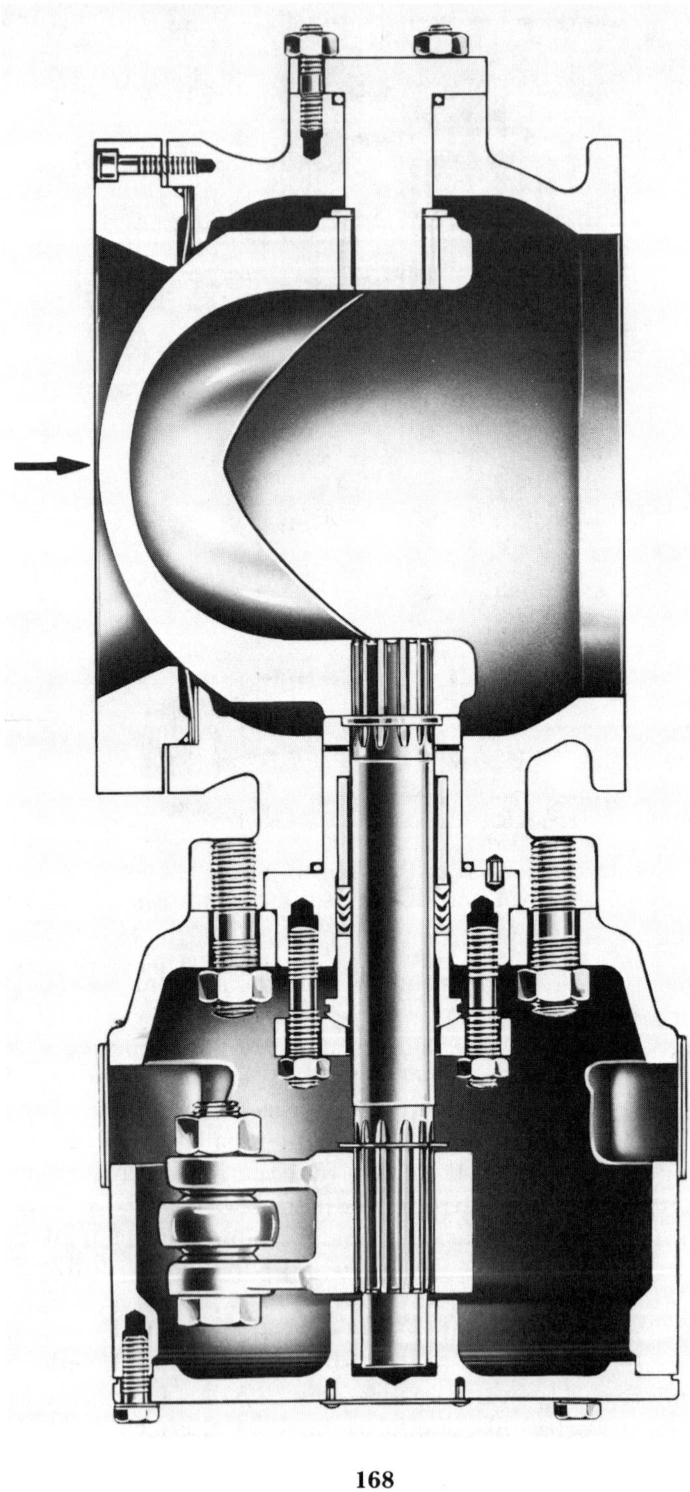

**Fig. 2-66.** Single-seated flangeless (wafer body) characterized ball valve for throttling service. *(Courtesy of Fisher Controls Intl.)*

**Fig. 2-67.** Some single-seated characterized ball valves are also available with end flanges. (*Courtesy of Valtek, Inc.*)

accurate control when the percent open is very small and to provide a constant increase in flow rate with a constant rotation of the actuator (equal-percentage opening). Some are U-shaped, to provide a somewhat faster opening (quick opening).

Some high-performance ball valves, the metal-seated types for severe service, have a notch that roughly parallels the edge of the ball but is set in slightly, known as arcuate cut (Fig. 2-68), which has the effect of opening the flow passage by a large percentage all at once and then gradually. This is good for erosive flows and high differential pressures, so as to minimize the tendency for high opening velocities to cut into the ball. The arcuate cut is also used in a number of high-performance ball valves intended for on-off service, because of its improved performance under high differential pressures and erosive services. Another technique used in high differential services where cavitation could be a problem is to provide internal baffles to break up the flow stream so that cavitation cannot occur (Fig. 2-69).

The eccentric ball valves used for flow modulation are almost all wafer types, installed between two flanges with a set of bolts that passes from one line flange to the other, and only centering the body and not actually bolting into it. This is not really a requirement for this type of valve, but the design is well ingrained in standard practice. These valves are end-entry design just like many others, and as with most eccentric butterfly valves, their preferred orientation is with the flow into the disc from the tailpiece end. These valves are often designed to rotate only 60 degrees from full open to full closed, since there is only one seat and these valves are often not designed for tight shutoff anyway, just like

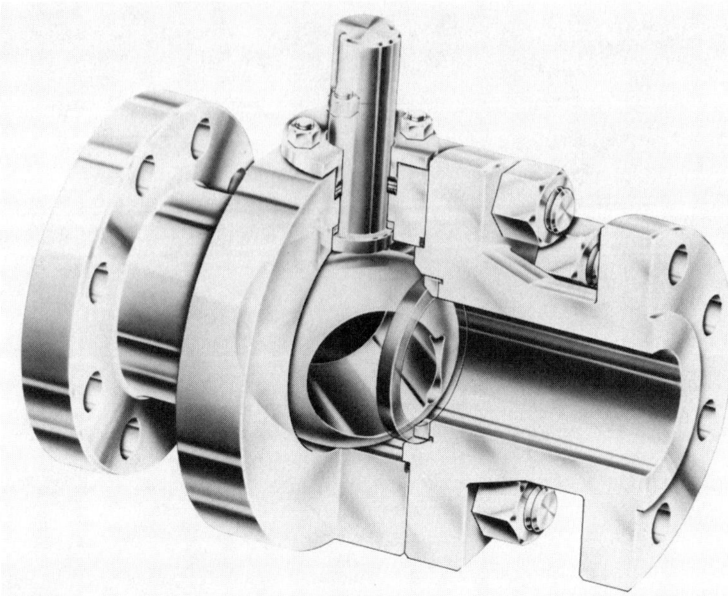

**Fig. 2-68.** Arcuate cut contour in leading edge of metal-seated high-performance ball valve for improved performance with high-pressure drops and quick opening. (*Courtesy of Mogas Industries, Inc.*)

many other modulating valves. For proper control, wafer-type valves almost always have spline connections between the ball and shaft and between shaft and actuator linkage bracket for minimal slop or hysteresis, when the ball and stem are not machined as one piece.

In common with butterfly types, eccentric-type valves are almost always designed with the shaft in the horizontal plane with the ball rotating upward to open. This presents an aspect to oncoming fluid very much like that of a gate valve, and means that debris or solids traveling with the fluid will be swept rapidly through the seat area as the ball closes and will be less likely to be trapped there and damage the ball and seat. These valves are always built as trunnion type. In general, the rangeability (ratio of maximum flow to the smallest flow that can be adequately controlled) is quite a bit higher than that of a globe-type control valve; in addition, these valves are much more able to handle entrained solids such as fibers and slurries.

The wafer-type balls are commonly available in sizes from 1½-in. or 2-in. (DN 40 or DN 50) up to 18-in. (DN 450) or larger pipe size. They can be used with manual lever-lock or handwheel actuators, and in fact are used as manual valves (Fig. 2-70) more often than many globe-type control valve bodies are. But, for smaller sizes, there is another design that is more like a modified manual valve that is commonly used for control (Fig. 2-71). This type, which is generally a three-piece bolted

**Fig. 2-69.** Ball valve for throttling or high noise service, equipped with anticavitation baffles. (*Courtesy of Neles-Jamesbury, Inc.*)

**Fig. 2-70.** Flangeless (wafer) end-entry reduced-port ball valve. (*Courtesy of Worcester Controls Corp.*)

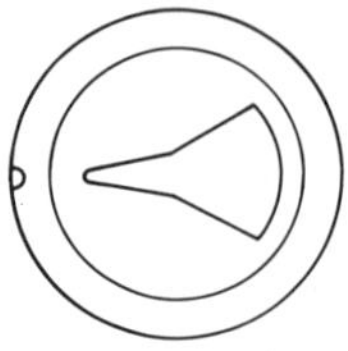

FIG. 2-71. Characterized seat for ball valve in modulating service. (*Courtesy of Worcester Controls Corp.*)

body construction, is also primarily unidirectional but has a characterized seat instead of a characterized ball, and the ball itself is a standard design. Many of these valves have been available for only a few years. They are generally available for the same range of pressure and temperature conditions that similar conventional ball valves are capable of withstanding. They are primarily ½-in. to 2-in. size (DN 12 to DN 50), sometimes 3-in. (DN 75), like the corresponding manual valves.

Unlike the characterized ball types, these valves are designed with a characterized seat, where the throttling takes place, on the downstream side just as in a conventional ball valve. They are, in fact, a conventional floating-ball design with slotted ball and tongue shaft. The fit between ball and stem has to be much closer than standard, to cut down on hysteresis, and in fact the ball and stem are usually matched and lapped together for near zero fit. They are capable of tight shutoff like other standard ball valves. For accurate control, the actuator bracketing and actuator internals have to be constructed tighter, with better alignment and less possibility for mechanical slop, than is needed for a simple actuated on-off valve.

The seat can be cut in a number of patterns to provide equal percentage or linear characteristics or a special characteristic. The equal-percentage pattern is the most common. By changing the included angle of the seat opening, the slope of the flow curve can be changed somewhat, but its general behavior remains equal percentage. Seat material is metal, generally powder technology sintered metal, which is required for the unsupported area that sees a fairly high force at its edges in throttling service. The use of sintered metal instead of solid metal allows the "alloying" or impregnation with graphite or fluoropolymer powder to yield a self-lubricating material that can take high unit seating stresses without galling or needing external lubrication.

### *Displacement Valves*

There are a few valve types collectively known as displacement type (Fig. 2-72). These valves are commonly called the ram valve, the catalyst sampling valve, or the tank bottom valve, depending on their applications. All of these valves are basically a piston that travels inside a cylinder. The valve is attached to a fluid space, either a pipe line or a tank, that the piston acts against. When the piston is withdrawn, there is

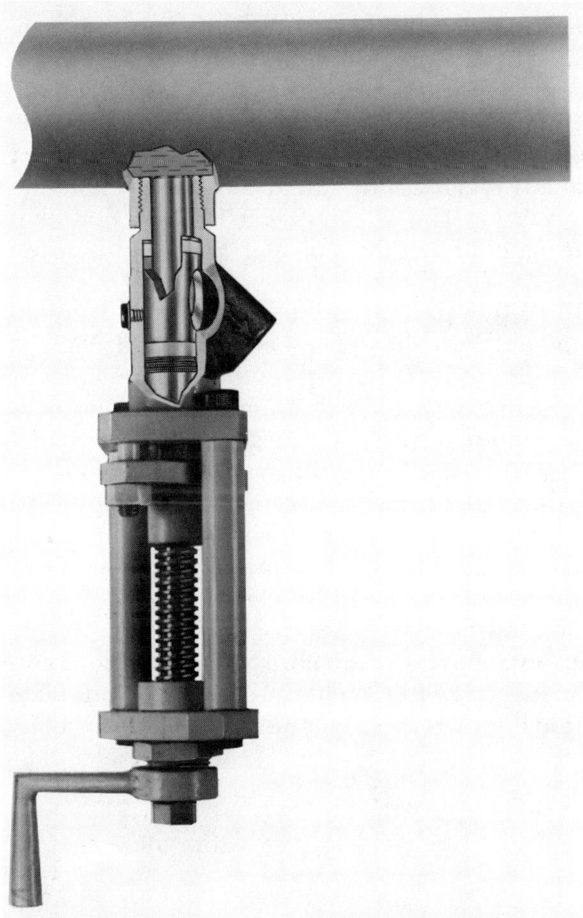

**Fig. 2-72.** Piston or ram-type flush bottom displacement valve for sampling or draining. (*Courtesy of Fetterolf Corporation.*)

an opening to one side that the fluid can flow through. The ram valve got its name from the fact that the piston looks and behaves like a hydraulic ram.

These valves are on-off in the truest sense of the term. The piston fills the fluid area inside the valve, often extending all the way to the point where the valve attaches to the pipe or vessel. Opening the valve is done by retracting the piston until the inner face of the piston slides past the opening that leads downstream. Many of these valves are also equipped with a seat that the rim of the piston can seal against at the end of its travel. Although these valves are by and large built as drain valves, with slight modifications they can be configured as in-line block valves. The in-line use of these valves represents a relatively new application, though, and at this time there are not many installed examples of this type.

The advantage of this valve for draining, as opposed to a simple gate valve mounted in a drain line, is that the piston can rise to the end of the line, completely filling the drain line and presenting a flat surface to the inside of the space being drained (Fig. 2-73). In the case of pipe lines, the end of the piston can be contoured to match the contour of the inside of the pipe. This has several advantages. One, the drain line presents no pocket for material to collect in. This is significant for any flow that includes solid particles, or material that can polymerize or set up inside the line.

With an extension of the piston past the end of the line, the piston in the closed position can extend into the space being drained. This is useful in vessel or tank drains, so that if a layer of accumulated or polymerized material is present, the piston extension provides an opening through this material when it retracts. The piston can even sometimes be used to ram its way through a layer of material inside a tank. Valves designed to break through a layer of material must be designed with a small stem and a larger seat area, and rise up into the vessel instead of retracting down into and through the valve. A variation of this design is the tank spray valve (Fig. 2-74), which can be inserted into a closed reactor for automatic washdown.

Because most of these valves have a relatively small production, and because of their geometry, they are often built as weldments from plate and forgings. This additionally increases their usefulness, since many

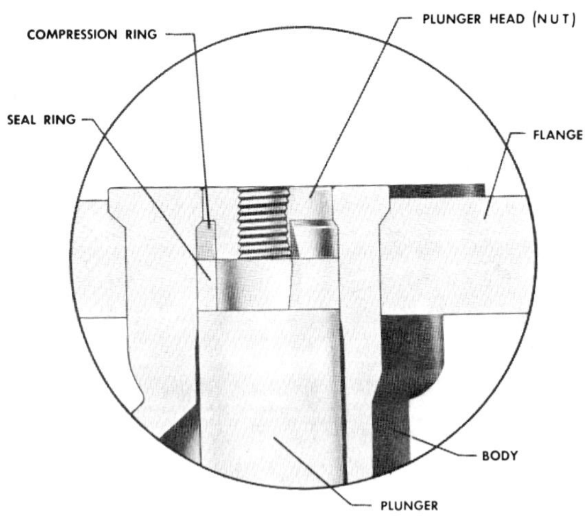


THE "RAM-SEAL" PRINCIPLE

**Fig. 2-73.** Detail of sealing mechanism and flush-surface configuration of ram valve. (*Courtesy of Fetterolf Corporation.*)

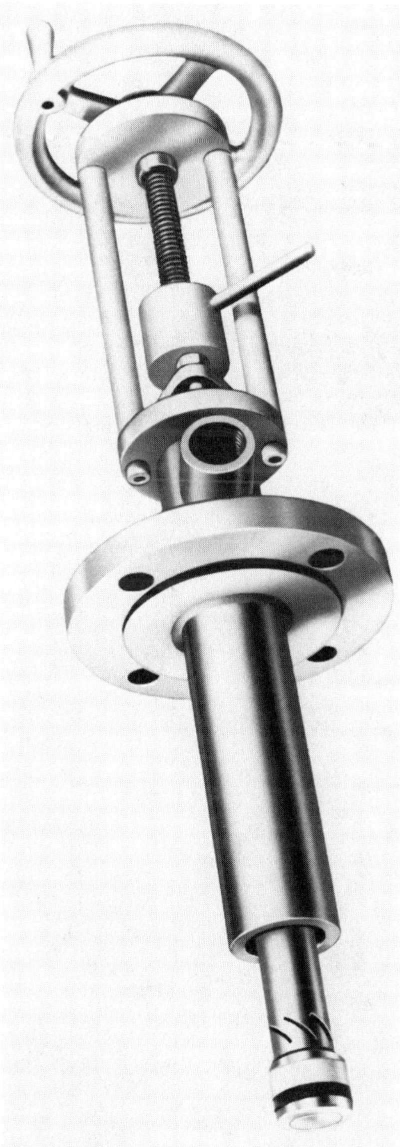

**Fig. 2-74.** Tank spray valve adapted from ram type, designed to be inserted into reactor for internal washdown between batches. (*Courtesy of Fetterolf Corporation.*)

exotic materials tend to be available more as plate, pipe, and forgings (or only as plate), rather than in the form of castings. There are a number of specialized designs, in part because so many of the valves are fabricated rather than made from cast bodies. This makes it easier to customize the design.

The outlet of the valve is at an angle to the inlet. The inlet can be welded, threaded, or flanged (Fig. 2-75) to the header or the tank, and

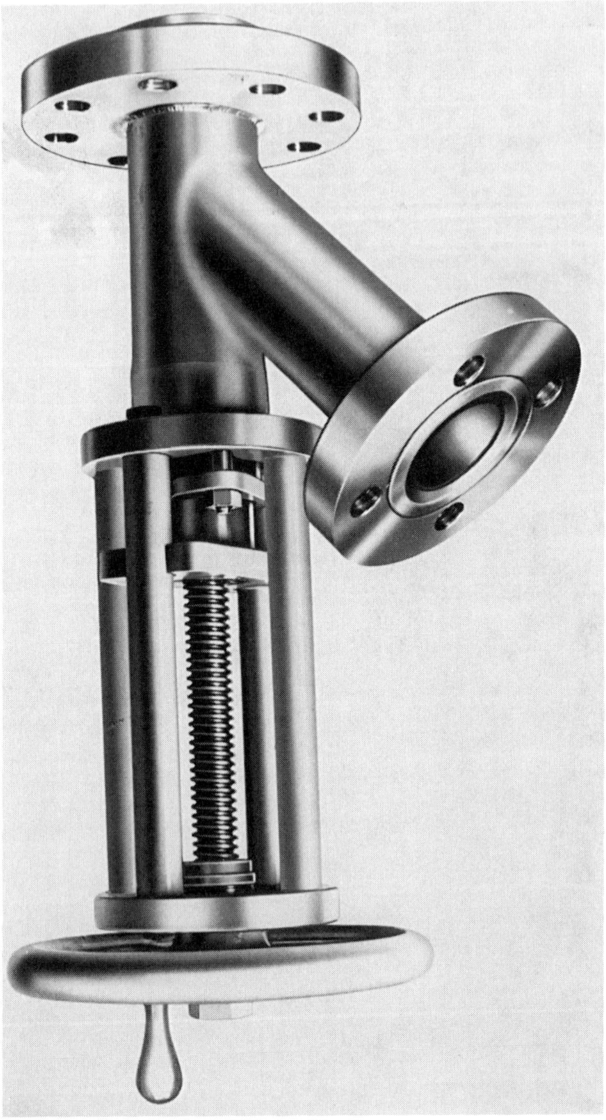

**Fig. 2-75.** Flanged-end ram sample valve. (*Courtesy of Fetterolf Corporation.*)

the outlet is often threaded. The outlet can be a smaller size than the inlet. When these valves are used for sampling, a sample container can be connected directly to the valve or an additional valve can be connected to the outlet of the sampling valve for convenience and additional shutoff ability. (The valve can also be used for injection, in which case the flow is opposite from the sampling mode but the valve is mechanically the same as the sampling configuration.) The entire valve can be

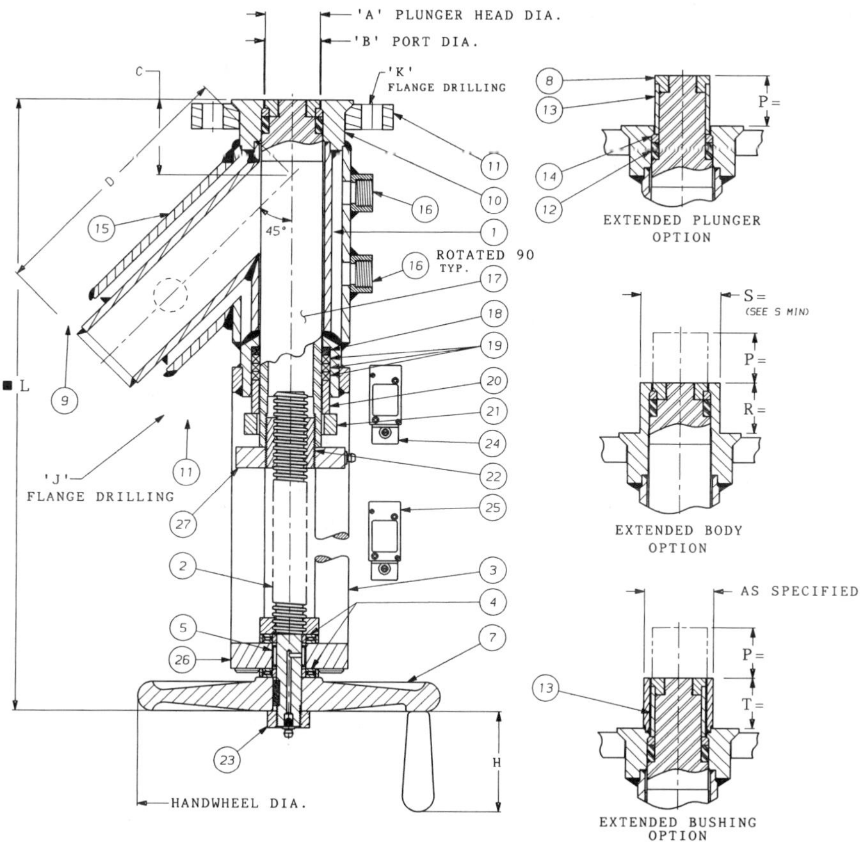


**FIG. 2-76.** Jacketed ram sample valve. (*Courtesy of Fetterolf Corporation.*)

jacketed for heating the body (Fig. 2-76), and sometimes the valve is supplied integral with a short section of header pipe, which then can be integrally jacketed. Very often these designs are used with high-alloy materials. Although the valve can be designed to be flanged up to the pipe it is intended to be attached to, these custom configurations can save a great deal of expensive material and in some cases can even cut down on the number of field welds, which is always desirable.

Because the piston is occupying the entire volume of the branch or drain pipe, when a soft seat is used the seat is relatively protected. In other words, when the face of the piston travels across the outlet opening, that action effectively shuts off most of the flow through the valve. Then, to seat, the piston contacts a circular seat somewhat like a globe valve seat but more angular. This provides a tight shutoff, and since very little flow is passing through the valve when it seats, very little high-velocity flow occurs across the seat, so it is not prone to damage or cutting. In general, these valves have very good shutoff abilities.

The displacement valve has a threaded stem like that of a gate or globe valve. Generally, the stem drives the piston directly, but some early models had a compound arrangement whereby the stem traveled twice as far as the piston did. This had the advantage of greater force to push material out of the way but was more mechanically complex. The stem length is another incentive to consider integrally mounting or close coupling the valve to the line or vessel, since every additional inch that the piston must move means an additional 2 in. that is required on the operator end for clearance. The displacement valve generally has a crank in small sizes, or a handwheel in larger sizes, and can be equipped with gears or motor or piston actuators just like a gate valve. Sizes range from ½-in. pipe size (DN 12) up to 6 in. or 8 in. (DN 150 or DN 200), although large ones are generally used for special purposes.

When the inlet side is built at an angle and the outlet side is moved closer to the end of the piston, the displacement valve becomes a Y-pattern block valve like a Y-pattern globe. Its shutoff ability remains the same. Since this valve is already fabricated from plate and pipe, it becomes a natural for block valves in piping systems made of exotic materials. Since generally the use of high alloys is a forced choice based on high corrosivity or some other extreme condition, and since valves often represent the single most expensive item and the single most exasperating item to source in a piping system, this is an active solution.

### *Other Block Valves*

There are a number of other valves that, while important for their specific area of interest, are not used in large numbers or are used only in very specific situations. This section will attempt to touch on these valves briefly.

One specialized design is the hydrant (Fig. 2-77). Basically, a hydrant is a valve that is in an underground line but whose operating mechanism and outlet are aboveground. The seat and disc assembly resembles that found in a globe valve, except that the orientation of the seat and disc is backward or upside down from that of a globe valve relative to the position of the operator. The vast majority of hydrants are in municipal water systems or industrial fire-water systems, but a few are used for utility water in places where the ambient temperatures are low in winter and the frost line is deep.

Most hydrants are designed as dry barrel, where the volume inside the hydrant is normally dry. Since the valve seat is at the bottom of the hydrant barrel, keeping water out is easy. Getting the water out after the hydrant is opened is accomplished by providing small drain openings that are exposed or pushed open by the closing of the disc. The water then seeps into the ground around the hydrant. To allow properly for

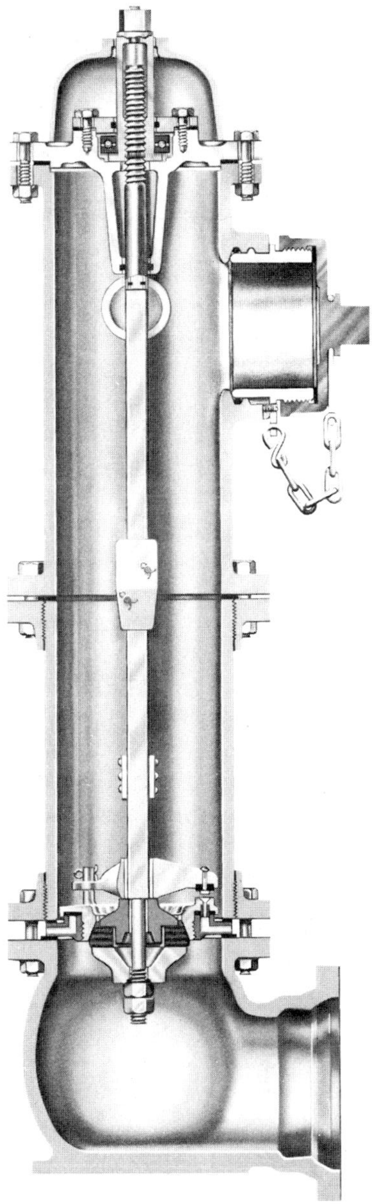

**Fig. 2-77.** Cross section of hydrant valve with automatic drain feature. *(Courtesy of American-Darling Valve, division of American Cast Iron Pipe Company.)*

this drainage, a volume of about a cubic yard of gravel should be provided below and around the base of the hydrant for the water to drain into.

The conventional hydrant design calls for the valve disc to be below the seat. (In hydrant terminology, the seat and disc assembly is often referred to as the valve.) This way, opening the hydrant will push the disc down, away from the seat, and water pressure will tend to close the valve, which is the opposite of the way that conventional globe valves are designed. This design means that the valve should fail closed if a traffic accident should damage or break the upper portion of the hydrant. Since the possibility of traffic accidents is always present, some hydrants are designed with a breakaway connection at grade such that the hydrant will break in half when struck. The design of the valve, with water pressure tending to seat the disc and close off flow, is supposed to prevent the gushers that commonly appear in the movies when hydrants are the victims of traffic.

Since the main valve is below grade, the connections above grade are generally not valved in municipal water systems. In industrial hydrants, they often are (Figs. 2-78 and 2-79), even when the hydrant is the dry-barrel type. Regardless, these connections are normally designed to be compatible with fire hoses and incompatible with anything else. The most common hydrant outlet connections are a pair of 2½-in. National Hose Thread (NHT) outlets and one 4½-in. NHT, known as a pumper connection for the 4½-in. hose that fire engines carry to connect to hydrants. The odd sizes and incompatibility with National Pipe Thread (NPT) dimensions are designed to make it difficult for anyone to withdraw water from hydrants for purposes other than fire fighting. This is the same reason that the operating nut on a municipal hydrant is generally five sided, on the theory that not everyone owns a pentagonal wrench. Industrial hydrants, though, often have handwheels on top to make operating the hydrant easier in an emergency.

Fire-fighting equipment could be a whole book in itself, but what follows is a brief background on the odd sizes mentioned above. During the nineteenth century, as cities on the East Coast developed their

**Fig. 2-78.** Gate valves for independent shutoff of hose nozzles. *(Courtesy of American-Darling Valve, division of American Cast Iron Pipe Company.)*

**Fig. 2-79.** Ball valve for industrial fire hydrant installation. (*Courtesy of Elkhart Brass, Inc.*)

infrastructure, water systems were installed to provide both drinking water and fire-fighting water. Each city developed its own, often built with components manufactured in town by the small foundries and manufacturing plants of the day. Many hydrants were designed with oddball connections so as to make it difficult for the citizens to get unauthorized water from the fire hydrants. Great, uncontrollable fires were still common in urban areas because of the time it took to report fires and extinguish them with manual and steam pumpers. One of the last such fires occurred in Baltimore on February 7–8, 1904. When the conflagration got out of hand, the city requested assistance from other fire departments, which sent men and fire engines in by rail from surrounding states. When this equipment arrived, virtually none of it was usable because the hoses and couplings all had thread standards that would not interchange with those in Baltimore. This and other incompatibilities eventually led to some standardization in thread designs while still making it unlikely that the average citizen could withdraw water from city hydrants.

In addition to the standard hydrant designs, there are a few hydrants available for other purposes. Where heavy freezes and deep frost lines are a problem, many outdoor water lines are buried and water is available through hydrant valves (Fig. 2-80). These valves are often 1-in. to 2-in. pipe size (DN 25 to DN 50), contrasted to the common fire hydrant size of 4-in. or 6-in. inlet size (DN 100 or 150). (Fire hydrant sizes stated as 5¼-in. [133 mm] or other oddball sizes refer to the valve size, not the inlet size.) In areas where freezing is so rare as to be disregarded, a wet-barrel design is common since it is cheaper. Here, the valves are on the individual outlets and there is no single underground inlet valve. The body of the hydrant is always filled with water.

Another important set of valve types that are often not considered part of the piping system, but that are certainly crucial to operating a process plant, are the manifold valves and gage valves that are attached to the process piping to connect and isolate the instrumentation. Basical-

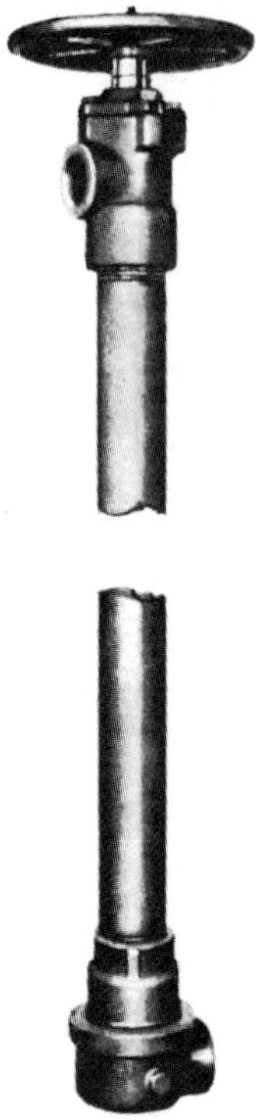

**Fig. 2-80.** Yard hydrant valve. (*Courtesy of American-Darling Valve, division of American Cast Iron Pipe Company.*)

ly, a manifold valve (Fig. 2-81) is used to connect an instrument that requires two connections to the line at a single location, to allow for one or both to be blocked off independently of the other. Examples are differential pressure instruments or orifice flanges. The manifold valve, which is available in three-valve and five-valve combinations, is designed to allow each of the process connections to flow through or to flow across the manifold valve internally. Many manifold valves now are machined from extruded sections, normally stainless steel, that have a large flat surface on one or both sides for mounting to other instrument items. On the side that connects to the line, female threads or unions are provided. Each flow line and valve are machined into this extrusion. Standard barstock-style instrument valve stems, packing, bonnets, and handles are provided for each of the three or five valves in the manifold.

There are two different types of valves commonly referred to as gage valves. One is a valve for pressure gages (Fig. 2-82a,b), which can also be used for flow instrument connections and other purposes, and the other is a valve for liquid level gages. The valve that is designed for level gages is also occasionally used for pressure gages.

The liquid level gage valves are designed to support level gage glasses, with adjustable outlets to make up minor differences in orienta-

**Fig. 2-81.** Manifold valve (left, on mounting bracket) and gage valves (right, mounted on orifice flanges) with bleed valves installed. (*Courtesy of Hex Valve Co.*)

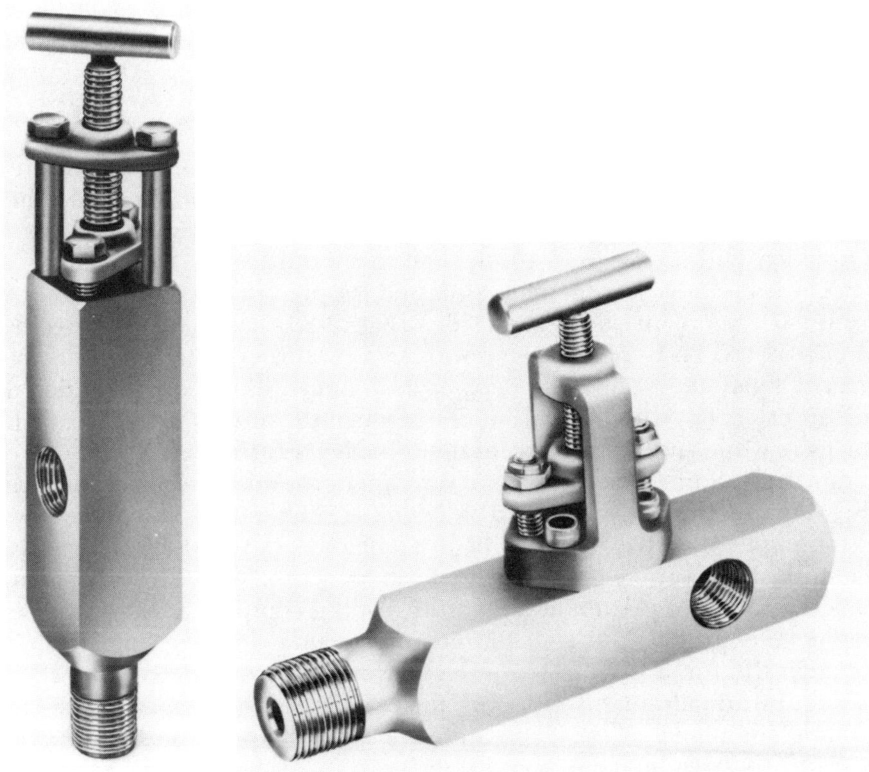

(a) (b)

**Fig. 2-82.** Gage valves with male inlet and multiple outlets, for orifice flowmeters and for pressure gages. (*Courtesy of Hex Valve Co.*)

tion from one valve to the other. They are usually designed to be used in pairs, with a left and right hand combination which refers to the orientation of the outlet, or gage connection, as viewed while looking toward the vessel or pipe that the gage is attached to. One right hand and one left hand are required for each gage segment. The shutoff valve is built like a needle valve, with a handwheel to open and close the valve. The stem is in line with the main axis of the valve body. On the outboard side of the valve seat is the gage connection itself, at right angles to the axis of the valve body, plus a drain or vent connection that is usually directly opposite the gage connection. The gage connection almost always has a ½-in. (DN 12) female threaded outlet, to accept a gage with male threads, but an integral union is often provided on the gage valve. This allows easier makeup of the gage assembly, allows the entire glass to be rotated for better visibility, and allows slight irregularities in the orientation to be compensated for without damaging the fragile gage glass. On the inboard side of the valve seat there is often an excess flow check valve, machined as part of the gage valve, which consists of a slightly larger diameter segment of the flow path with a ball in it, like a ball bearing. The principle of the excess flow check is that very small flows, such as would be caused by the vessel level fluctuating and causing liquid to flow in and out of the gage glass, will not disturb the ball, but a very large flow with high differential pressure, as would result from the glass breaking and pressurized liquid pouring out of the vessel, would move the ball into its seat and shut off the flow. The liquid level gage valve is often forged steel or stainless steel, but barstock is occasionally used. The size of the vessel connection varies, but 1 in. (DN 25) is common. Threaded and socket-weld connections are both used.

The same valve is sometimes also used for pressure gages, but a more common type of pressure-gage valve (Fig. 2-83) has the stem at right angles to the body of the valve with two or three connections downstream from the seat. This body style is more like the conventional needle valve, but with multiple outlets. One of the outlets serves to attach the gage while another is the vent used for depressuring the gage before removing it. Sometimes one of the vent openings will have another valve threaded into it, sometimes sold as a complete assembly. This valve may also be a needle valve, or it may be just a bleeder valve, which is basically like a bushing with a plug threaded into it (Fig. 2-84). This plug may have a retainer to prevent it from unthreading completely, but it always has a hole drilled into the bottom intersecting another hole drilled through the threads from one side, so that as it is unscrewed the flow passage gradually opens. Its only purpose is to depressure the small volume contained in the gage and downstream from the block valve prior to unthreading the gage itself.

The common gage valve normally has a male connection on the side that attaches it to the line. Generally this connection is threaded, but

**Fig. 2-83.** Gage valve shown in typical installation. (*Courtesy of Hex Valve Co.*)

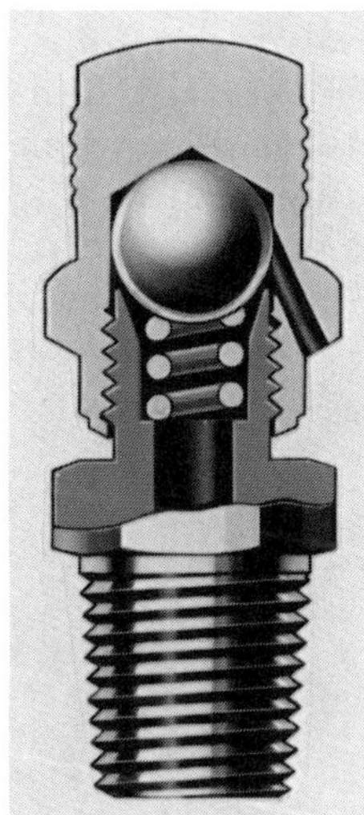

**Fig. 2-84.** Bleed plug with integral check valve. (*Courtesy of Swagelok Co.*)

socket-weld or other connections are available. The size of the male connection is often ½ in. (DN 12), but for mechanical strength and resistance to breaking or fatigue failure, many users prefer ¾-in. pipe size (DN 18). The standard sizes for the pressure-gage and vent connections are almost always ½-in. pipe size (DN 12) and female threaded. In a few cases, the internals of the pressure-gage valve are a ball type

(Fig. 2-85), which is used when the entire assembly must be rodded out. These are unusual, with the needle type predominating at this time.

Previously, the almost universal practice was to use a standard piping valve as the root valve, or main valve coming off of the pipe line, with an assembly of nipples, fittings, and needle valves attached to the piping valve. The gage valve replaced this whole assembly, but originally the gage valve was just attached behind the root valve. Many users now are abandoning the separate root valve and using the gage valve as it was designed, to be its own root valve (Fig. 2-86).

Relief valves are another subject that would require a book of their own to describe fully. Here our concern is with the block valves used to isolate them. Normally, standard block valves such as gate valves are used. But the ASME code has some very specific rules on allowable pressure drop and flow area in the lines connecting the relief valves to the equipment they protect. The most germane features of these rules are that a valve with significantly reduced port area is not permitted—which excludes standard plug valves and API 602 pattern reduced-port gate valves, for instance—and that the valve must be equipped with a locking device to prevent its being closed while the relief valve is supposed to be in service. Many operating companies make their own rules that exceed the ASME requirements, especially in the hydrocarbon industries with their high degree of fire danger. Sometimes two relief valves are required, such that one is in service at all times (relief valves are also required to be removed and recalibrated every so often). Sometimes this is done by providing two vessel nozzles, one for each relief valve and its associated block valve, and at other times it is done by providing two relief valves, one nozzle, and a valve that is constantly open to one or the other and cannot be positioned completely closed.

Relief valve inlet block valves may be three-way ball or plug valves with transflow pattern, such that one flow passage starts to open before

**Fig. 2-85.** Three-piece bolted ball valve configured as gage valve. (*Courtesy of Swagelok Co.*)

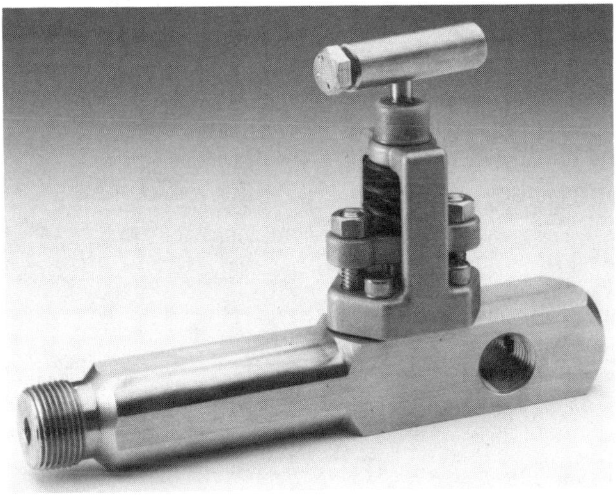

**Fig. 2-86.** Gage valve with extension for mounting on insulated lines, with oversized line size connection for strength. (*Courtesy of Anderson, Greenwood & Co.*)

the other one completely closes, and they may be diverter valves (Fig. 2-87a,b) with plugs that slide back and forth between the two outlet ports. These diverter valve designs are based on and operate in the same manner as the ram sample valves described in the previous section. A very different type has a rotating flow passageway almost like a three-way ball or plug valve (Fig. 2-88).

Valves used to control solid flows, which are either air blown or simply falling through vertical pipe or ducting, are quite different from valves for liquid or gas flows. The knife gate valves discussed in the gate section are one example. Others would include the lockhopper valve, which is a block valve and a flow-control valve in a sense. It consists of a rotating chamber, which is mounted with its axis horizontal in a vertical line. Flow of solid particles is into the top of the chamber, which is basically a bucket. In other words, the bottom of the chamber is solid and solids pile up in the chamber. To allow flow to continue past the valve, the chamber is rotated, by hand or by a motor, 180 degrees until it is upside down, at which time the solids fall out of the chamber and continue on their way. Meanwhile, on the upstream side more solids pile up on the outside of the chamber until it rotates around again, where they fall into the chamber and the cycle repeats.

Another type of solid-flow block valve is the swing gate (Fig. 2-89), which serves the same function as the knife gate but operates in a different manner. The swing gate does what its name implies, that is, the gate is pivoted and swings into the flow path to shut off flow, rather than sliding laterally. Levers and cylinder operators can be used on these

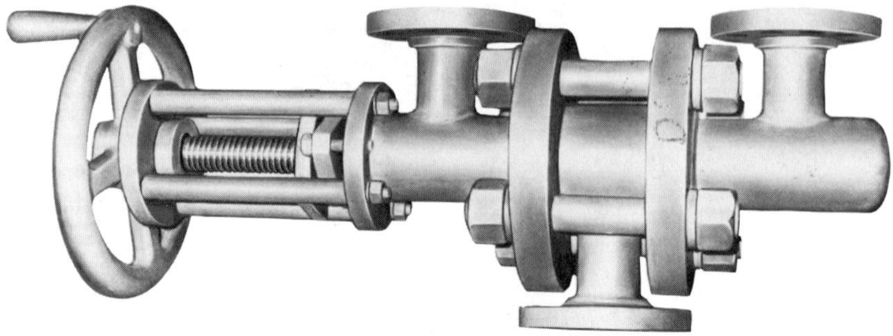

(a)

(b)

**FIG. 2-87.** Changeover valves designed for relief valve inlets with full flow area capability. (*Courtesy of Fetterolf Corporation.*)

valves, which have a slightly smaller overall dimension than knife gate valves do.

Other types of solid-flow control valves include the diverter valve, which is not the same as the diverter valve described above for relief valve inlets. The diverter valve used for solid flows is a flow-switching valve, with a Y-pattern body where flow is into the trunk of the Y and flow is out either of the legs of the Y. There is a vane or blade that is pivoted at the crotch of the Y and seats across one or the other branch, diverting all flow into the other branch. The shaft that the vane is attached to has a lever or some type of power actuator on the outside. Diverter valves could be built as flow-splitting valves, but such applications are unlikely since, for one thing, in most solid-flow systems there is not always an even distribution of flow across the flow area, unlike a liquid flow stream.

There are many other specialized one-of-a-kind valve designs, probably as many as there are machine shops capable of building valves, but these are the ones that an average user is most likely to encounter in a career spent specifying valve types.

**Fig. 2-88.** Relief valve inlet changeover valve that operates by rotating a section of flow passageway. (*Courtesy of Anderson, Greenwood & Co.*)

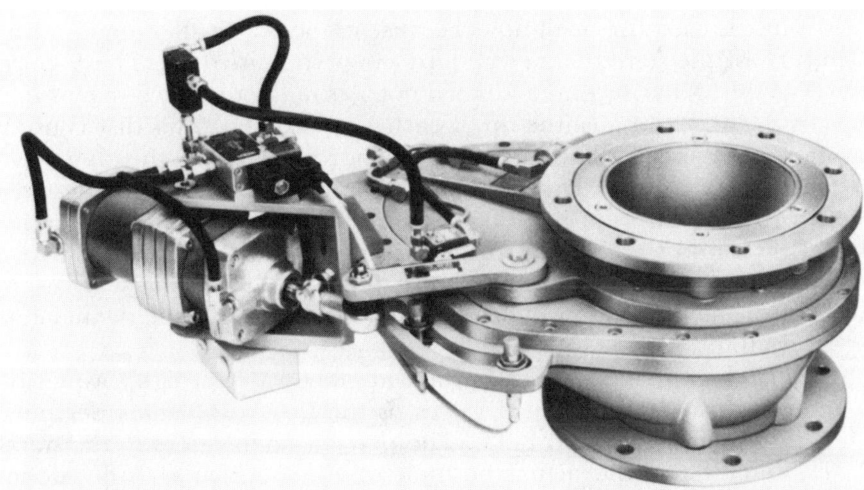

**Fig. 2-89.** Swing disc valve for solids handling, equipped with cylinder actuator. (*Courtesy of Schütte-Meyer GmbH.*)

## Throttling and Control Valves

### *Globe Valves*

For many years, the most common valve type used for throttling flows has been the globe. The principle of the globe valve is found in household faucets and in control valves capable of throttling down several thousand pounds (several hundred kg/cm$^2$) of differential pressure. When the design of the seat (or seats) is varied, the response of the valve, in terms of how much the flow changes when the disc is moved, can be changed. These two factors make the globe very useful for automatic control. Although the valve that we commonly refer to as a "control valve" looks very different from the valve that is usually referred to as a "globe valve," the two basically are the same.

The globe valve got its name when it was first used to throttle steam in industry and on steam locomotives. In the nineteenth century and before, valves were built by foundries and many of the early castings for globe valve bodies were almost spherical. This holds true to this day on many low-pressure steel and iron globe valve castings.

The globe valve (Fig. 2-90) has a circular seat with a disc, or plug, that moves away from the seat to open and into the seat to close. The plug has a contour of some sort in the area directly opposite the seat. The most common contour is the plug-type disc (Fig. 2-91a), which has a straight tapered or cone-shaped section. This type is almost the only one found on low-rated manual valves such as bronze or iron bodies.

There are a number of other types of plug available. The ball-type disc (Fig. 2-91b) is commonly used in high-viscosity services because it opens a greater flow area faster for a given movement of the stem. The needle disc is used for small flows because it opens a smaller amount for each turn of the stem than a standard plug-type disc does. The V-port disc (Fig. 2-91c) has several V-shaped notches in the disc such that as the valve is opened, first just the tip of each V is exposed. This disc is used for more precise throttling since it opens more slowly than the plug-type disc, and since it can be controlled more closely, it is one of the designs used for actuated, or control-type, throttling valves. For bubble-tight shutoff, it is also possible to provide soft inserts in the disc (Fig. 2-91d). Many other special plug designs, known as characterized plugs, are used for control valves. These will be discussed in more complete detail later in this chapter.

One concept that is very important in globe valve design, primarily but not exclusively for control valves, is the flow-rate versus percent-open curve (Figs. 2-92 and 2-93). The percent open refers to the stem travel, and relates the actual position of the stem to its fully open position. Since any actuator, be it a handwheel or an automated actuator of some type, actually moves only the stem and disc, this is a way of

**Fig. 2-90.** Typical globe valve. (*Courtesy of Newmans Inc.*)

relating the input (signal, number of turns, etc.) of the valve to its output (flow through the valve). There are three common curves. Each of these is also known as a flow characteristic curve. One is known as linear, and it is the easiest to graph but not so easy to physically obtain. In the linear characteristic, flow increases by the same absolute percentage for every percentage increase in the stem travel, except as the valve comes very close to full open. A variation of this curve that is much more useful for automatic controls is the equal-percentage curve, where flow increases by a constant percentage of the percent open. In other words, every

**Fig. 2-91.** Globe valve discs, plug type, ball type, V-port, and soft seated with insert. (*Courtesy of Pacific Valves a unit of Mark Controls.*)

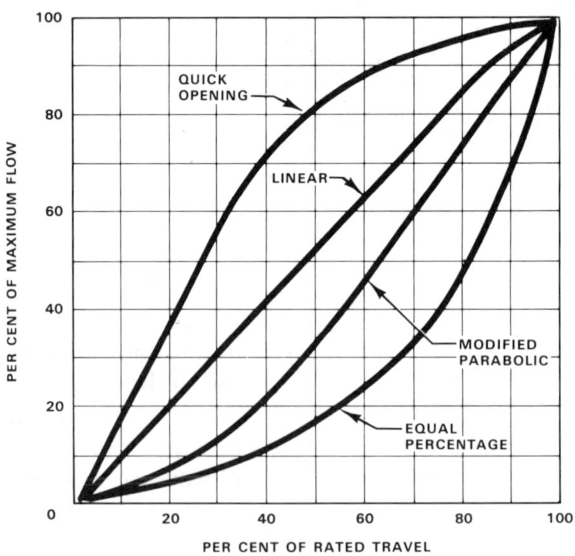


**Fig. 2-92.** Flow characteristic curves. (*Courtesy of Fisher Controls Intl.*)

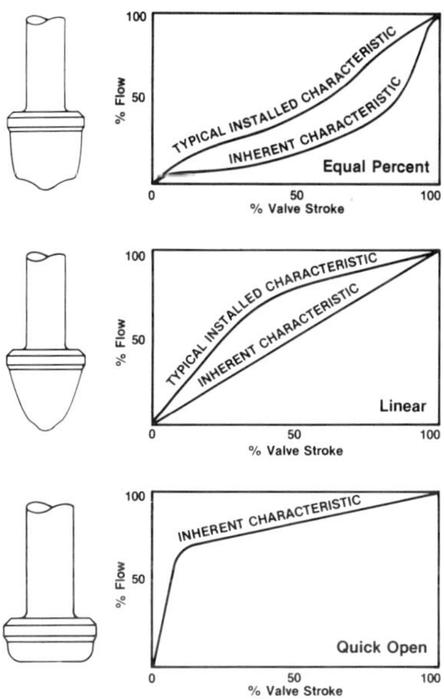


**Fig. 2-93.** Flow characteristics of various plug shapes are influenced somewhat by the surrounding piping. (*Courtesy of Valtek, Inc.*)

2 percent increase in the stem travel increases the flow rate by 2 percent *of what it is at that point on the curve.* The equal-percentage curve thus looks more like a parabolic or geometrically increasing curve than a straight line. The third type is the quick-opening type, which is the easiest to physically obtain, in which the first half or so of the curve is linear, but with a steeper slope than the true linear curve, followed by a curve that asymptotically approaches the maximum flow line.

The most common body configuration for globes is the straight pattern, also known as the T-pattern. The term "straight" comes from the fact that this body design can be placed in a straight pipe line, although the flow through this valve is anything but straight. Since the flow passes on all sides of the disc, and since the stem has to pass through the body wall, the disc opens perpendicular to the flow direction. This means that the fluid flow has to make a right-angle turn, flow around the disc, and then make another right-angle turn to return to its original direction. In fact, in most globe valves, the flow turns down at about 30 degrees, changes direction about 120 degrees and then another 120 degrees, and finally turns another 30 degrees back to its original direction.

One consequence of this is that there is a significant pressure drop. This is perfectly acceptable in many applications, since the whole purpose of the globe valve is to reduce line pressure, either for its own sake or for the corollary purposes of reducing flow velocity or fluid temperature. In many cases, however, the pressure drop is not wanted, and so some other types of globe valve design have been developed.

One of the earliest alternate body designs was the angle valve (Fig. 2-94). In this configuration, since the flow has to approach the disc from the bottom anyway, the valve inlet was simply placed at the bottom of the valve. This meant that the outlet piping had to be at a 90-degree angle from the inlet piping. This design is naturally suited for a number of piping applications in which the original flow is upward. The most obvious of these was steam, exiting a boiler at the top (Fig. 2-95). The first throttling valve on an oil or gas well, called a choke (Fig. 2-96), is also an angle type.

Another variation, which has existed for many years but has been common only in the last 30 years, is the Y-pattern globe (Fig. 2-97). In the Y-pattern globe the inlet and outlet piping are on the same centerline, but the stem and disc are at an angle to the inlet and outlet, normally 45 degrees. This considerably reduces the twists and turns that the flow stream sees, which reduces the pressure drop by up to half. Decreased pressure drop also means that the $C_v$ goes up. The Y-pattern

**Fig. 2-94.** Angle body globe valve. Flow is into the bottom and out the side. (*Courtesy of Newmans Inc.*)

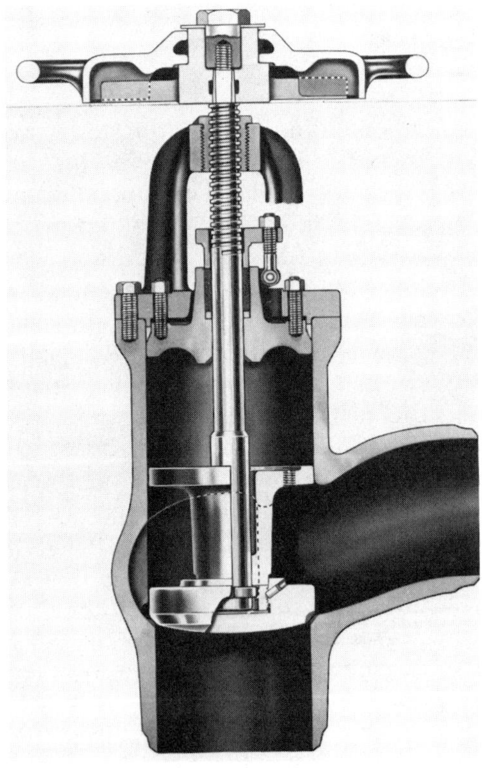

**Fig. 2-95.** Cross section of butt-weld-end angle globe valve. (*Courtesy of Edward Valves, Inc., formerly Rockwell Edward.*)

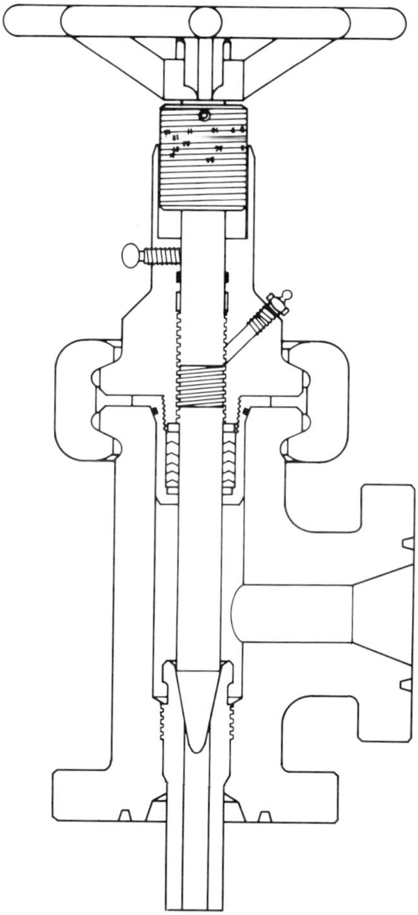

**Fig. 2-96.** Choke valve used on wellheads. Markings on collar below handwheel are used to calibrate flows. (*Courtesy of Vetco Gray, Inc.*)

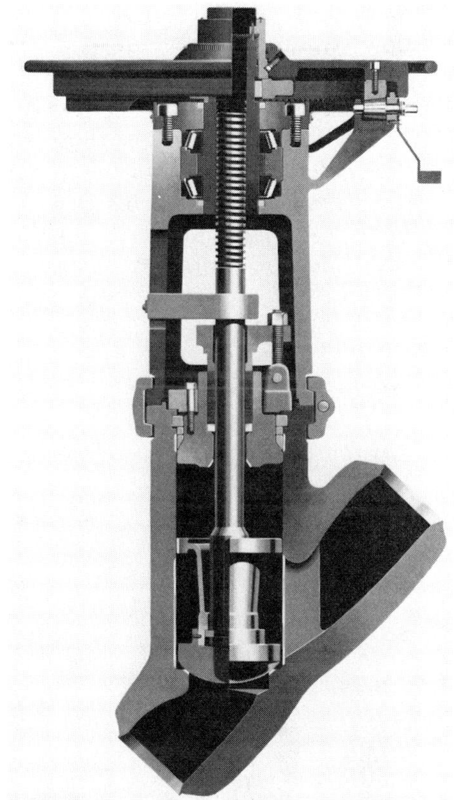

**Fig. 2-97.** Y-pattern globe valve. (*Courtesy of Edward Valves, Inc., formerly Rockwell Edward.*)

valve seat opening is almost in line with the inlet and outlet, which means it is possible to rod out this valve to a limited extent. The common uses for Y-pattern valves are in very high pressure services. Because of the asymmetrical design of the Y-pattern valve, it does sometimes make layout of piping systems more difficult. This is generally a minor factor when weighed against savings in pressure drop that it offers.

The Y-pattern is also sometimes used in very large actuated valves (Fig. 2-98) for the purpose of increasing $C_v$ through the lower pressure drop. Unlike standard Y-pattern globe valves, these valves are often oriented with the actuator vertical for ease of access and proper support and the valve body at 45 degrees to the horizontal. Other specialized Y-pattern valves can be built in fabricated-steel or cast bodies for applications in which there is virtually no seat at all, sometimes with a rubber insert in the disc for sealing, and whose purpose is for complete drainability. These valves have an almost or completely smooth bottom inner surface, and the disc raises almost completely out of the way of the flow stream.

A couple of other variations on the globe valve are used very extensively in steam boiler work. In fact, the ASME code for fired pressure vessels (boilers) requires these valves be used under circumstances that are spelled out very specifically in the code. One is the first valve out of the boiler, which in many cases is required to be a stop-check valve (Fig. 2-99). Other names for this valve are the nonreturn valve or the boiler stop valve. The stop-check basically consists of a globe-type valve, in any

**Fig. 2-98.** Large Y-pattern globe control valve with double-acting piston actuator and geared handwheel override. (*Courtesy of Fisher Controls Intl.*)

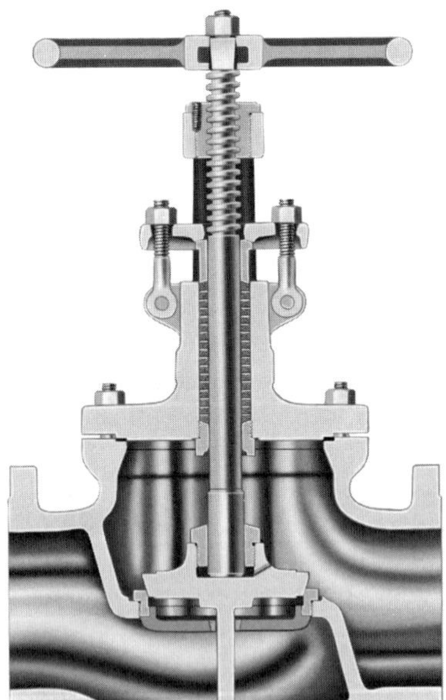

**Fig. 2-99.** Globe-style stop check in the closed position. Disc is guided from above and below. (*Courtesy of Pacific Valves, a unit of Mark Controls Corp.*)

body configuration, with the stem-to-disc attachment redesigned into a guide and not a connection. The stem can be lowered into the valve, causing the disc to seat and shut off steam flow. However, when the stem is raised, the disc does not follow it unless there is pressure under the disc. The disc then acts like the disc or piston in a lift check. The reason for this is to prevent reverse steam flow (as from a higher pressured boiler or in the case of a rupture in the boiler being protected) by pushing the disc closed and thus shutting itself off as if this were a check valve, and in addition to permit manual closure of the valve.

The second type of valve is the blowdown valve, also known as the tandem valve (Figs. 2-100 and 2-101). The purpose of the blowdown valve is to allow certain portions of the water in the boiler to be blown down or discharged from the boiler every so often, to remove minerals that will have concentrated in the boiler water after a period of operation. Usually two blowdown lines are required, one to blow down from the area near the water-vapor interface, and another to blow down from the area just above the bottom interior surface of the boiler. These valves are known as tandem because the code requires two valves, one to take the brunt of the pressure drop and a second to provide the tight sealing that the other cannot provide. The two are generally designed differently, although this is not a code requirement (Fig. 2-102). The one that takes most of the pressure drop (the one that is closer to the boiler, opened last, and closed first) is often a sliding-plug type while the other valve has something like a needle-seat design. Sometimes these two valves are built into one body, with the stems at right angles to each other. The blowdown valves have to be very carefully sized.

The stop-check valve is also used in applications other than on boilers. For building service systems, such as chilled water, these valves

**Fig. 2-100.** Hard-seated valve for tight shutoff in boiler blowdown line. (*Courtesy of Yarway Corp., a subsidiary of Keystone International, Inc.*)

**Fig. 2-101.** Seatless valve for throttling of boiler blowdown. Flow is into the side of the valve and out the bottom. (*Courtesy of Yarway Corp., a subsidiary of Keystone International, Inc.*)

**Fig. 2-102.** Pair of valves in tandem, a Y-pattern for tight shutoff, and an angle for throttling, for boiler blowdown. (*Courtesy of Edward Valves, Inc., formerly Rockwell Edward.*)

are often used on pump discharges as a combination block-check (Fig. 2-103), thus getting the actions of two different valve types in a single body. This valve is usually equipped with an indicator, to allow for precise settings, and an O-ring resilient seat.

Several other special valves have been designed for steam services. All are basically throttling valves, but they have been intended for some very specific services. One is the desuperheating valve (Fig. 2-104), which is designed to lower the temperature of superheated steam by spraying water into it and desuperheating it. The water flows down around the valve stem to the plug region, which controls the flow of water into the steam line by uncovering nozzle-shaped flow passageways. Another is the conditioning valve (Fig. 2-105a,b), which is a high-pressure angle pressure letdown valve with the desuperheating feature built in. This allows the direct conversion of, say, a typical 600-lb steam service into a 300-lb steam service, with lower pressure, temperature, and degree of superheat, in one valve.

The bellows seal configuration is used on globe valves also (Fig. 2-106). Its first applications were probably on globe valves, because of the relatively short stem travels and consequently relatively short bellows lengths required. An interesting variation of the bellows seal that is possible only on a globe valve uses a metal diaphragm to completely seal off the stem opening (Fig. 2-107). The stem is located above the diaphragm, with its bottom end being rounded off and bearing on the

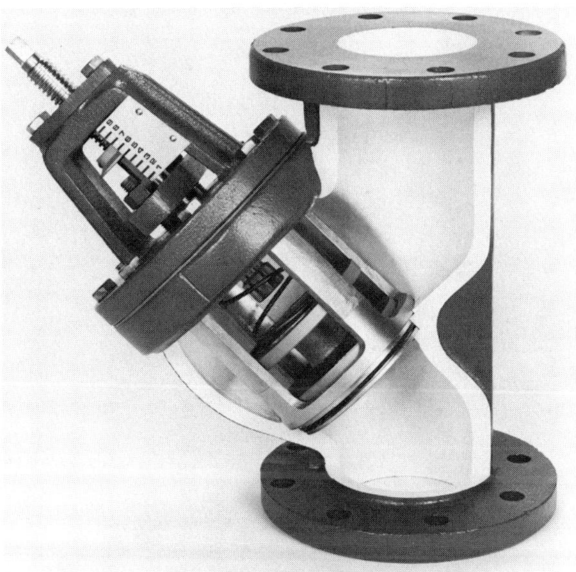

**Fig. 2-103.** Combination globe-check (stop-check) valve for water service. (*Courtesy of Mueller Steam Specialty Co.*)

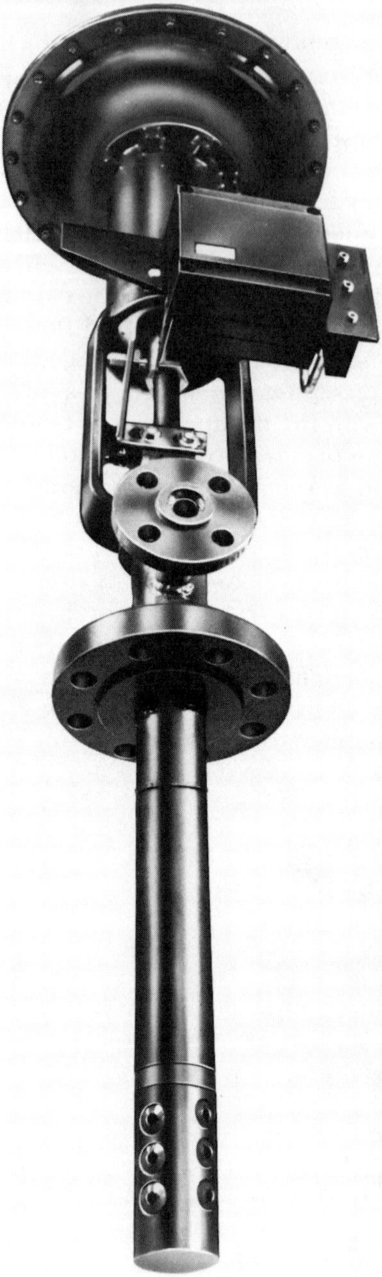

**Fig. 2-104.** Desuperheating valve, with water inlet connection at center and outlet nozzles at bottom. (*Courtesy of Yarway Corp., a subsidiary of Keystone International, Inc.*)

(a)

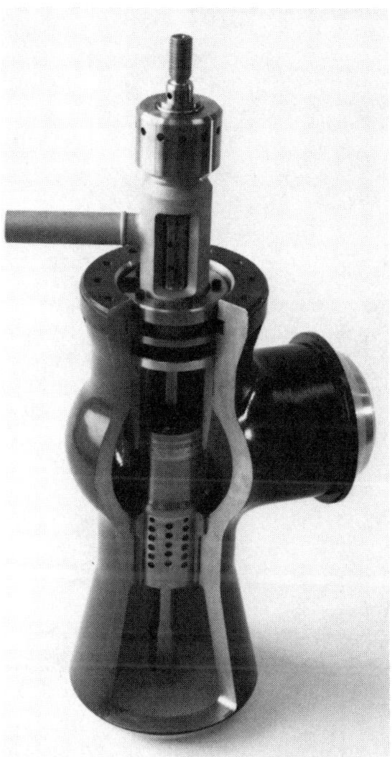

(b)

**Fig. 2-105.** Exterior and cutaway of conditioning valve, with water inlet at the top, high-pressure steam in from the side and lower-pressure desuperheated steam out the bottom. (*Courtesy of Yarway Corp., a subsidiary of Keystone International, Inc.*)

diaphragm. Under the diaphragm is a standard disc like one that might be used in a check valve, with extra guiding and a spring that is positioned to push the disc upward so as to want to hold it open. When the stem is threaded down into the valve, it pushes on the diaphragm, which in turn pushes down on the top of the disc, which then closes. When the

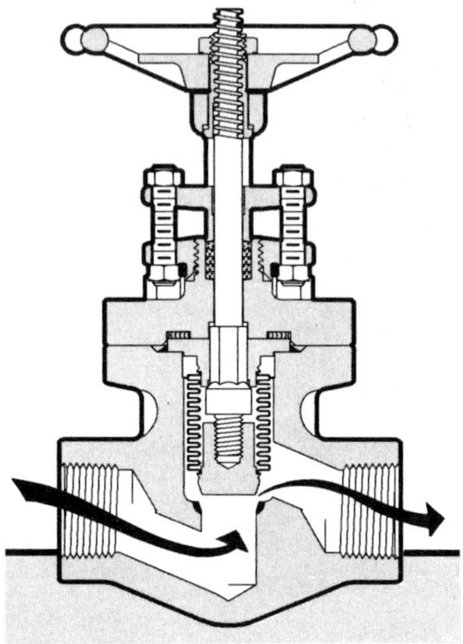

**Fig. 2-106.** Bellows seal globe valve. (*Courtesy of Henry Vogt Machine Co.*)

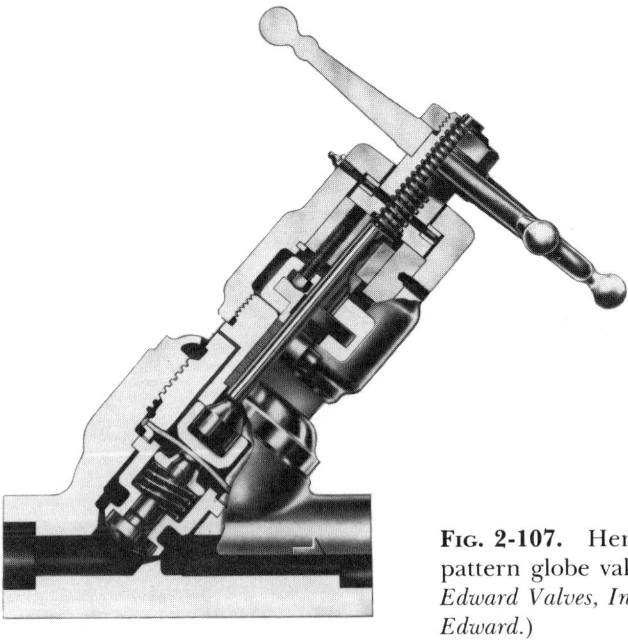

**Fig. 2-107.** Hermetically sealed Y-pattern globe valve. (*Courtesy of Edward Valves, Inc., formerly Rockwell Edward.*)

stem is backed off, the spring lifts the disc back to the open position.

Globe valves are manufactured in very small sizes in barstock or tubing type valves, and in these size ranges (½-in. pipe size [DN 12] and smaller) they are the valve of choice for almost all shutoff and throttling applications. Most of these are the variant called the needle valve (Fig. 2-108), in which the disc is cone or needle shaped, and often machined integrally with the stem, that is most common in these sizes. Needle valves are also built with the same construction as standard globe valves (Figs. 2-109 and 2-110), in sizes up to 2-in. pipe size, with a restricted size seat ring. Many of these valves are equipped with special stem designs (Figs. 2-111 and 2-112). Many of the needle valves have a finer thread or a micrometer-type adjustment to allow ultra-fine control for metering (Fig. 2-113). Many of them have indicators to show the exact percent open and stops that can be set to allow the valve to be opened to the same position time after time.

Various disc shapes are used in the barstock valves, some of which are almost a cross between a globe and a plug valve because they have a soft sleeve-shaped seat into which the disc seats. They are still rising-stem, though, not quarter-turn valves. Some barstock globe valves are designed for throttling; some are not. Those that are not are often simply referred to as block or on-off valves, and their internal construction is secondary. It simply happens to be that the globe design is the cheapest and easiest to build in those small sizes. Most barstock valves have stem threads above the packing to protect the fine threads from line fluids, and a very rudimentary bonnet design that is often threaded in and then pinned to make it secure from backing out under vibration or operation.

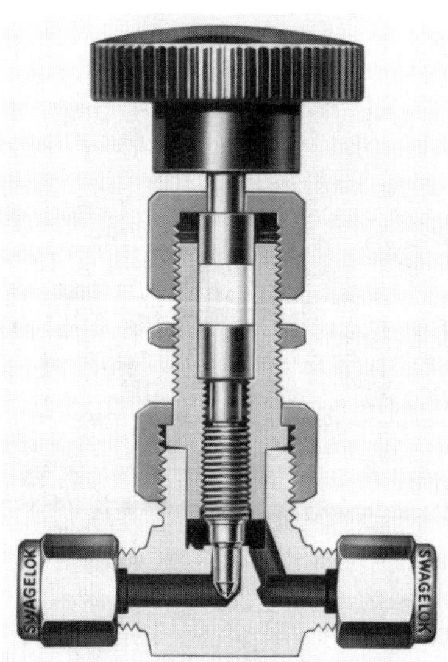


**Fig. 2-108.** Forged-body needle valve with union bonnet and union gland, for shutoff or throttling service. (*Courtesy of Swagelok Co.*)

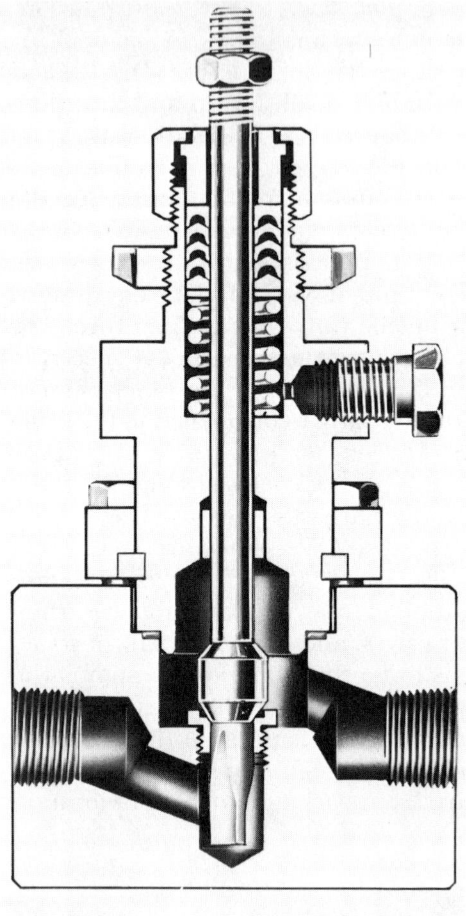

**Fig. 2-109.** Cross section of barstock control valve with fluted plug for close control of small flows. (*Courtesy of Fisher Controls Intl.*)

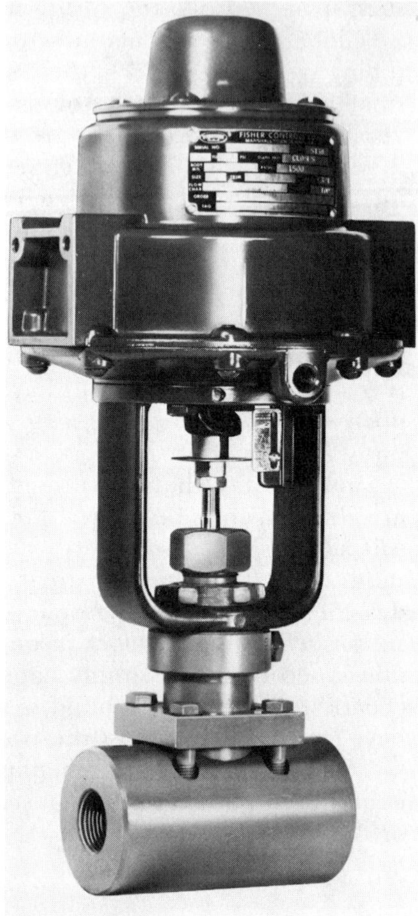

**Fig. 2-110.** Control valve with barstock body. (*Courtesy of Fisher Controls Intl.*)

In larger sizes the globe valve is available in any common material and type of construction, but globes are not normally available in very large sizes. This is primarily because of the higher stem force required to operate a globe as compared to a gate or other on-off valve, and the increased bulk of a globe as compared to the same size block valve. Manually operated globe valves, even when equipped with gears, become very difficult to operate in sizes larger than about 12 in. (DN 300) or so in moderate pressures and in sizes larger than about 4 in. (DN 100) in high pressures. Actuated globe-type valves (either control valves or power-actuated by virtue of the force required to operate them) are

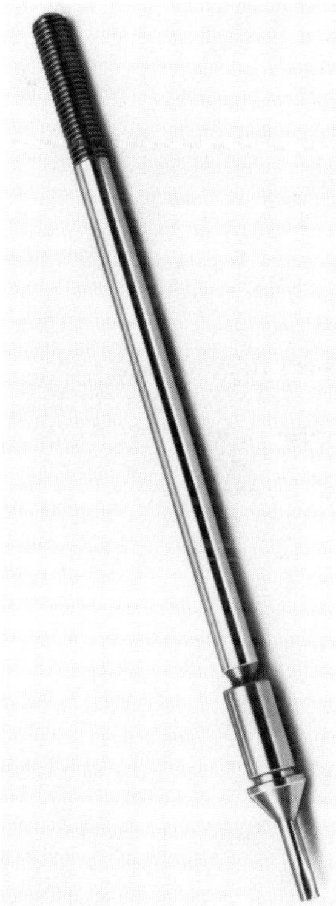

**Fig. 2-111.** Fluted plug with stem. (*Courtesy of DeZurik, a unit of General Signal.*)

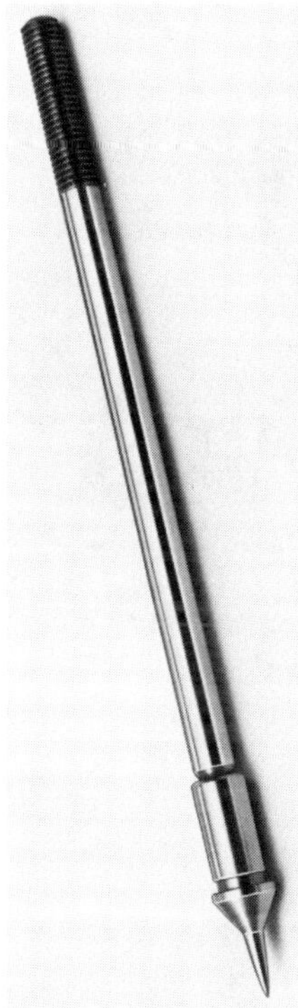

**Fig. 2-112.** Equal-percentage contour needle plug for very small flows. (*Courtesy of DeZurik, a unit of General Signal.*)

available in larger sizes, but a 20-in. or 24-in. (DN 500 or DN 600) throttling valve is a very large valve indeed. When control of a very large flow is required, it is more common to split the flow into two or more smaller lines, each with its own throttling valve. The other choice is to use a butterfly or ball valve, which do not throttle with exactly the same characteristics as a globe but which are more feasible for very large lines. These are most common in services such as large water and air lines.

In addition to its throttling abilities, the globe valve has one other valuable property. Because the disc moves down directly on the seat, the closing force applied to the stem translates more or less directly into seating force, rather than a highly oblique wedging seating force as in the gate. Thus the globe valve can be made to seat more tightly than the gate, which is of some significance in high-pressure services where the fluid forces are very high. It is quite common to use globes as *block* valves in high-pressure services for this reason, especially where leaks would be

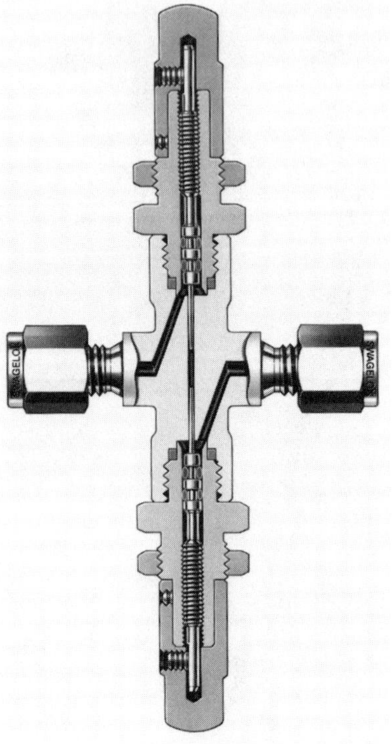


**Fig. 2-113.** Two needle valves in one body, for extremely fine throttling down to cc/hr flowrate. (*Courtesy of Swagelok Co.*)

hazardous, as in drain valves discharging to atmosphere. To reinforce this ability, it is also common to use two valves in series, such that if the upstream valve develops a leak there is a backup already in place. The use of two valves also allows the pressure drop to be split between the two, although operating personnel may not be very likely to take advantage of this feature. By splitting the pressure drop between the two valves, the overall noise level can be reduced somewhat also since noise levels are related (in a very complicated way) to the "suddenness" of pressure loss.

Virtually all globe valves have a preferred flow direction. Generally, this flow direction is marked on the body with an arrow to aid in installing the valve correctly. (Most of the time, we are speaking of the preferred direction of the *differential pressure,* and not, strictly speaking, the flow direction, but during flow conditions the two are one and the same.) The flow direction is either from the bottom of the disc to the top, or from the top of the disc to the bottom, depending on circumstances. The two cases are described below.

The first case, where the flow is into the bottom of the disc, is almost universal for manual, conventional globe valves. When a globe valve is viewed from the outside, the inlet end is the one that leads to the bottom of the disc and the outlet end is from the top of the disc. There are several reasons for this, one being that the disc is free to swivel on the stem such that as the disc approaches the seat, it will "land" in one orientation while the stem turns behind it. This means that the disc is most stable when the fluid force is applied to the underside. Fluid flow is the smoothest when the disc is approached from the underside, because of the contours, and control of closing is the best when the fluid force opposes the disc seating force rather than trying to assist it. Some amount of backflow is tolerable through a conventional globe valve, but the throttling and closing behavior is unpredictable. Sometimes a guided disc is provided, wherein a guide stem reaches down from the bottom of the disc and is held by a ring that it slides through. The ring is attached with two or three arms that reach out to the body just below the seat ring. Sometimes there is a separate ring just below the seat ring that these arms are an integral part of.

There is an exception to the first case flow rule, and that is when the globe is used as an on-off valve in very high pressure applications. In this case, the arrow marked on the body really indicates the preferred *pressure* direction. When the valve is opened fully, the flow characteristics in the "reverse" direction are not very different from those in the "forward" direction, and the best closing performance under high pressure is with the disc movement opposing the direction of differential pressure. An example would be the block valves on a pump line, with one in the suction line and one in the discharge line. When the pump is blocked in and depressured for maintenance, the differential pressure in the suction line is in the same direction as the differential pressure created when the valve is closed against fluid flow, but in the discharge line, the depressuring differential pressure is in the opposite direction. Therefore, if the main purpose of the valve is to allow blocking in and removal of something upstream, it is perfectly reasonable to install the valve with the arrow pointing upstream.

A word of warning—this is a very difficult concept to explain adequately to maintenance and operations personnel. If it creates severe confusion, it might be as well to forget the whole thing.

The second, or opposite, case is most often applied to actuated valves for control applications (Fig. 2-114). Here, the valve body and trim are designed quite differently from those of a manual globe valve, to reduce the required opening and closing forces as much as possible, to guide the disc from its outside edge, and to eliminate the disc instability that is present in a conventional globe where the disc is guided only by its loose attachment to the stem. In many cases, the bodies can be installed perfectly well in either direction, and sometimes the flow into the top of the disc is more appropriate.

Although it is possible to take a conventional globe valve designed for manual actuation and install a power actuator on it, the valves that are used for constant modulating services are generally built quite different-

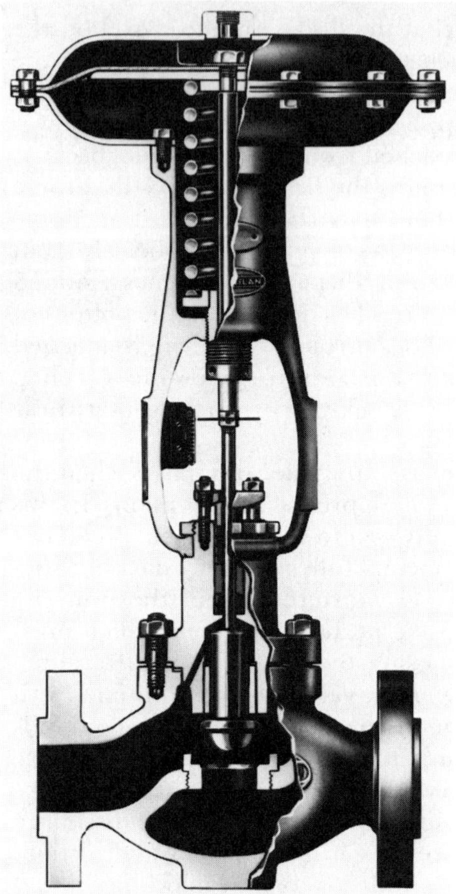

**Fig. 2-114.** Cutaway view of typical globe-style control valve—in this case, a post-guided balanced plug type with direct-acting fail-open diaphragm actuator. (*Courtesy of Masonielan/Dresser Valve and Controls Div.*)

ly. The terminology is also quite different. In control valve language, the "disc" is generally known as the "plug." "Rising stem" is known as "sliding stem" or "reciprocating stem." Quite a few more parts are employed also. Two different types are used, both being guided more securely than the manual globe type.

In one type known as the top-guided or post-guided type (Fig. 2-115a), a disc or plug is used that is very much like the one in a manual valve. The difference is that, because of the necessity for continual motion back and forth over sometimes small increments, the disc has an upper extension that is surrounded by a guide in the upper part of the body just below the bonnet. The guide keeps the disc from moving around as much as a standard globe valve disc would, and especially prevents osculation or the wobbling of its axis around that of the stem. The disc may be drilled for pressure balancing between the top and bottom (Fig. 2-115b). The stem, also, is much stiffer and fixed to or integral with the disc, since these stems do not rotate when powered by

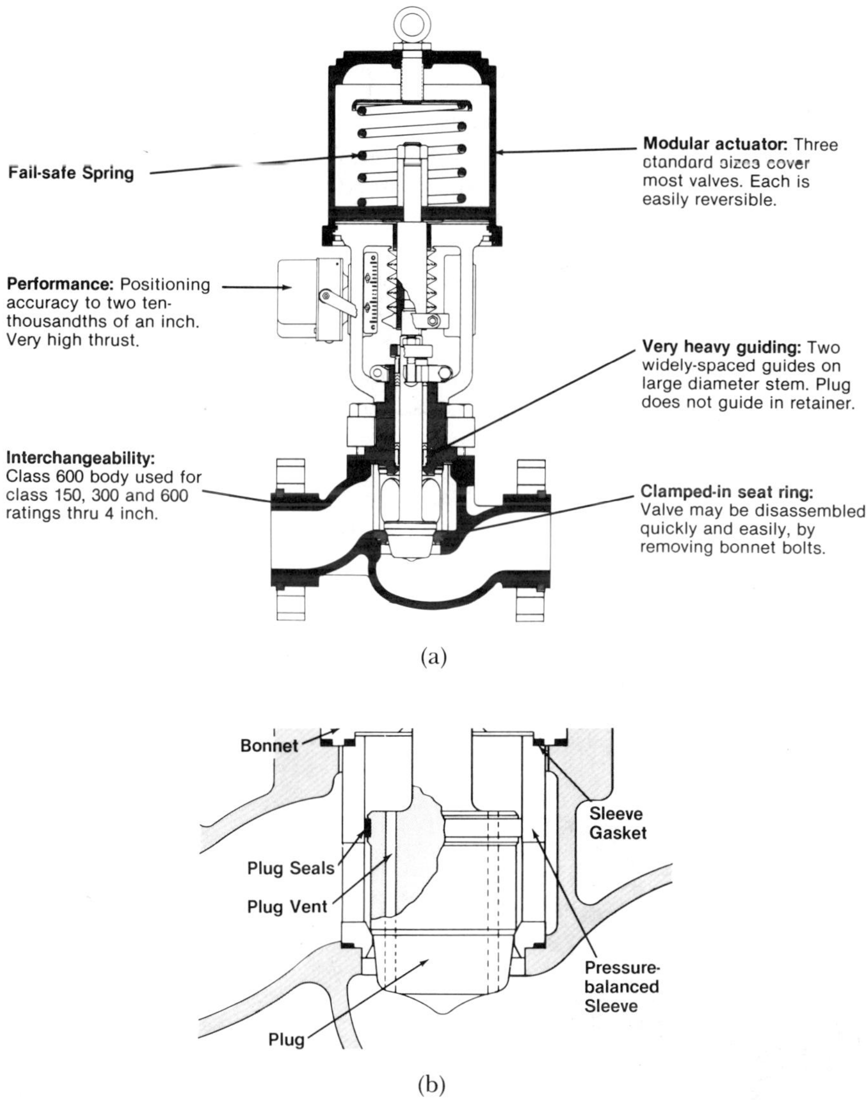


**Fig. 2-115.** Top-guided globe-style control valve. (*Courtesy of Valtek, Inc.*)

an actuator. The port opening can be almost full sized, which is the normal case, or it can be very small with a needle-type plug. These needle plugs can then be characterized further to allow control of ultra-small flows. A variation on this design is the port-guided type, which is most common with V-port plugs. Here the seat ring has a guide surface extending down from the inside edge of its seating area that engages an extended plug surface.

In the other, more common type, known as the cage-guided type,

(Fig. 2-116), the plug is generally equipped with upper and lower guide rings. These set the surface of the flow passageways in the plug back slightly from the inner edges of the body-flow passageways. Multiple, discrete flow paths are common, rather than a continuous 360-degree flow opening. Standard control valves have three to eight openings, depending on size, and valves designed for low-noise applications or for cavitation control will have hundreds of openings. The plug has straight threads to accept the end of the stem. An angled hole is then drilled through plug and stem for a pin to hold the two together. If a seat ring is used, the bottom of the plug is shaped to seat the way a manual globe would. Sometimes separate soft-seat rings of elastomeric or fluorocarbon plastic are provided, where very tight shutoff is needed. The degree of seat tightness is described by classes I through VI, each one a decreasing amount of permissible leakage.

The plug is surrounded by a cage (Fig. 2-117a–e), which guides the plug and also characterizes the flow as linear or some other type. The bottom end of the cage is seated against a flat surface in the valve body, and the top surface is seated against the underside of the bonnet. Often

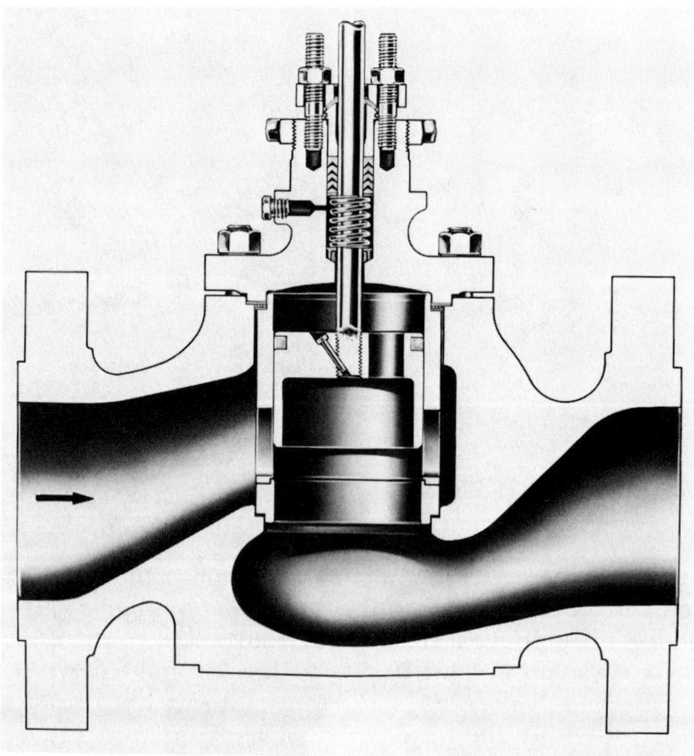

**Fig. 2-116.** Cage-guided globe-style control valve with balanced plug. (*Courtesy of Fisher Controls Intl.*)

(a) (b)

(c) (d) (e)

**Fig. 2-117.** Common cage designs—quick opening, linear, equal percentage, low noise, and anticavitation. (*Courtesy of Fisher Controls Intl.*)

a seat ring is provided as well, in which case the bottom of the cage seats on the back of the seat ring, which is then seated into the valve body. Sometimes top and bottom are metal-to-metal seated, but usually gaskets are employed to help prevent leakage around the cage that would bypass the controlling action of the plug. The gasketed configurations are more common with high differential pressures and low-noise valves.

Body configurations are also quite different. In the quest to produce low operating forces, one early design that is still common today (Figs. 2-118 and 2-119) splits the flow into two, with two plugs on one shaft, such that in the upper flow passage the flow comes under the disc, and in the lower flow passage the flow is over the lower disc. The result is a net force of close to zero on the stem. This is known as a double-ported type. Any trim can be provided (Figs. 2-120 and 2-121). The double-ported types are almost always built reversible, with top and bottom ends both having a bolted bonnet. The bottom cover is blank or with a guide built in to it. When a push-down-to-open actuator is used, the valve body

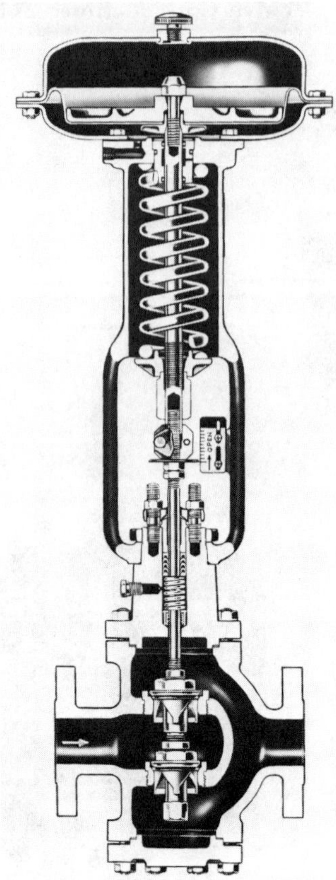

**Fig. 2-118.** Double-ported balanced control valve with V-port plugs. (*Courtesy of Fisher Controls Intl.*)

**Fig. 2-119.** Cutaway of double-ported top- and bottom-guided balanced control valve with equal percentage plugs, showing actuator mounting configuration. (*Courtesy of DeZurik, a unit of General Signal.*)

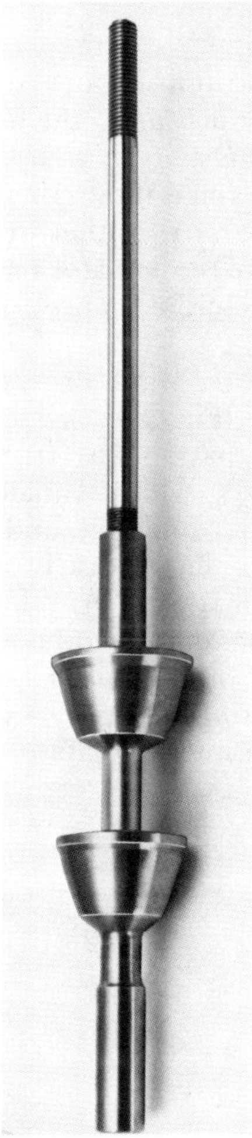

top- and bottom-guided double-ported control valve with linear contour. (*Courtesy of DeZurik, a unit of General Signal.*)

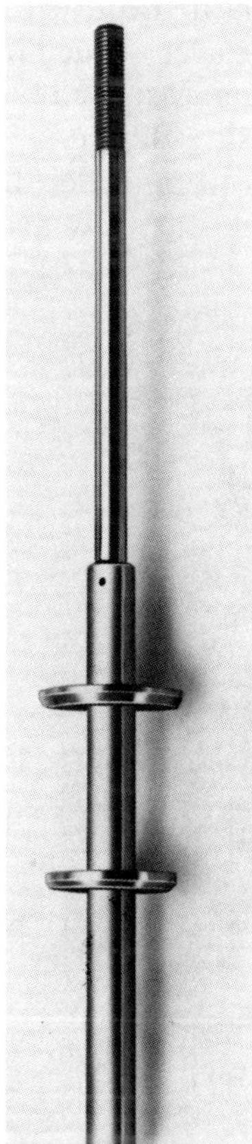

with quick-opening contour. (*Courtesy of DeZurik, a unit of General Signal.*)

is turned "upside down," with the seats on the bottom side of the body port openings. When a push-down-to-close actuator is used, the body is "right side up" with the plugs assembled on to the shaft in an opposite way from the type previously described. In cases like these, there are other reasons for selecting one type of actuator over the other that are more significant, so the body position just adapts to the actuator requirements. These actuator selection problems are described more fully in Chapter 3, on actuators.

The double-ported type is also suitable for three-way valves, either for diverting flows or for mixing or splitting flows (Fig. 2-122). The only body modification required is to block off the passageway from the bottom port to the side outlet. In the diverting type, which is analogous to an on-off style of operation, flow enters from one side and exits through either the other side or the bottom. The plug or disc is double faced, with one seat in each of the ports, each facing the other. When the plug is seated against the bottom seat, the flow goes out the top port and out the other side flange. The flow through the bottom port is directed through the bottom outlet. In the mixing or splitting configuration, two plugs are used and they are controlled to some intermediate position

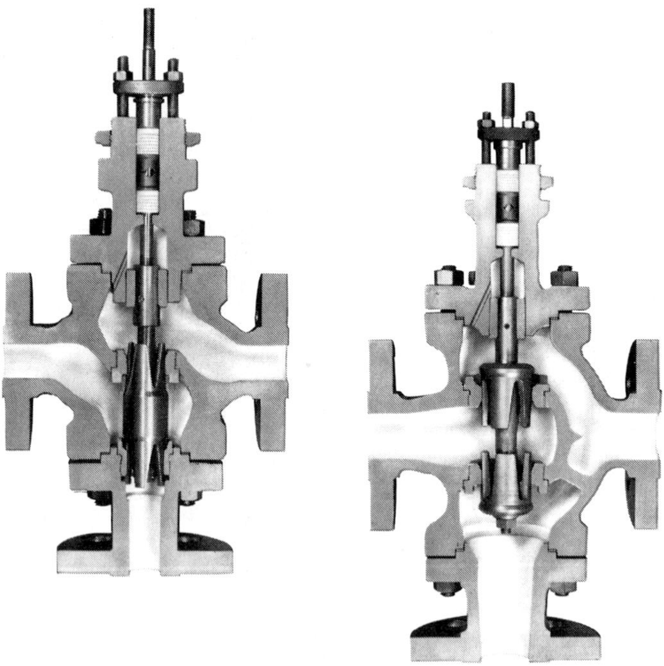

**Fig. 2-122.** Three-way double-ported valves designed for flow mixing (left) and flow splitting (right). (*Courtesy of Neles-Jamesbury, Inc.*)

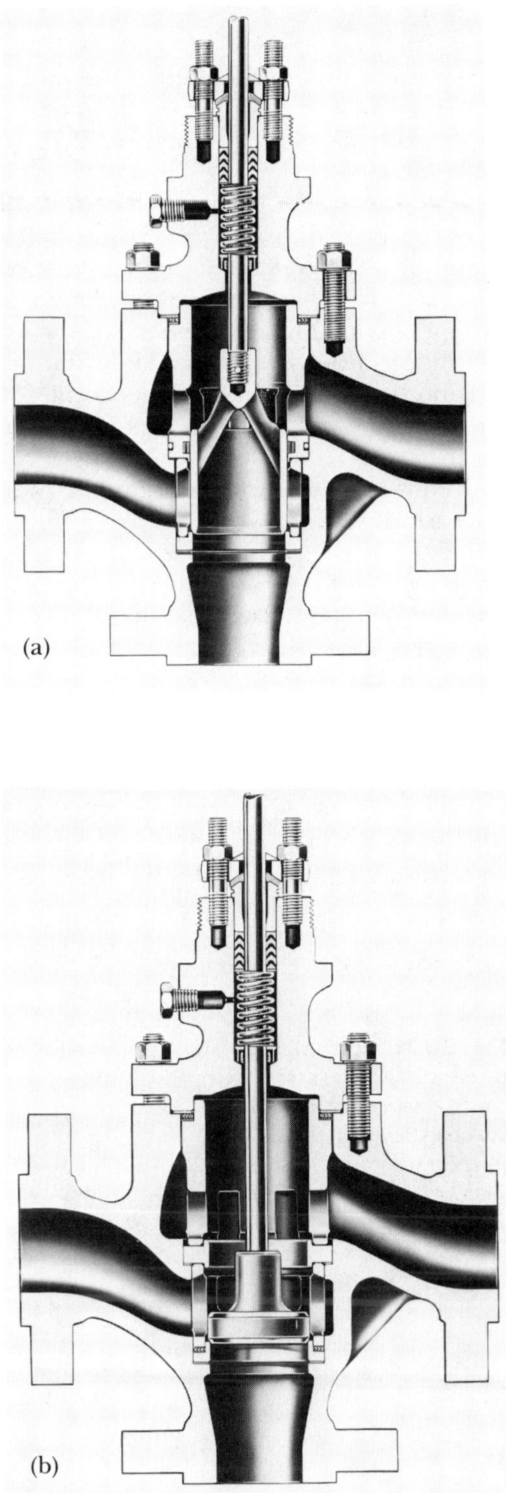

(a)

(b)

**Fig. 2-123.** Three-way single plug valves designed for flow mixing (with full proportioning capability) and flow splitting or diverting (with limited proportioning capability). (*Courtesy of Fisher Controls Intl.*)

between the two seats to get a proportion of one flow versus the other. Mixing and splitting are the same principle, but with flows in opposite directions. In some cases, single-ported types with two seats (Fig. 2-123a, b) can be used for three-way operation.

Configurations like these are the norm for valves to be used in sanitary service (Fig. 2-124), where complete disassembly is required for cleaning and tight shutoff is required. Although these services can be fairly corrosive (tomato juice is an example), the primary requirement is for absolute cleanliness, which if not obtained by disassembly and manual cleaning is done by steam or solution cleaning in place. Tight shutoff requirements mandate soft plugs and the cleaning requirements call for minimum of bearing surfaces and small parts. Even the actuators are subject to these requirements.

Sometimes the flow rate required is quite a bit smaller than the pipe size is capable of delivering. For various reasons, it is not easy to force the actuator to control more finely. For one thing, it is not possible sometimes to move in small enough increments. A better design choice is to reconfigure the valve with reduced trim (Fig. 2-125) so that the plug and cage are smaller but the body size stays the same. Sometimes it is

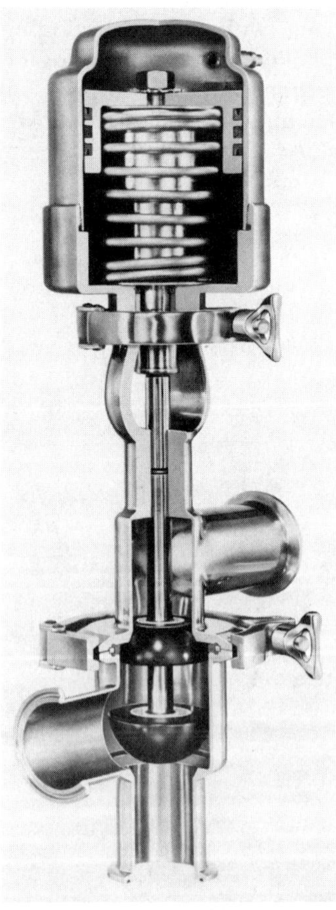

**Fig. 2-124.** Cutaway of three-way sanitary service control valve with polished surfaces, clamp ends, soft plugs and stainless steel sanitary actuator. (*Courtesy of Tri-Clover, Inc.*)

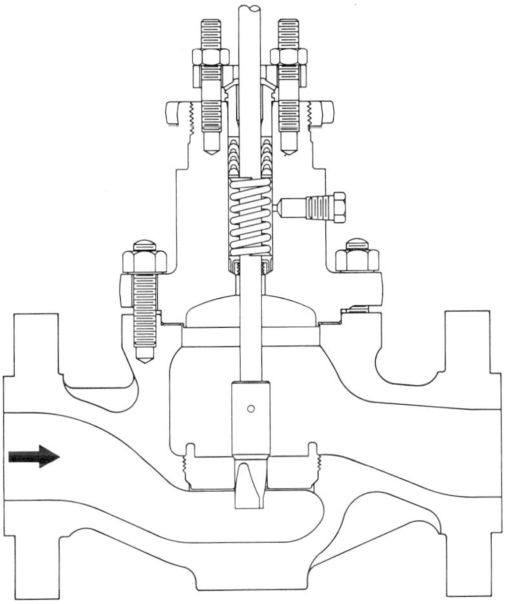

**FIG. 2-125.** Globe-style control valve with reduced trim and fluted plug in standard body. (*Courtesy of Fisher Controls Intl.*)

possible to use a trim that might ordinarily be installed in a 1-in. size (DN 25) valve in, say, a 3-in. size (DN 75) valve. (Trims are really rated in terms of their $C_v$, but the principle is the same.) All that is required is an adapter that holds the smaller cage and is in turn held by the valve bonnet (Fig. 2-126). Sometimes, but not often, the stem is downsized also. It usually depends on what size actuator shaft is being provided.

Dedicated control valves equipped with diaphragm actuators, which represent at least half of all actuated valves, are built with an adapter bonnet design that consists of a short plain, tapered, extension with threads above it, usually from 2-in. to 4-in. (50 to 100 mm) diameter. The actuator yoke has an opening this size, which fits over this extension. A large locknut then threads down over the yoke to hold it in place. In place of the locknut, a two-piece bolted clamp is also used.

The various types of flow characteristic required are created by various configurations of plug openings. Almost any type of custom plug configuration is possible, but the standard ones are most available. Often the contours are all in the cage, and the plug just has a flat bottom. As it moves upward, it exposes more of the flow passageway. In addition to the contours used to produce flow characteristics, several designs are available for lower flow noise. The simplest of these breaks the flow stream up into about 50 separate streams. Its design provides a somewhat linear characteristic. Other more sophisticated designs are specifically for the control of cavitation, a phenomenon in which the pressure

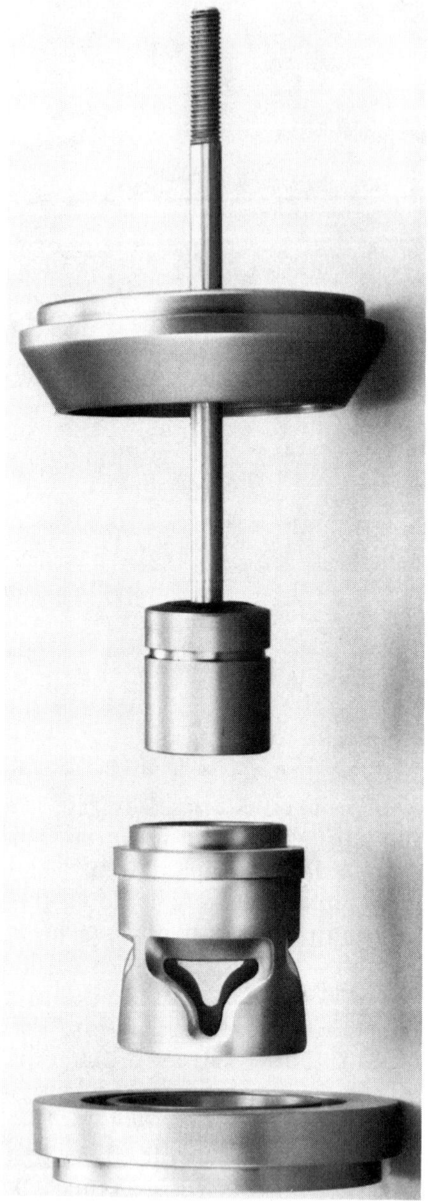

**Fig. 2-126.** Reduced trim set with equal percentage cage. (*Courtesy of DeZurik, a unit of General Signal.*)

in a very small volume of the fluid drops so low as to be below the vapor pressure of the fluid, which causes bubbles to form and then be imploded as the unit pressure rises again. This causes localized shock waves, which are very damaging to the valve surfaces nearby. Anticavitation trims are often linear flow. Almost always the plug is either open at the top or has a flow passage drilled in the top to prevent entrapping of

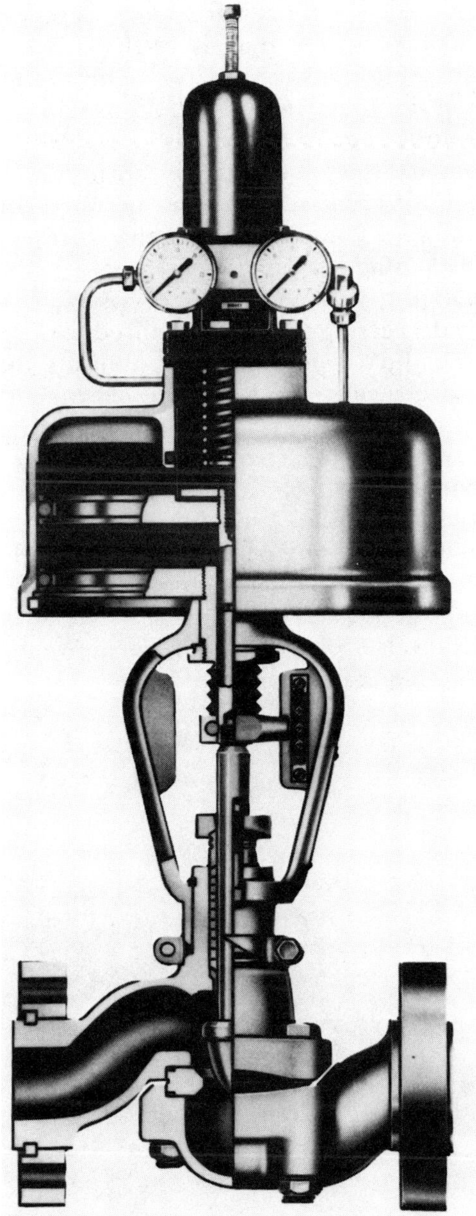

**Fig. 2-127.** Two-piece control valve body with seat ring, shown with double-acting piston actuator and positioner. (*Courtesy of Masoneilan/Dresser Valve and Controls Div.*)

fluid above it, which would introduce significant opposing forces during rapid plug movement. Some top-guided types have threaded-in seat rings very much like those in standard globe valves, and some have a separate seat ring that is clamped into place by a two-piece body (Fig. 2-127).

For large valves with very high differential pressures, such as power plant steam shut-off valves, sometimes a pilot plug is used. This is a two-piece plug within a plug with springs between the two. This design requires flow to be from the top of the plug. In the seated position the upstream pressure is applied around and on top of the plug. As the valve starts to unseat, first the stem lifts the pilot plug. The springs hold the main plug down. Only a small differential movement between the main plug and the pilot plug is allowed. When the end of this differential movement is reached, a flow passageway is opened between the downstream side of the plug and the volume above the plug. This depressures this area, significantly decreasing the force required to open the main plug. In the closing mode, the opposite happens, allowing the volume above the main plug to pressure up and assist in closing the valve.

Other techniques used for very high pressure drops (Fig. 2-128a,b) are similar to those used for cavitation control—namely, to break the flow stream down into many smaller streams and, in addition, to reduce the pressure as gradually as possible. A different concept (Fig. 2-129a,b) uses several stages, each with its own seat and plug, connected together to open simultaneously but forcing the flow to pass each seat one after the other. A few high-pressure-drop valves pass the flow from above the seat into a very hard venturi port (Fig. 2-130).

(a) (b)

**Fig. 2-128.** Cage designs for high-pressure drops. The toothed or expanding-wave pattern is for gases and the multiple perforated-plate cage is for liquids. (*Courtesy of Valtek, Inc.*)

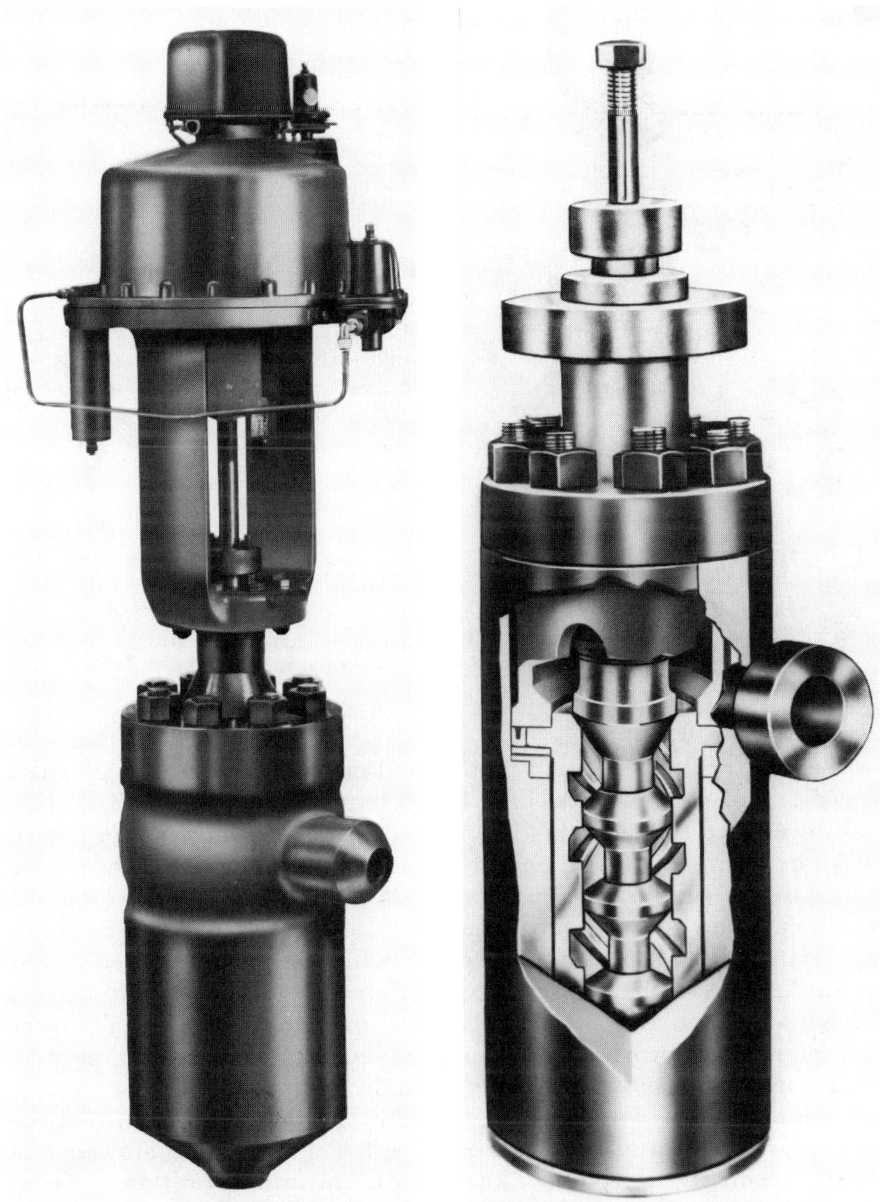

**Fig. 2-129.** Angle-pattern throttling valve designed for cascaded pressure control. (*Courtesy of Yarway Corp., a subsidiary of Keystone International Inc.*)

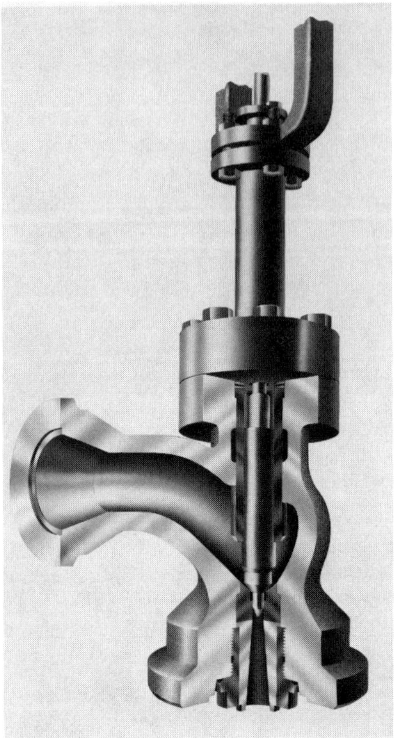

**Fig. 2-130.** Sweep-angle control valve with oversized post and tungsten carbide venturi liner for very high differential pressures. Flow is into side and out bottom. (*Courtesy of DeZurik, a unit of General Signal.*)

Because of their frequent motion and often momentarily high shaft speeds, many control valves have had better packing systems than manual valves used (Fig. 2-131a–c). Actuated valves were among the first to make extensive use of graphite foil packing such as Grafoil, at a time when it was significantly more expensive than asbestos packing. In addition, a number of positive-loading, or active, packing designs are in use. Many of these contain a coil spring as the component located closest to the body inside the packing box, which pushes up as the gland bolting pushes down on the packing. The use of lantern rings is also very common. These provide a space inside the packing chamber opposite a grease injection point, to allow lubrication from outside.

The size of valve required is generally determined by the size of the seat or trim required, in the case of both manual globes and control valves, when the purpose of the valve is throttling or flow control. Oversizing the valve—for instance, using a full-size valve in the same size line—means that the valve may not control the flow effectively, and it certainly means that you have bought too much valve. There are many formulas for sizing valves, but as long as some attention is paid to the correct valve for each location, great savings can be achieved and great pains can be avoided.

(a) (b) (c)

**Fig. 2-131.** Packing stacks commonly used in control valves. (*Courtesy of Fisher Controls Intl.*)

There are three solutions when it is determined that the valve size can be smaller than that of the line that it is being installed in. The first, applicable to control valves, is the reduced trim. This also means that if the flow rate increases later, the trim can be replaced without affecting the body. The other two solutions, which apply to manual and control valves equally, are reduced flow area either in the valve body or in the piping outside of the valve body. Circumstances will dictate which of these two will be easier and cheaper. If space is not a limitation, and the piping is standard weight carbon steel or some other low-cost easily available material, then the preferred solution is to install reducers in the piping. If the material is exotic, or space is a limitation, it is sometimes possible to obtain a venturi or reducing valve body (Fig. 2-132). This does make for very slightly smoother flow since the reduction is taking place at the same spot that the directional and shape transitions are taking place as the flow stream approaches the throttling point.

The same design principles described here can be used with valves up to 100,000 psi (7000 bar) or greater (Fig. 2-133). Most commonly, these very high pressure valves are also small. The flow streams are small, that is. Often, in the very high pressure ranges, the body is basically a big block of metal with small flow passageways drilled in it.

### *Butterfly Valves*

The butterfly valve, or rather the various types of valves that are collectively known as butterfly valves, will be discussed here as members of the family of throttling valves. Butterfly-type valves are indeed fully capable of flow control, and that ability will be discussed. However, it will be apparent from even a casual survey of valve installations that the majority of butterfly valves are used as block valves only. It is also true that some butterfly valves are more suited for throttling than others are, just as some are more suited for severe services than others are. It happens to be that the less expensive ones are among the ones that are not that suitable for throttling services.

There are two broad families of butterfly valves. Briefly, the difference is whether the stem is concentric with the centerline of the valve body or eccentric or slightly off-center in two or three planes. This geometric difference is responsible for the differences in the way that various butterfly valves seat and shut off. These two groups overlap somewhat in rating and performance, but in the operational characteristics that really matter, such as pressure rating and shut-off ability, these two types of valve are quite different.

Needless to say, because the two valves look quite alike (that is, they both are often very short face-to-face and often do not have separate flanges of their own, and they both have a disc that turns inside the pipe) and they share the butterfly valve name, they are often confused. Both

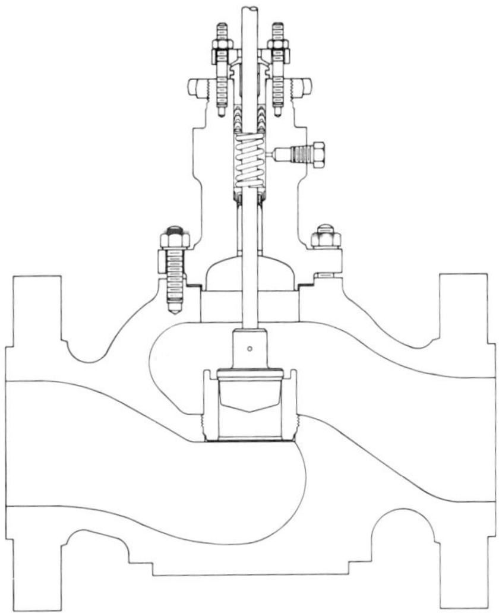

**Fig. 2-132.** Venturi body control valve with top-guided plug. (*Courtesy of Fisher Controls Intl.*)

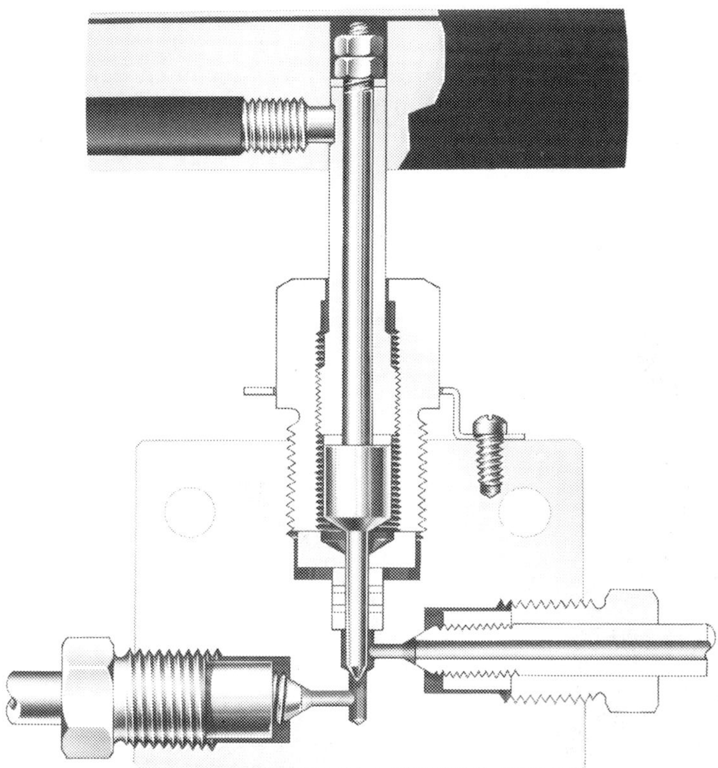

**Fig. 2-133.** High-pressure needle globe valve for block or throttling service. (*Courtesy of Autoclave Engineers, Inc.*)

valves have certain characteristics that make them worthy of choice over, for instance, a gate valve. But both have certain very definite shortcomings, and not the same shortcomings for each type.

To illustrate the point that confusion between these two has been rampant, a company that was one of the pioneers in the high-performance butterfly valve market spent 10 years and many advertising dollars trying to get their valve referred to as something other than a "butterfly" valve. They tried referring to it as a "trunnion" valve, which is an apt name but not very distinctive. They tried referring to it as an "eccentric disc" valve, which is more descriptive but longer. Other manufacturers put forth their own nomenclature, as well as a sizable amount of effort to demonstrate the difference between the two valves, and sometimes the similarities, depending on their strategy at the time.

At this writing, it appears that the name "high-performance butterfly" is almost universally used for these valves. The other branch of the family is not "low-performance," but is generally referred to as "rubber-lined" or "concentric" butterfly valves. Some of these, especially the ones that are not actually rubber lined, are quite special valves in their own right.

To start with the low end first, the typical rubber-lined valve (Fig. 2-134) is made of cast iron or other fairly low-cost body material, with a bronze disc and steel or stainless steel stem, and a rubber lining. The

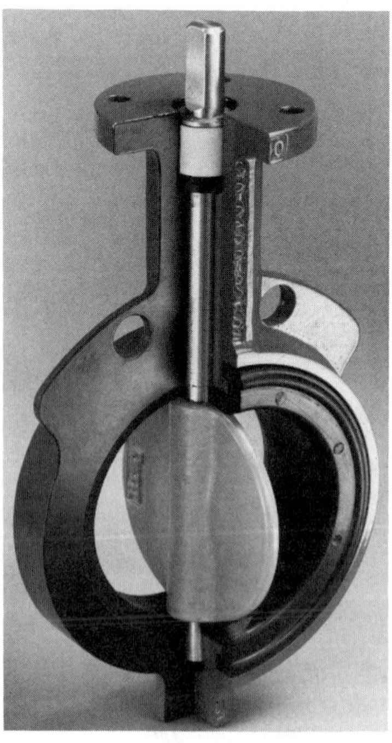

**Fig. 2-134.** One-piece-body resilient-seated wafer butterfly valve with epoxy coated disc. (*Courtesy of Bray Valve and Controls.*)

lining is the key to the entire valve. The lining is one piece, and seals against the disc and against the mating flanges. The concept employed here is unique. In effect, the flanges that the valve mates against are used to hold the lining in place so the disc can seal against it more firmly. As a payoff for this assistance, the liner seals against the end flanges without requiring the normal gaskets that every other flange set requires (Fig. 2-135). Surprisingly, if you try to use separate gaskets on a rubber-lined butterfly valve, the flanges are *more* likely to leak than if no gaskets were used. This unusual situation has been proved time and again, much to the surprise of individuals who had not encountered it before. The reason that gaskets do not work as well against the rubber lining is probably that the liner is designed with ribs that seal with a high local seating force in a narrow area, and the addition of another soft gasket probably spreads the seating force over a broader area.

As for the design concepts of the rubber-lined valve, the liner also normally has a small molded rib around the stem at top and bottom, or a tight-fitting area of greater compression (Fig. 2-136) to seal off leakage that would go up and down the stem surface, and sometimes an extratight fit on the disc just next to its top and bottom surfaces next to the stem. Because the stem and disc are normally concentric to the body and liner centerlines, the disc rotates around without any translational motion in the area at top and bottom. These areas of the liner are subject to wear from the rubbing of the disc, and closing the valve does not force the disc into the seat any tighter here than when it is open. Therefore, not much sealing takes place in these areas. If you examine a leaking rubber-lined valve, the areas at top and bottom are where you would normally look first to find evidence of wear or erosion. In contrast, the remaining 150 or so degrees of arc on either side of the stem penetration seal quite effectively because the liner is designed to be compressed radially outward as the disc closes and pushes into the seat.

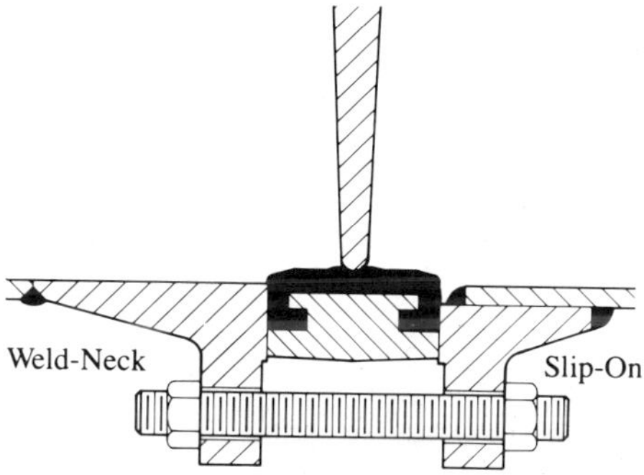


**Fig. 2-135.** Retention of resilient seat by the mating flanges. (*Courtesy of Bray Valve and Controls.*)

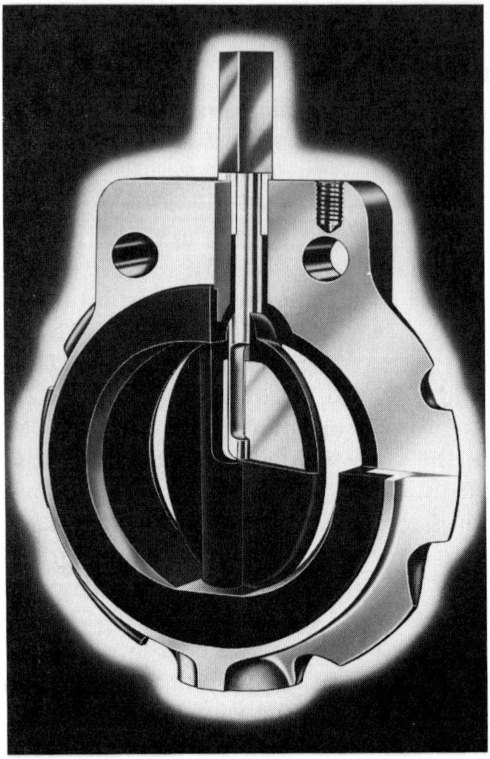

**Fig. 2-136.** Fully lined resilient-seated butterfly valve with encapsulated disc. (*Courtesy of Media Valve Co. Inc.*)

It should be kept in mind that the rubber lining is normally held in place by the force from the mating flanges. There are two ways of installing the lining. One is by machining a lip, or dovetail, into the body below each seating surface and mechanically locking the seat in (Fig. 2-137). This type of seat is forced into place during assembly by flexing the rubber. In some designs, the body is actually made in top and bottom halves (Fig. 2-138) and bolted together around the seat. Sometimes the seat is adhesive bonded to the metal body. The second basic assembly method uses a cartridge—usually phenolic or some other material that is rigid but somewhat less rigid than cast iron—that the rubber seat is bonded to, and then the whole affair is slipped into the iron body. This design is easier to repair, and because there is more resiliency in the seat-cartridge system, the seat is not as vulnerable to gross distortion or as likely to be pulled entirely out of the body, which of course makes the seat and the valve useless.

The butterfly valve is normally designed as a "through-bolting" body, either a wafer or lug (single-flange) design. These body styles do not actually have true flanges on the ends, and they both depend on bolting to pipe flanges on each end (Fig. 2-139). In the wafer type, stud bolts or machine bolts are used, inserted through the bolt holes in one flange,

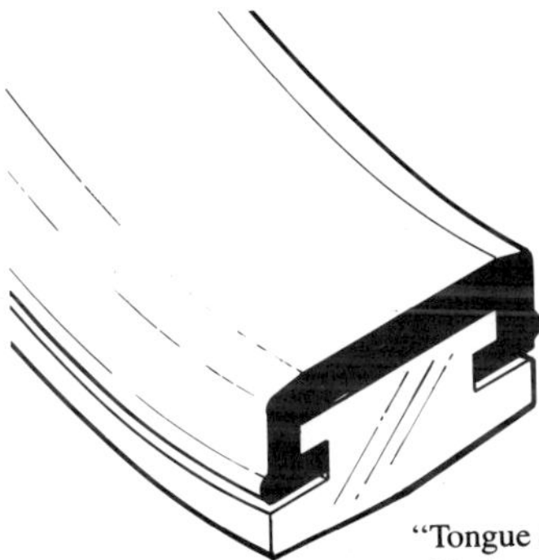


**Fig. 2-137.** Resilient seat locked into the body. (*Courtesy of Bray Valve and Controls.*)

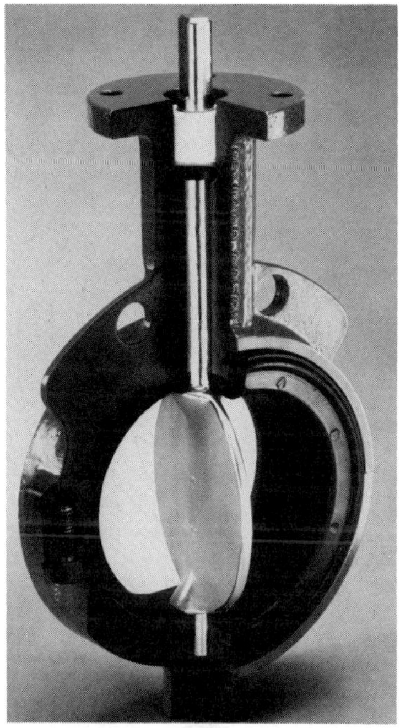

**Fig. 2-138.** Two-piece-body resilient-seated wafer butterfly valve with one-piece disc and stem. (*Courtesy of Bray Valve and Controls.*)

past the exterior of the valve, and into the bolt holes of the opposite flange. The bolt length required is thus the length required for a standard flange-to-flange joint, plus the thickness of the valve. The lug type, which is also called single-flange because the valve body behaves somewhat like a flange, is attached to both flanges by cap screws that go through the bolt holes on one flange into tapped holes in the valve body. The same hole in the valve body receives two cap screws coming in from opposite ends. The cap screw length required is the thickness of one flange plus slightly less than half the thickness of the valve body, since the two cap screws must not interfere with each other. A different, rarely used technique is to thread a stud bolt completely through the tapped holes in the body, long enough to extend through the bolt holes in each end flange, with a nut behind each flange.

The essential difference between wafer and lug styles is that the wafer type requires the presence of both flanges at all times in order to retain the valve in the line. The lug type, on the other hand, can have the piping on one side removed while the valve remains attached to the piping on the other side. This is valuable in locations where disassembly

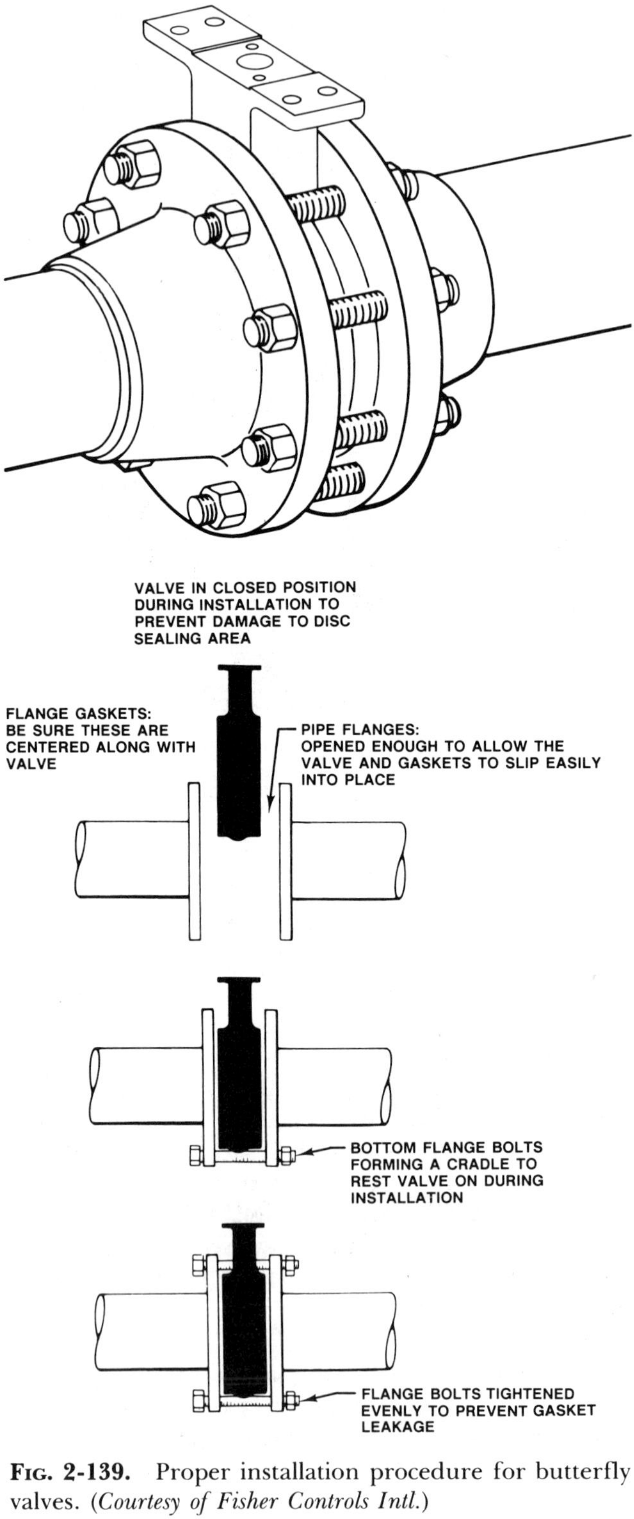


**Fig. 2-139.** Proper installation procedure for butterfly valves. (*Courtesy of Fisher Controls Intl.*)

is required, such as cooling water piping to heat exchangers where the piping must be disassembled in order to remove the tube bundle.

Note that even in cases where the valve can be attached to piping on one side only, the valve is not necessarily capable of holding full line pressure when only one flange is present. Sometimes this is possible, sometimes not, depending on the valve design, size, and rating. If there is any doubt, the user should specify "dead-end shutoff" so the manufacturer can assure that the seat is capable of being adequately retained with only one flange in place.

In larger size valves, the number of bolt holes and the size of the stem make it so that a few of the bolts cannot pass outside the valve body in the positions next to the top and (sometimes) bottom of the valve. In these cases, some quantity of cap screws will be required even though the valve body is basically a wafer type and through bolting is used elsewhere around the periphery of the body. Sometimes, for purposes of helping to orient the valve with respect to the flanges, a few of the bolt holes are drilled through the body or are indented into the side of the body. These designs still require through bolts and not cap screws.

There are several double-flanged designs also. Some larger size valves, from about 24 in. (DN 600) and up, are built as double flanged (with a separate true flange on each end, with the ability to use stud bolts on each joint) just for convenience and ease of bolt-up. They still have the same seat design that wraps around to form the end seal, at least up to moderately large sizes (Fig. 2-140). In some cases, it is still possible to use through bolting where the valve body flanges do not really do anything but guide the bolts, but it is unusual to want to do this. The main reason for it is when there is insufficient room to install an extra set of nuts on the two sets of short studs that would be used for double-flange bolting. Valves built to AWWA C504 standard (Fig. 2-141) for municipal water service are designed with end flanges even down to 3-in. size (DN 75). These valves have flanged or mechanical joint or push-on ends just like gate valves, rather than the wafer or single-flange configuration. The resilient seat is retained by internal metal seat retainers, rather than being retained by the body seating surfaces. Another exception is very large butterfly valves, ranging from 24 in. (DN 600) up to 84 in. (DN 2100) or 108 in. (DN 2800) or 144 in. (DN 3600) or even larger, depending on the capacity of the manufacturer. These valves are so large that the seat retainers are usually assembled in segments, although the resilient seat itself is normally in one piece, usually not molded but made from an elastomer extrusion and fused into a continuous piece. Other valves where the seat is retained internally and not as part of the end connection are some very small butterfly valves (down to 1-in. size [DN 25]) with bronze body and threaded ends, and a number of sanitary service butterfly valves (Fig. 2-142a,b) with special clamp ends for easy disassembly. Here, the seat is retained in the body joint of a two-piece body that is itself capable of being taken apart for

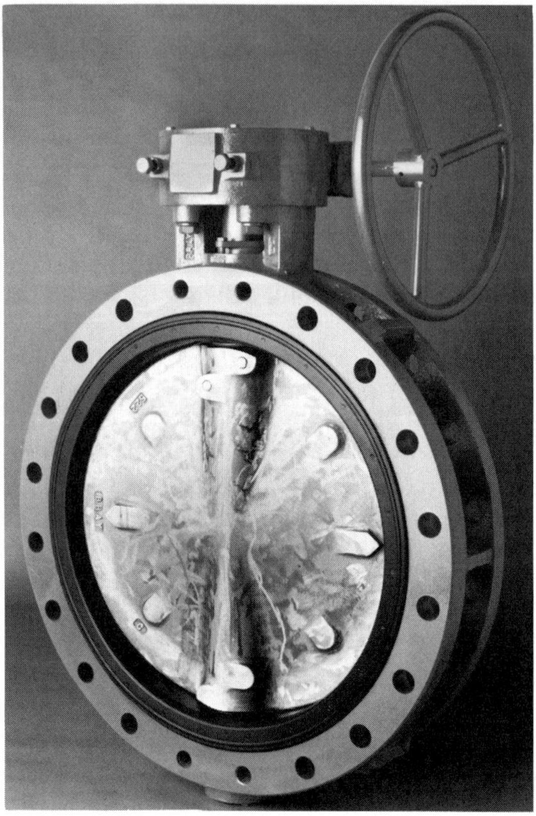

**Fig. 2-140.** Large-diameter resilient-seated double-flanged butterfly valve. (*Courtesy of Bray Valve and Controls.*)

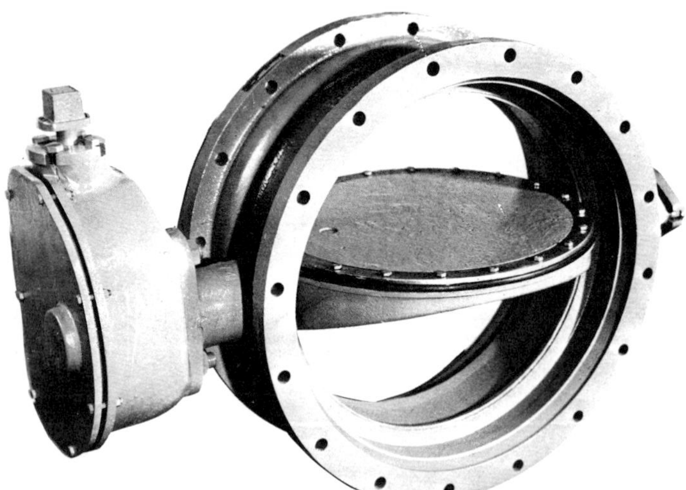

**Fig. 2-141.** Resilient-seated AWWA standard butterfly valve with flanged ends. (*Courtesy of American-Darling Valve, division of American Cast Iron Pipe Company.*)

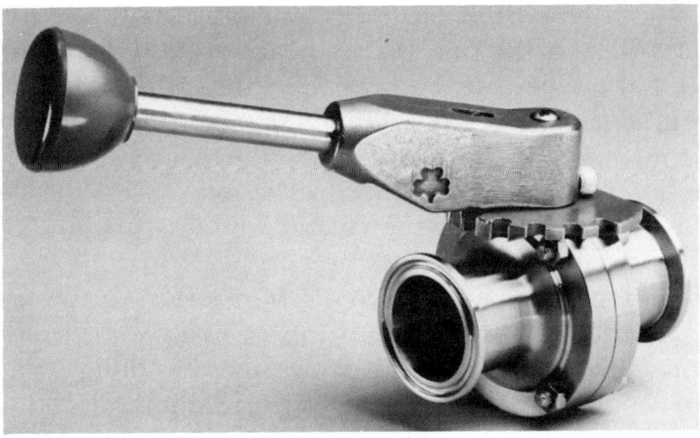

(a)

(b)

**Fig. 2-142.** Exterior and cutaway views of sanitary service butterfly valve. (*Courtesy of Tri-Clover, Inc.*)

cleaning. Butterfly valves are also made with grooved coupling ends and sanitary connector ends, and are even available in oval body models for duct work.

The primary markets for resilient-seated valves are uses related to water and air, meaning utility water services, HVAC, and other low-pressure, low-severity services where there are lots of valves and where, if the valves leak a little, it is not too serious. Although this is a blunt assessment of the market, it is the one that is heard frequently from valve users. It is true that rubber-lined valves will shut off tightly, but they will not shut off tightly over a period of many years or under severe conditions (sand in the lines, higher pressures), unless great care is exercised. Great care involves installing the valves exactly according to the manufacturer's instructions so that the rubber lining is not under undue strains and that the seating surfaces are properly loaded, and generally that the valves are not operated too often or left to sit in one position for too long. Exercise is beneficial to valves, like people.

The most common body material is cast iron, since by design the body itself is not under a lot of force from the adjacent piping and the liner theoretically protects the body from the line fluid. Other body materials are occasionally available, including ductile iron and steel. Aluminum bronze is often used where the external corrosion problem is severe, such as in marine environments or brackish water services. It is also common to apply a highly resistant paint system to the body, such as a multipart epoxy coating, or to use a chrome or nickel plating on the exterior of the body, also for corrosion resistance or for cleanliness.

The disc is most commonly aluminum bronze for water services, but it can be ductile iron with a plated or electroless metal-coated surface, or occasionally stainless steel. The stem is steel if no corrosion resistance is required, such as in air service in HVAC lines, or more commonly a 300 series or 400 series stainless steel. The top of the stem is normally machined with two or four flats, and the operator is generally a lever handle where torque permits. In larger sizes a worm gear drive with a handwheel is usually supplied, and of course any rotary actuator that can be set for 90 degree rotation can be used (Fig. 2-143). Normally the stem extends through the disc, and the disc is pinned to the stem with taper pins going through the stem or with cap screws going through the stem and into the disc on the other side, or with cap screws and nuts going through the stem and out the other side, or occasionally with cap screws going on to a flat in the stem. It is also possible to build butterfly valves in an all-plastic design (Figs. 2-144 and 2-145), usually in PVC, with a plastic body and disc, standard rubber lining, and coated-steel stem.

Some concentric disc valves are not rubber lined at all, but are designed for more corrosive services with a polymer lining material such as PTFE or PFA or polypropylene. These valves generally have to have an elastomer backup behind the more rigid plastic lining, for resiliency in order to get good sealing. Many of these valves have either a stainless disc or a disc coated with the same material as the lining. These valves normally have to have a two-piece body, since the lining is not resilient enough to be assembled into a one-piece body. Since the body has to be assembled around the disc and stem and lining anyway, it is then feasible

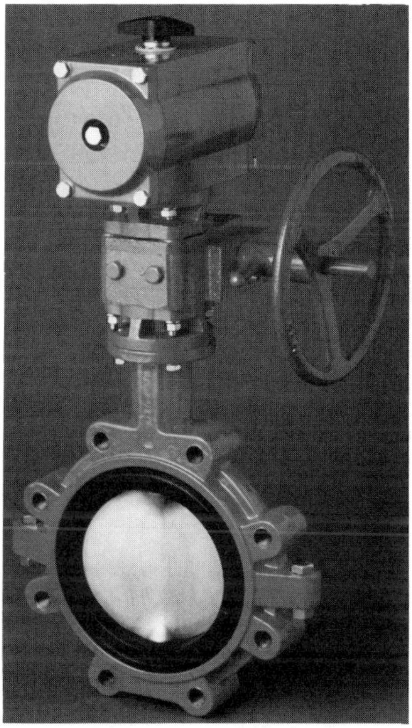

**Fig. 2-143.** Two-piece-body lug-type butterfly valve with air actuator and manual gear override. (*Courtesy of Bray Valve and Controls.*)

**Fig. 2-144.** Butterfly valve with molded PVC body. (*Courtesy of Nibco Inc.*)

to make the disc and stem one piece, which eliminates the area between disc and stem as a site for corrosive attack. The two-piece body also permits the disc to be made thinner, since the shaft does not have to pass through the center of the disc. This makes this valve even more attractive in the smaller sizes. The split in the body is at the horizontal centerline, usually with one cap screw on each side to hold the top and bottom halves of the body together.

Some rubber-lined valves are not truly concentric disc valves. Some types have the disc somewhat offset, so there is some extra force applied to the liner in the areas near the top and bottom to get better sealing in those areas. Many butterfly valves are built for duct work in HVAC services. Some are just standard rubber-lined butterfly valves such as would be used in process water services. In fact, chilled-water services in a building have many of the same requirements as cooling water in a process plant. However, many valves that are marketed purely for building or commercial services, since the governing codes and standards are less strict, are a simplified or lightened version of the valve used in process service. Often the pressure rating of the valve is reduced by

**Fig. 2-145.** Butterfly valve with body cut from PVC sheet, typical of larger sizes. (*Courtesy of Nibco Inc.*)

using a lighter disc and stem. The seat compound used is likely to be nitrile rubber, such as Buna-N, since it is low cost and no particular chemical resistance is required.

Other valves used in duct work generally look quite different from rubber-lined butterfly valves, often being fabricated of light steel or cast iron or aluminum, and built to the contour of the duct. Most of these valves are rectangular or square, and many do not have the ability to shut off tightly. Some valves used in duct work that passes through fire walls are required to shut off sufficiently to prevent the spread of a fire. Most of these valves are actuated, and designed to close down the air supply system in a building when a fire is detected, in the same manner as emergency shutdown valves in a process plant.

Most concentric disc or rubber-lined valves have a design pressure rating somewhat lower than that of the adjacent piping. Because most of these valves do not have true flanges, the rating is determined by the ability of the liner to shut off pressure. Many valves are rated at "150 lb," but this is not ANSI class 150 (PN 16) rating. Normally in a butterfly valve this means exactly 150 psig at 100°F (7 bar at 38°C). Various valves made by the same manufacturer will be rated at anywhere from 100 to 250 psi (7 to 17 bar), and there are some (not many) that are rated at 285 psi (20 bar), which is full ANSI class 150. There is not as much margin for overpressure in a rubber-lined butterfly as there is in some other valves. While most metal-seated valves can be expected to take 150 percent of rated pressure in the closed position, the most that the rubber-lined butterfly is required to be tested to is 110 percent of rated pressure. This is reflective of the actual performance of these valves.

The other half of the butterfly valve family is the high-performance valve (Fig 2-146). Here, again, there are several variations, but the characteristic features of the standard high-performance butterfly are a stem offset from the body centerline, a relatively thick disc with a seating surface contoured like a spherical section, and a body with removable end piece to hold a seat that contains a circular spring.

A high-performance butterfly valve closes by pushing the disc into the seat, in the same way as the rubber-lined butterfly, but instead of compressing the seat by line contact, the high-performance valve wedges into the seat and spreads it uniformly 360 degrees around. The spherical seating surface is centered at a point somewhat behind the disc centerline (Fig. 2-147), so that in the open position no part of the disc is in contact with the seat. This is the principal difference between the two valve types. In the rubber-lined type, the pressure force does not con-

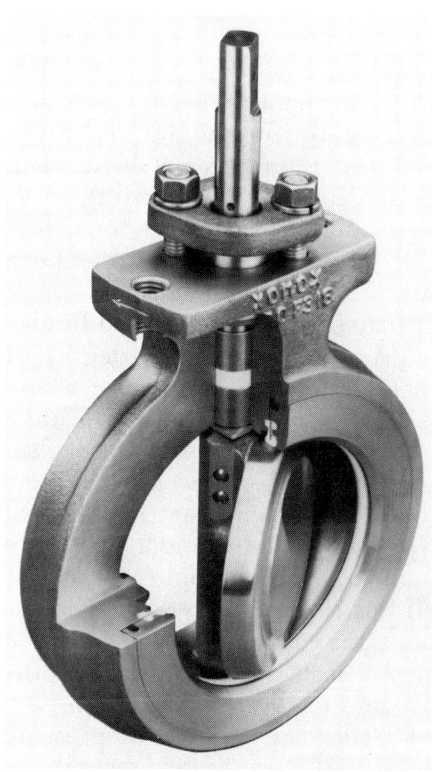

**FIG. 2-146.** High-performance wafer butterfly valve with preferred flow direction shown (disc on upstream end of valve). (*Courtesy of Xomox Corporation.*)

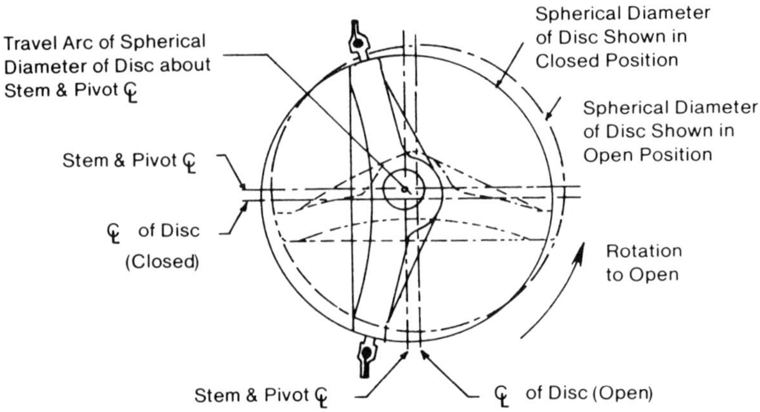


**Fig. 2-147.** Principle of the eccentric shaft design in high-performance butterfly valves. (*Courtesy of Flowseal, a unit of Mark Controls.*)

tribute much to the sealing quality; in fact, it is more likely to work against good sealing. But in the high-performance type, the pressure force is employed to push the seat more tightly into the disc, at least in one direction. By geometry, the best sealing force is obtained when pressure is coming from the side with the seat retainer toward the disc, opposite the side of the disc that the shaft is on. The pressure force does act on the disc also, to push it away from the seat, but that movement in a well-designed valve is minute compared to the movement of the seat.

The seat in a standard high-performance valve is a semirigid fluorocarbon polymer such as PTFE or PFA, and since a fair degree of deformation is required during seating (as opposed to a ball valve, where the ball is forced directly onto the seat), some means of providing resiliency is required (Fig. 2-148a). This is normally provided by an elastomer backup, enclosed by the polymer so as to protect it from direct contact with the line medium, or by a spring enclosed by the seat material. The spring can be a flat-type or a coiled toroidal spring. In virtually any valve intended for manual block service, some capacity for bidirectional sealing must be provided for. In a butterfly valve, because the same seat must seal in both directions, the concept of forcing the seat against a solid backup the way a ball valve seals is not practical and the spring action is one of the two mechanisms involved in the sealing act. The other is the geometry of the seat itself, in relation to the pressure force. By careful design, it is possible to configure the seat such that a pressure force operating on it in either direction pushes it into the disc surface. Of course, sealing is almost always better in one direction than the other. Very few valves are really 100 percent bidirectional, in that the leak rate is just slightly less in one direction than the other. But in many cases, the leak rate with the pressure force coming from the stem

**SOFT SEAT DESIGN**

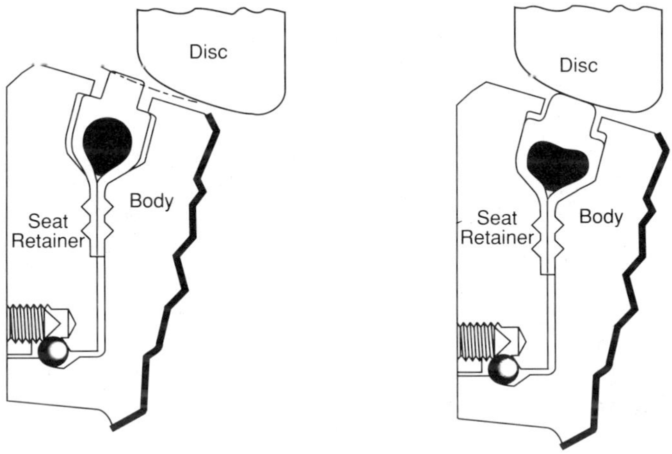


**DISC OPEN–**
Soft Seat is retained in cavity. Shoulders of seat and cavity abut.

**DISC CLOSED/UNPRESSURIZED–**
Mechanical preload seal between disc and seat. Outward radial forces stretch and flatten o-ring.

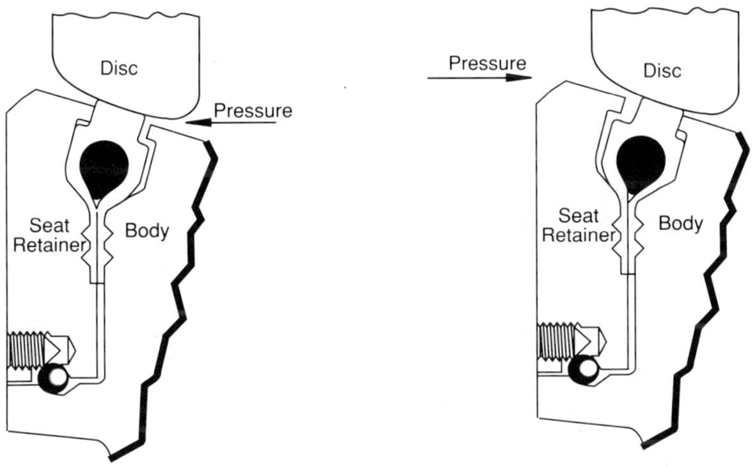


**DISC CLOSED/SEAT DOWNSTREAM–**
Line pressure moves axially and directs forces inward to create energized seal.

**DISC CLOSED/SEAT UPSTREAM–**
Line pressure (opposite direction) moves seat to opposite sidewall for identical energized seal effect.

(a)

**Fig. 2-148.** Sealing mechanisms of the three basic styles of high-performance butterfly valve seats: soft seated, fire safe soft seated, and metal seated. *(Courtesy of Flowseal, a unit of Mark Controls.)*

## FIRE-FLOW SEAT DESIGN

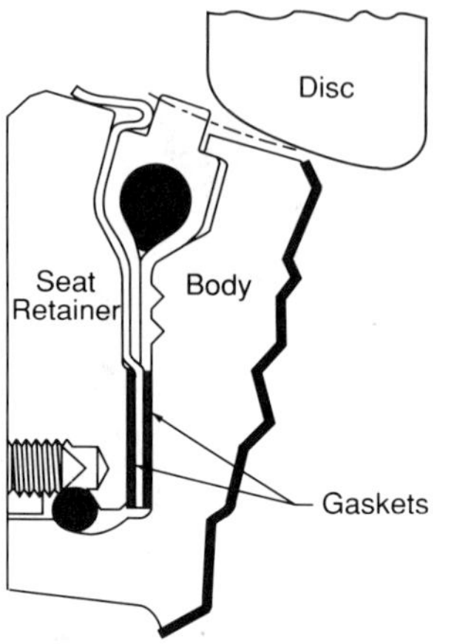


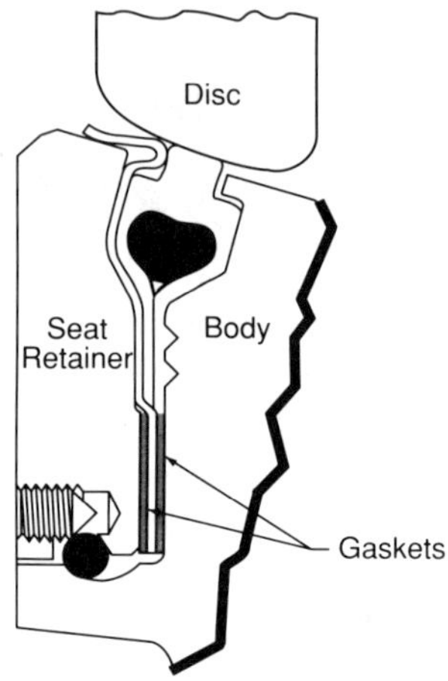


**BEFORE FIRE/DISC OPEN–**
Soft Seat and Metal Seat are retained in cavity by seat retainer ring. Graphite gaskets seal the Metal Seat tail.

**BEFORE FIRE/DISC CLOSED–**
Soft Seat and Metal Seat are in contact with disc, producing low pressure mechanical seal and high pressure energized seal.

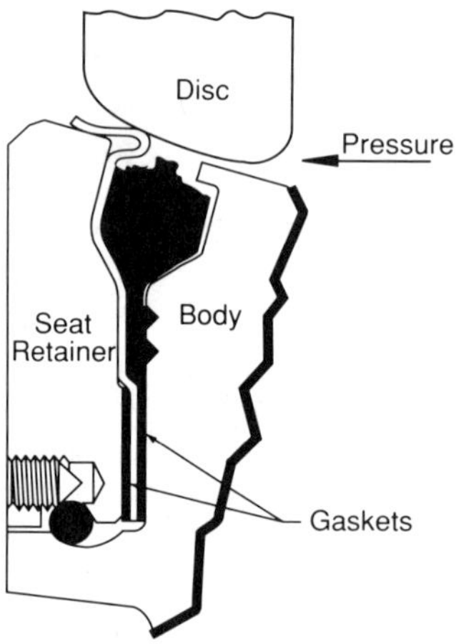


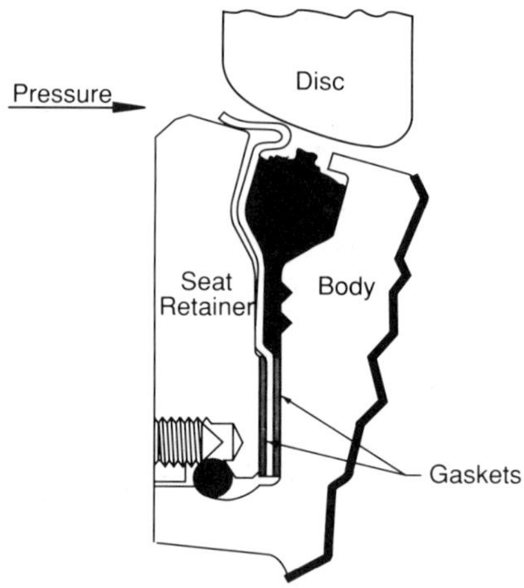


**AFTER FIRE/SEAT DOWNSTREAM–**
With full or partial Soft Seat destruction, the Metal Seat maintains metal-to-metal seal.

**AFTER FIRE/SEAT UPSTREAM–**
As with "seat Downstream", the Metal Seat provides a mechanical and pressure-energized seal with loss of Soft Seat.

(b)

**Fig. 2-148.** *(continued)*

**METAL SEAT DESIGN**

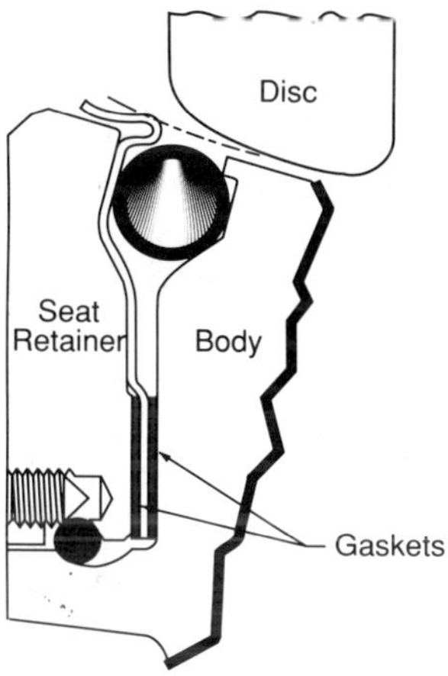


**DISC OPEN–**
Metal Seat is retained in cavity by seat retainer ring. Graphite gaskets seal the Metal Seat tail.

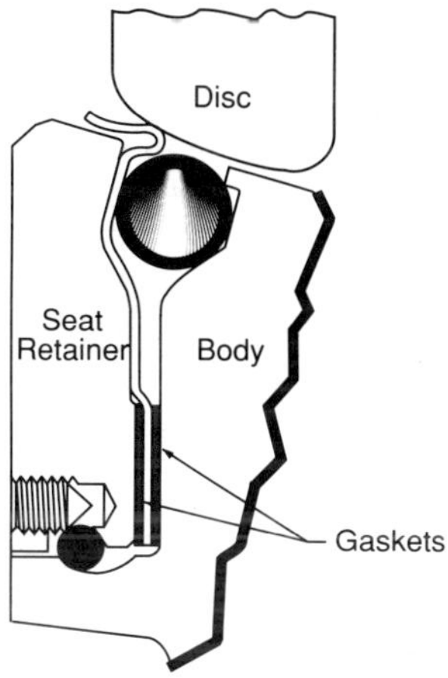


**DISC CLOSED/UNPRESSURIZED–**
Seat spring force applies mechanical loading against disc edge for metal-to-metal seal.

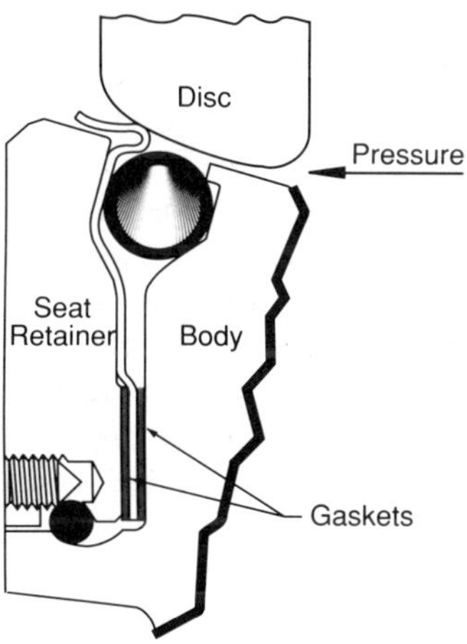


**DISC CLOSED/SEAT DOWNSTREAM–**
Line pressure in cavity applies radial inward force to seat while disc is deflected towards seat, creating energized seal.

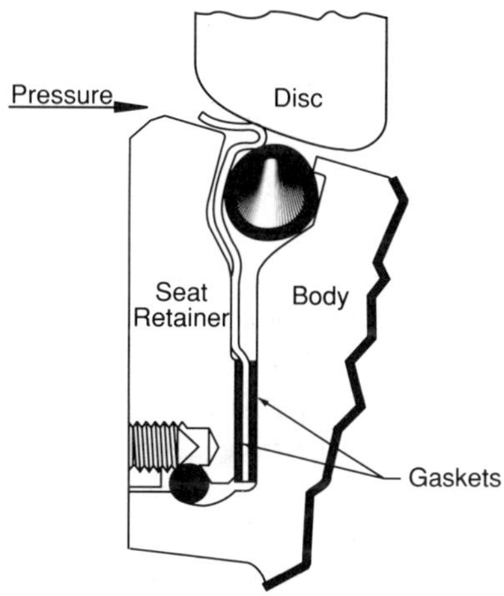


**DISC CLOSED/SEAT UPSTREAM–**
Line pressure (opposite direction) moves seat axially and dynamically loads seat against disc for energized seal.

(c)

FIG. 2-148. *(continued)*

side can be almost as low as the leak rate with the pressure force coming from the disc side.

As with a ball valve, it is necessary to prevent leakage behind the seat, but generally this can be accomplished in a high-performance butterfly valve just by extending the seat material back a distance into the body. Like a ball valve, the high-performance butterfly can be made fire-safe. The simplest way is to provide a metal lip such that when the seat is melted out, the metal lip will provide some sealing (Fig. 2-148b). However, since the disc cannot move farther down into the body, the metal backup seal is also designed as a spring. Normally, in soft-seated valves with fire-safe backup, this metal lip is mounted behind the soft seat so that it is helping to push on the soft seat. Butterfly valves can be built for high-temperature or abrasive service with a metal-only seat, using basically the same design (Fig. 2-148c). Generally speaking, the all-metal seat is less capable of tight shutoff than a soft seat is, except with very special designs.

Although butterfly valves can be and are used for abrasive or erosive services, such applications are rare and require special design. In general, the high-performance valve stands up less well to abrasive material in the fluid than the rubber-lined valve does because the rubber tends to deform around small particles caught in the sealing area and tends to bounce particles off rather than suffer scratches as a metal surface does. However, for both valves the damage caused by unavoidable abrasion such as sand in water can be minimized. If the valve is oriented with the stem horizontal, rather than the more obvious vertical position, and the particles tend to hug the bottom of the flow stream, then as the valve begins to close the local flow velocity near the seat tends to increase. This washes out particles rather than settle them near the seat. In the vertical position these particles would tend to congregate near the lower stem and bearing area, where there is little disc movement to dislodge them and where they could drift downward between the stem and its bearing surfaces. The disadvantage to this orientation is that normally there is more room for stem and operator above a group of pipelines than alongside, especially between two lines. However, the savings in life extension of valves are such that it is worthwhile to make room for horizontal orientation wherever the presence of solid particles is known.

High-performance butterfly valves tend to be available in a wide selection of materials, including stainless steels and high alloys, but the most typical is a steel body (either cast or cut from thick plate), with a stainless disc or a steel disc with electroless metal coating, a stem made of precipitation-hardened steel such as 17-4PH for greater strength in torque, seat springs of Inconel 718 or similar high-strength heat-resisting material, and stem bearings top and bottom. Many high-performance valves have either a packing gland at both top and bottom or at least some means of accessing the bottom stem bearing from outside. The disc attachment mechanism is normally either taper pins or bolts. Often the stem runs completely through the disc, and to withstand the seating torque the stem and disc are somewhat larger and heavier than in the rubber-lined valve. It is possible to provide steam jacketing around the exterior of the high-performance valve, and even possible to bore through the stem and pipe steam or purge gas up the stem and through the inside of the disc (Fig. 2-149).

A high-performance butterfly valve does the same job as a gate valve with significantly less metal mass, so in high alloys there is often a considerable cost savings in using a high-performance butterfly. In

**Fig. 2-149.** Steam-jacketed high-performance butterfly valve with steam provided both external to the valve body and through the stem to the disc. Valve also has a double-reduction worm gear operator. (*Courtesy of Xomox Corporation.*)

carbon steel the difference is less pronounced. Speaking in general terms, the butterfly is more expensive than the gate in sizes up to 20-in. to 24-in. pipe size (DN 500 to DN 600) in carbon steel, while in stainless steel the size break is about 6-in. or 8-in. pipe size (DN 150 or DN 200). If both valves can be considered to do the job equally well, the price break can be used to determine which to use. There are other factors than price to be considered, however.

One factor is that the gate valve, being metal seated, is basically intrinsically fire-safe. Both valves leak, as all valves do to some extent, and if the seat is damaged, generally no amount of force will get either valve to shut off. However, the butterfly valve is more likely to shut off bubble-tight in normal use. The butterfly valve is certainly smaller and it weighs less than the gate valve, which can be very significant in locations like ship and offshore platform installations. The butterfly closes faster, which makes it better both for control applications and garden-variety block valves if ease and speed of shutoff are important. The butterfly valve does have the ability to throttle. Generally, throttling butterfly valves have a slightly different design than on-off butterfly valves, but any butterfly valve will throttle better than any gate valve can.

Since the two ends of the high-performance valve are not identical, it does naturally seat better in one direction than the other. Early high-performance valves were marketed as one-directional only, and some older designs are still primarily capable of seating in only one direction. Most modern designs will seal equally well in either direction, especially at lower pressures, but where there is a choice the valve should be installed with the stem end downstream, that is, with the blank side of the disc facing the flow. The thick disc does not deform significantly from pressure force, but with the blank side of the disc facing the flow, the fluid force helps to push the seat more tightly against the disc while in the opposite direction the fluid force tries to oppose the sealing force. Metal-seated valves are more decidedly one-directional than are soft-seated valves.

Most high-performance valves have a travel stop located just behind the seat on one side to keep the disc from overtraveling and lifting back off the seat. The travel stop is not possible with rubber-lined valves. The actuator normally has some type of stop in addition, usually a lever-lock device in the case of a handle operator. Generally, the seat end is clearly marked on the outside of the body to indicate which way the valve should be installed.

As with an end-entry ball valve, access to the seat for assembly and maintenance is through the end of the valve. Unlike those in some ball valves, this end piece almost always comprises the complete gasket seating surface on one end of the valve. This end piece is normally retained by set screws or cap screws driven into its face. When the screws are toward the outside of the seating surface, normally no disruption of gasket seating occurs. Some designs have the screws more toward the inner edge of the seating surface, sometimes directly in conflict with the primary seating area. Spiral-wound gaskets, especially, seal over only a narrow band toward the inner edge of the flange face, and disruptions in the seating surface such as screw holes are very injurious to the ability of spiral-wound gaskets to seal. This feature is being phased out by many manufacturers of high-performance valves, but if there is any possibility of a problem the user should still specify the cases when spiral-wound gaskets will be used.

The end piece also provides an opportunity for leakage to the outside, if the design has the end piece extending completely to the outside (which is better for end flange sealing purposes). This requires an additional gasket just as any other valve with a body joint does, although often the seat itself can act to seal off this leak path. For services in which there must be as few leak paths as possible, there are some valve designs in which the end piece joint does not lead to the outside of the valve. These generally are special order designs.

An improved version of the high-performance valve has the stem offset not just toward one end of the body, but offset slightly (about 1/16 in.; 1 to 2 mm) to one side also (Fig. 2-150). In addition, the center of the

**Fig. 2-150.** Triple offset eccentric butterfly valve with double-acting air cylinder, positioner and air reservoir for fail-safe actuation. (*Courtesy of Adams Valve.*)

disc is offset slightly downward from the vertical center of the stem. This triple-offset design, which goes by various proprietary names, has an even better shutoff ability than does the standard high-performance butterfly. This valve is often capable of sealing bubble-tight in very large sizes. As a rule, sealing ability drops as valves get larger, mostly as a result of accumulated manufacturing tolerances, but this valve is the exception. The sealing area in this valve (Fig. 2-151) often consists of laminated metal and soft-seat components much like a spiral-wound gasket, and is located in the edge of the disc. These valves can also be built with a significantly offset stem (Fig. 2-152a,b) with much the same seat geometry as a tilting-disc check valve, and the stem can be arranged to allow the valve to behave like a globe-type stop-check valve. These valves are even available with refractory lining for service at very high temperatures, such as steel mill hot gas lines. Valves of this type are generally fabricated steel, although cast bodies are becoming more common, and this type is of course somewhat more expensive than a standard high-performance butterfly. But it does deliver even better performance, where such performance is needed.

Because this valve often has an even thicker disc and longer seat area than a more conventional high-performance valve, it is often built as a double-flanged design with overall length quite a bit longer than the common butterfly valve dimensions. Other special application butterfly valves are often built as double flanged, generally in the larger sizes. There are also butterfly valves, designed primarily for cryogenic service,

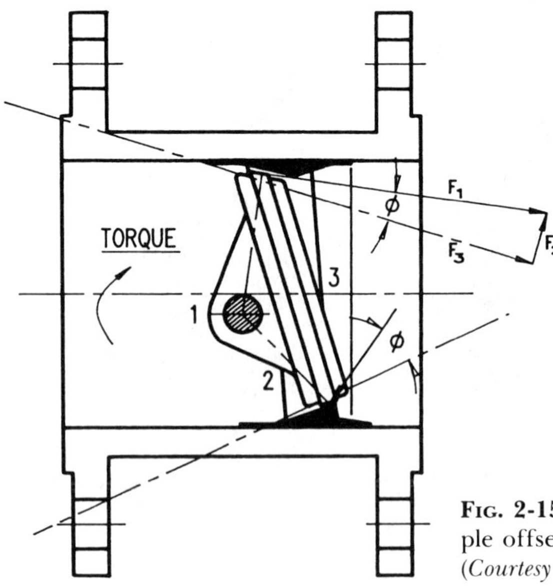


**Fig. 2-151.** Sealing principle of triple offset eccentric butterfly valve. (*Courtesy of Adams Valve.*)

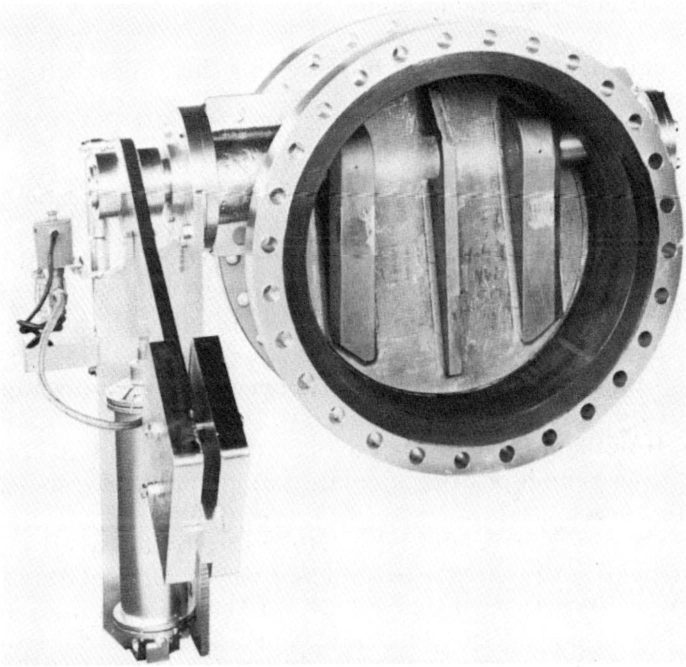

(a)

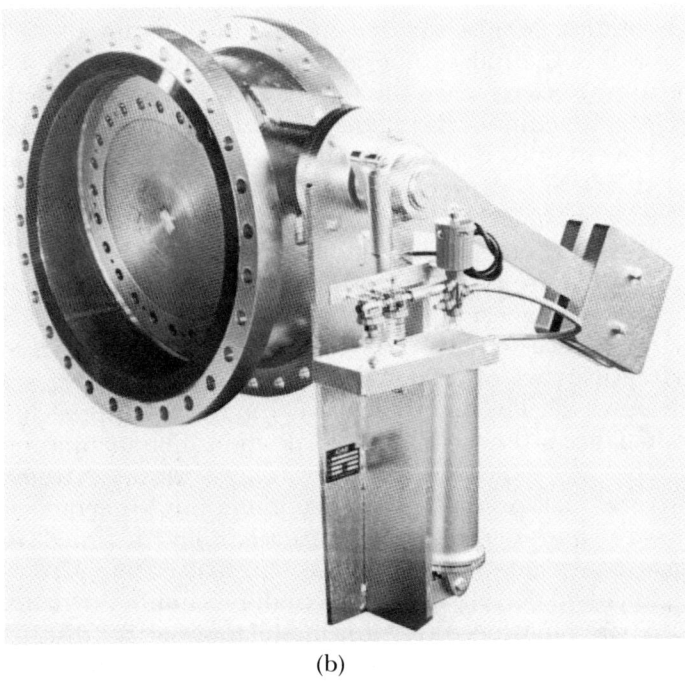

(b)

**Fig. 2-152.** Triple offset butterfly valve configured as stop-check. (*Courtesy of Adams Valve.*)

that have butt-weld ends and a bonnet for accessing the internals from the top. This is a very rare configuration.

The butterfly valve concept is a fairly common one. The choke in your carburetor is a butterfly valve. Its application for industrial piping, though, began only in the 1930s. For years one of the main applications of butterfly valves was in oil field use. Eventually their use spread to a number of industries, but as the various types of butterfly appeared the attempts that were made at standardization were not very successful. For a long time the butterfly was one of the least standardized of all valves in the fluid-processing industries. There was a standard for flanged waterworks valves, and then later a standard for rubber-lined lug and wafer valves. Finally, in 1982, the high-performance valve joined the world of standardization. It is still possible to buy valves that are not standard end to end to replace valves built prior to standardization, but it may not be so for many more years.

There are two types of actuated high-performance butterfly valves, just as there are major differences between globe valves and globe-style control valves. Many butterfly valves are built for remote operation in the shutoff or block mode, and these are generally standard high-performance butterfly types. The butterfly does have some points to consider, though. Almost all high-performance valves are already equipped with bracket mountings for actuators. The actuator sizing needs some consideration because the disc and the flow stream interact aerodynamically in ways that might not be expected. For instance, the greatest dynamic torque occurs when the valve is near full open (actually, at about 80 degrees) because this region is where the cross-sectional flow area is varied the most by disc movement. Seating torque occurs at the other end of travel, at 5 degrees of opening and below.

Butterfly valves used purely for throttling can be standard unmodified butterfly valves, but generally some design differences are found. One common feature is a more secure disc-to-stem attachment such as a spline, or sometimes a one-piece disc stem (which requires a two-piece body), to reduce or eliminate mechanical hysteresis. Since throttling butterfly valves are often not required to shut off, some are built with no true seat at all. The best these valves can do is to approach 5 percent or so of maximum flow in the closed position. Eliminating the seat makes the valve less expensive, and throttling with an extreme turn-down ratio (very small percentages of maximum flow) is erratic at best unless the valve and the actuator are specifically designed to accommodate such requirements. Valves of this type may be intended to operate in the region of 40 percent to 90 percent of maximum flow rate.

Some valves that are built specifically for modulating service discard entirely any attempt to seal in both directions, since control valves are rarely expected to see reverse flow, and are built with an asymmetrical disc that interacts better with the flow stream. These designs have a wider disc edge on the back side of the disc, the side that opens downstream. The purpose of the wider disc is to eliminate some of the hydrodynamic forces on the trailing edge of the disc, thus reducing the torque required to operate the valve, and in some cases raising the $C_v$ enough to permit the use of a smaller valve. In many cases standard valves no longer change the flow rate effectively after they open to about 60 degrees (two-thirds of the travel from closed to full open). Use of the wider disc edge extends this control closer to 90 degrees, which also makes the actuator capable of more precise regulation since it then can operate over a longer stroke. This design is still capable of sealing against a standard seat. Some other valves can be furnished without any seat, and act purely as a throttling valve with no capability at all of shutoff. This is perfectly adequate in many situations such as flow balancing. Other types that do not completely seat can be built with a reverse-tapered disc edge to help cut through deposits that accumulate near the seat area.

It is even possible to build a three-way butterfly valve (Fig. 2-153). Perhaps the term "three-way butterfly" is a misnomer since no one butterfly valve can actually be capable of possessing three ports. This term is usually applied to actuated butterflies, since the configuration includes two separate valves mounted on two sides of a tee. There is a common linkage that connects the levers of each valve such that as one closes, it opens the other one. This does permit flow splitting or mixing or diverting. Each of the different position combinations is possible, some requiring more complicated linkage than others. The whole point in doing this is allowing one actuator to operate both valves as if they were a single valve.

### *Diaphragm Valves*

The diaphragm or wier valve is another design that can be used for throttling, but is often used as a block valve (Fig. 2-154). The diaphragm valve was invented during World War II and is one of the class of valves that close by narrowing the flow passageway until it closes off. Like most valves of this class, it is completely bidirectional.

In the standard form, the diaphragm valve has a flow pattern that enters the valve from one side, passes over a wier that deflects it upward (toward the bonnet), then back down again to exit the valve body in the same direction and orientation that it entered. Above the wier, and below the bonnet, is a diaphragm which is pushed downward by the stem until it seals against the wier. The stem is a conventional threaded design, but attached to the stem is a device called a compressor that spreads the stem force over the entire upper surface of the diaphragm to give even pressure. The diaphragm is attached to the compressor by a

**FIG. 2-153.** Three-way butterfly (two butterfly valves and linkage mounted on a flanged tee) shown at far left. (*Courtesy of Bray Valve and Controls.*)

**Fig. 2-154.** Cutaway of plastic-lined diaphragm valve. (*Courtesy of ITT Engineered Valves, Inc.*)

stud that is imbedded in the diaphragm and allows the stem to lift the diaphragm upward.

The diaphragm itself is often a rubber compound, but in cases where no elastomer is resistant to the fluid being handled, it is possible to use a thin nonelastic fluoropolymer such as PTFE in contact with the process fluid and an elastomer behind it to give the diaphragm the flexibility that it needs. One of the virtues of the diaphragm valve is that the diaphragm itself seals the process fluid off from contact with the inside of the bonnet, stem, compressor, and other parts. Any leak paths to atmosphere occur by leaking under the diaphragm, or by the diaphragm itself rupturing. The diaphragm is molded with ribs along the line of contact with the wier and around the edge, to give higher contact pressure in those regions. It is molded as if it were in the closed position, so the most relaxed state of the diaphragm is when the valve is closed.

The body of the valve can be any material, but one of the reasons for using the valve is that the diaphragm blocks the process fluid from the inside of the bonnet and thus makes it more suitable for severely corrosive fluids. Therefore, the body and bonnet is often cast iron or some other low-cost material, with a full liner on the body to improve its corrosion resistance. In the early days, and even now, it was common to build bodies with glass linings. The soft diaphragm seals against the glass with very little likelihood of damage to the brittle lining. Other linings are more common, including PTFE, polypropylene, various rubbers, and almost any other of the linings that are commonly used in piping systems. All-plastic bodies are also used, including PVC, CPVC, PVDF, and PP. In very small sizes, the valve is built in a barstock design.

The operating mechanism used is normally a fixed handwheel. The stem rises through the handwheel threads as the handwheel is turned. A few very large diaphragm valves—they are available up to as large as 20-in. size (DN 500)—can require gear operators, but this is very rare. The position of the stem indicates the valve's position. A problem with the soft diaphragms is the tendency to be overtightened, so many valves are equipped with an adjustable travel stop on the top of the stem to limit the stem's downward travel, which also helps make the position a little more obvious. Most valves also have a finger plate in addition to the compressor, which fills the gaps between the compressor fingers and provides a soft stop for the diaphragm in the open position. In the full open position, pressure pushing up against the diaphragm is resisted by the springy action of the finger plate. In some small sizes, it is possible to equip the valve with a lever that lifts the diaphragm completely by means of a quarter turn, like a ball valve. The stem still rises, through a cam action.

Because the wier required in a conventional diaphragm valve obstructs the flow somewhat, there are some restrictions on orientation. In horizontal flow, the wier prevents free draining of the pipe line. If the valve is oriented with the stem at about 20 degrees up from the horizontal, free draining can occur. If solids are present, the wier may cause a buildup of particles that may block the line. Because of their geometry, standard diaphragm valves cannot be rodded out. Solids in the flow stream, though, is precisely one of the reasons for choosing a diaphragm valve in the first place since the diaphragm closes well against small solid particles. The diaphragm just flexes around them until it makes complete contact with the wier. It is also possible to build diaphragm valves in an angle pattern (Fig. 2-155), where the flow into the valve is from the bottom, making a slight turn and then encountering the wier, followed by another bend and exiting the valve at 90 degrees from its entering orientation.

There is a solution to the wier problem, known as the straight-through diaphragm valve. Here, the wier is not present and the bottom of the valve has a smooth interior. The diaphragm is somewhat longer (deeper) than in a standard valve, and it must deflect farther to close off the flow passageway. As a consequence, straight-through diaphragm valves are not available in sizes as large as standard valves, and are generally larger and more expensive. The best analogy is to a full-port plug or ball valve. The straight-through design does present a less restrictive opening, has no spot for settling out to occur, can be rodded out, and can free drain through the valve. It is available in most of the same materials as standard valves.

**FIG. 2-155.** Angle pattern diaphragm valve. (*Courtesy of ITT Engineered Valves, Inc.*)

Because of the diaphragm, any of these valves is limited somewhat in its pressure-temperature rating. Many are rated to 150 psi (10 bar) or thereabouts, some are rated to ANSI class 150 (PN 16), but very few are rated higher than that. Depending on the service and the diaphragm material, the maximum temperatures range from 175 to 350°F (80 to 175°C). Common process applications include water treating, corrosive chemicals, and food processing. The diaphragm is easily removable without requiring gaskets, and the smooth interior is easily accessible for cleaning, making it ideal for food and pharmaceutical services.

Valves used for food services are normally either polished stainless steel (Fig. 2-156) or lined with an inert material and epoxy coated on the exterior, with quick-release cleanable end connections. Because of the possibility of contamination if the diaphragm ever does leak, the bonnets of sanitary service valves are often required to be of noncontaminating materials and have a V-notch plug in the bonnet designed so that if there is a bonnet leak, the plug can be backed out until the bonnet starts to depressure to the atmosphere without causing a loss of bonnet pressure and a possible total rupture of the diaphragm.

Other special diaphragm valves are built for high-purity gas service (Fig. 2-157), which basically requires the minimum possible leakage, often under vacuum conditions (leakage from the atmosphere to the

**Fig. 2-156.** Diaphragm valves for sanitary service with polished surfaces and butt-weld tube ends (*left*) and quick-connect sanitary clamp ends (*right*). (*Courtesy of ITT Engineered Valves, Inc.*)

inside of the valve). These valves must be assembled under clean room conditions, tested for particle generation (the giving off of tiny fragments of elastomer or metal from the inside of the valve), and usually tested for leakage with helium.

Actuating a diaphragm valve can sometimes be done with standard air diaphragm actuators, but special ones are often furnished. The diaphragm is one valve that cannot be built to operate in reverse (push down to open), so in any case where the valve must fail open, the spring must be located under the actuator diaphragm. Since the valve opens fully with a very short stroke, it is common to direct-mount the actuator diaphragm to the valve diaphragm with a one-piece shaft. In fact, in a few cases, it is even possible to dispense with the actuator altogether and simply pressurize the bonnet with air to push directly down on the valve

FIG. 2-157. Small diaphragm valves with butt-weld tube ends for high-purity gas service. (*Courtesy of ITT Engineered Valves, Inc.*)

diaphragm. Such a method is actually used, although not often and not with much accuracy of control.

Any diaphragm valve can throttle perfectly well. However, for heavy-duty continuous modulating service, it is possible to build the compressor in two segments, so that for small-percent openings only the center section opens. After the center section opens, when further opening is needed the outer segment starts to open also. At full-open, the configuration looks like a standard valve in the full-open position.

### *Other Throttling Valves*

There are still other types of valves that are capable of throttling flows. Most of these valves are used primarily for shutoff, but can be used in partly open positions, just as the various types of butterfly valves can be. This section will briefly discuss these valves, most of which are designed to solve special problems. Some are locally very useful, and may occupy a big chunk of some of the smaller valve markets.

It would seem simple enough to throttle a flow—why not just pinch the line shut? That is basically what the pinch valve does (Fig. 2-158a,b). As in the diaphragm valve, all the moving parts of the pinch valve are outside the body and not exposed to line fluid. But unlike the diaphragm valve, the pinch valve does not really have a valve body at all. It consists of end flanges connected to a frame that surrounds a tube,

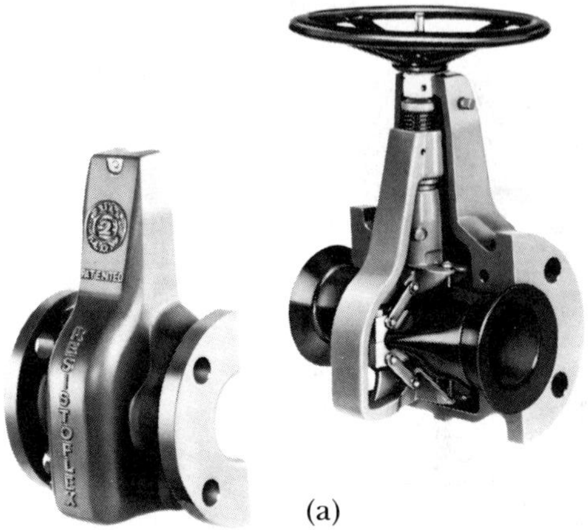

(a)

**Fig. 2-158.** Clamp valve cutaway and cross section showing operating arrangements. (*Courtesy of Crane Resistoflex Co.*)

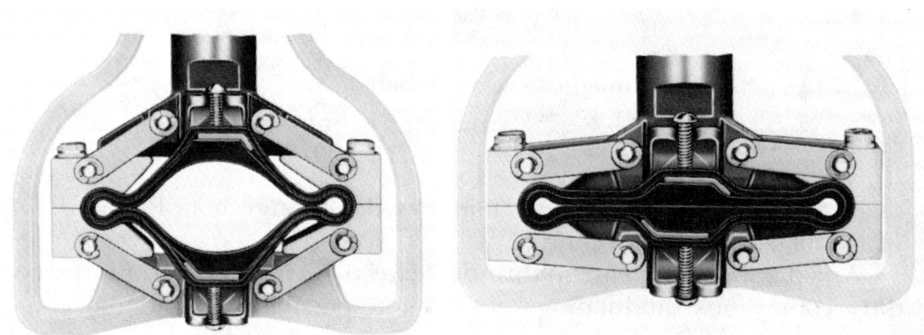

(b)

usually rubber with reinforcing wire. The operating mechanism is part of the frame, and simply pinches off the tube from each side until the tube is completely closed.

The pinch valve is thus capable of handling solids, slurries, and anything else that gums up the works of more conventional valves. It is often used in ash-handling systems, municipal sewer systems, and sand pumping. Rubber liners or tubes are inherently better at resisting abrasion than harder materials, essentially because harder materials "bounce off." The elastomer, of course, also deforms around whatever happens to be in the sealing area at the time that the valve is closing. Of course, the temperature and pressure limits of the pinch valve are determined by the limits of the rubber used for the liner. A number of rubber compounds are commonly used, including Buna-N, Neoprene, and Viton. Like other lined valves, the liner wraps around onto the faces of the end flanges to form the end seals.

The pinch valve is generally available in medium sizes, from about

2-in. to 12-in. pipe size (DN 50 to DN 300). Ratings are generally expressed as actual working pressure based on rubber hose ratings. Higher pressures are available by means of woven steel braiding much like that on flexible metal hoses. These valves generally have end flanges drilled to match ANSI 125/150 (PN 10/16) drilling templates.

It is possible to provide actuators for pinch valves. Generally, two diaphragm actuators are installed, one on each side, pushing toward each other. They can be directly installed, since the actuator travel approximates the movement available in the pinch valve. In fact, actuators are not even necessary, since the valve operates equally well by pressure differential from the outside. When an airtight shell is provided (which does not have to be suitable for the process conditions) (Fig. 2-159), the region outside the tube can be pressurized to provide throttling or shutoff force.

A variation on this principle is the iris valve, used in solid gravity flow systems (Fig. 2-160). Here the boot is twisted, with one end fixed and the other end free to rotate like the aperture in a camera lens. The valve is capable of tight shutoff, but practically no differential pressure. However, most gravity flow systems have no differential pressures to speak of anyway. The valve can be used for on-off or throttling (flow restriction) services.

**Fig. 2-159.** Pinch valve operated by air pressure. (*Courtesy of Red Valve Company.*)

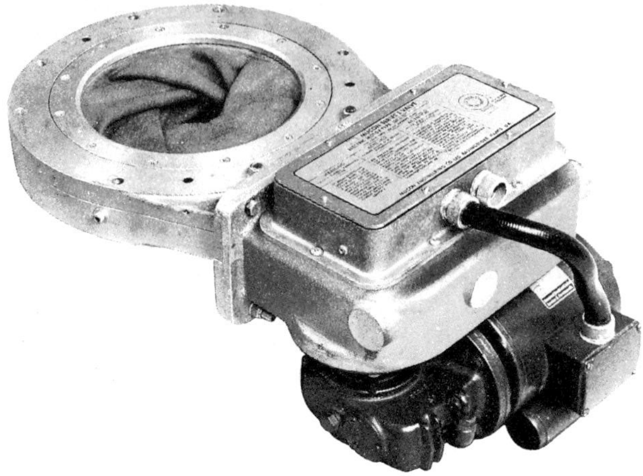

**Fig. 2-160.** Iris valve for throttling service in solids conveying lines. (*Courtesy of Kemutec Inc.*)

## Nonreturn Valves

### *Swing Check Valves*

This section deals with check valves, which are inherently different from almost all other types of valves in that no outside intervention is required for the valve to perform its designated function. Check valves, which are more generically known as nonreturn valves, are intended for control of fluid flow just like any other valve, but their contribution to the system is to keep fluid from returning in the direction from which it came. This principle is used widely in nature; the human body contains a number of check valves.

In piping systems, there are a number of ways to prevent backflow without using check valves. Any system that flows by gravity is automatically protected from backflow just by having one portion of the system higher than another. Having an air gap or a spot where the fluid has to fall freely from one part of the system to another is an even better isolation method. This method is often required where absolute separation is needed, for instance, in connections to potable water systems where the clean water may be used for other purposes but where it is unacceptable to have any possibility of contaminated water flowing back into the drinking water.

For the more usual industrial piping system where all parts of the system are pressurized, some type of valve is needed. The earliest type of check valve to be developed was based on the door or gate analogy. Like a door built to open in one direction but to come to a stop against a sill or jamb that prevents it from swinging both ways, the swing check valve has the ability to open freely when fluid pushes against it. In one direction the door, or disc, opens wide, and in the other direction the disc swings until it hits the seat.

Many check valves close dependably enough, but in the closed position they really are not closed tight. Fluid can get around a check valve disc a lot more easily than a person can get through a closed door. In most cases that is just the way it is, and there is not much you can do about it. The check valve is there to slow down the flow in the reverse direction, or to cut the mass flow rate down to about 1 or 5 percent in the backward direction of what the forward flow rate is. In some cases, a much better shutoff rate is required, and there are several varieties of check valve that can do that job.

The swing check still has a large portion of the check valve market, although there are many other types of check valves that are an improvement over the swing. Physically, a standard swing check valve looks like most other valves (Figs. 2-161 and 2-162). It has flanges or other standard end connections, and it has a cover that is removable. The cover or cap is just that—there is no stem or other mechanism to apply force or torque from the outside. The disc is attached to a swing arm or lever that is mounted on a bearing shaft. The arm pivots freely around this shaft, and the fluid swings the disc open or closed.

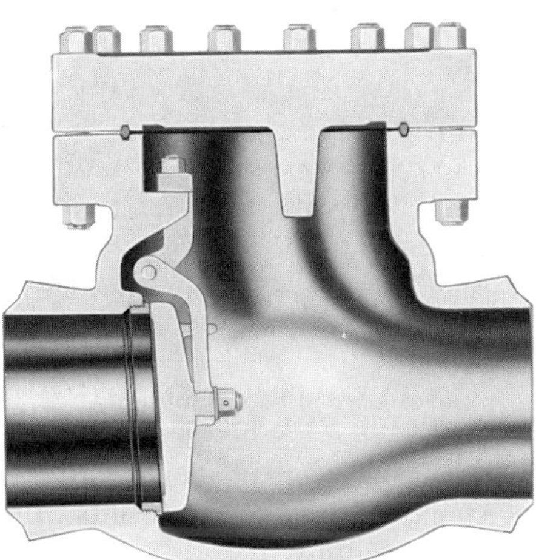

**Fig. 2-161.** Butt-weld-end swing check valve with internal swing arm. (*Courtesy of Pacific Valves, a unit of Mark Controls.*)

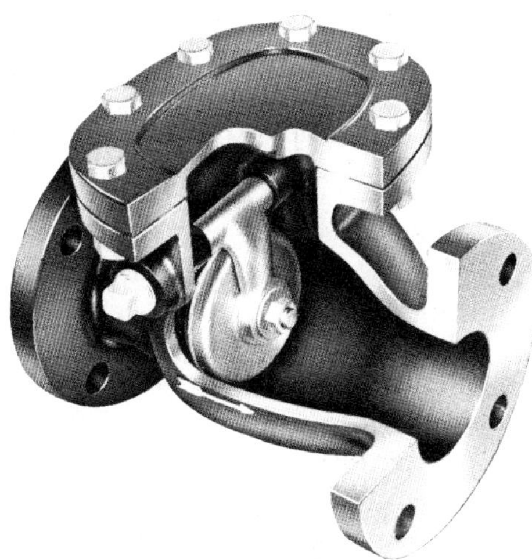

**Fig. 2-162.** Cutaway of flanged-end bolted-cap swing check valve with hinge pin retainer plug. (*Courtesy of Crane Co.*)

Actually, although this is the way that almost all swing checks are built, this simplistic picture is not entirely true. Since that shaft usually has to have been inserted from outside the body anyway for mechanical assembly reasons, the end of the shaft can be used for actuation just like any other valve. Often the reversal of flow is still required for the check to begin functioning, but a power actuator can be used to help in closing the valve. Sometimes a weight or a spring on the outside can also be used, as an external but still passive actuation aid. These springs or weights can be used either to assist in closing or to retard closing, depending on how they are connected and why they are needed. Theoretically, a valve built like a swing check could be actuated from the outside and used as a shutoff valve, but because of the torque required to close such a valve against line flow, such a scheme is almost never used.

The outside lever and weight, or lever and spring, requires a packing gland just like any other valve that has an actuation mechanism from the outside (Fig. 2-163). However, since no positive actuation is applied, the force available is still limited to that of the weight or spring. This means that sometimes the force that the packing gland exerts is great enough to bind or restrict the pivoting motion. The outside lever design, therefore, is generally practical only in large valves. It is available in sizes down to about 6-in. pipe size (DN 150). It is most common in transportation and building service piping. When a positive closure is required in refinery piping, it is more common to install a different type of check that inherently possesses better closure characteristics or, alternately, some type of actuated valve that closes on a flow signal, usually in addition to a check valve.

Note that the normal swing check has a pivot that is inserted through the body wall in the same location that this packing gland would be, but there is no gland because the retainer pin does not extend through the wall. There is normally just a standard pipe plug on one side to plug the opening used to assemble the retainer pin into the valve. It is possible to design a swing arm retaining device that is entirely within the body, installed from the top or attached to the cap and retained with some bolted-on or welded-in device (Figs. 2-164 and 2-165). These designs are slightly more expensive and potentially more complicated, but definitely do not provide the leak path to the outside that the plugged opening presents. It might be better if more swing checks were available with this feature, but they normally do not have it except in pressure-seal bonnet and other high-pressure designs. It is true that anything is available if you want it badly enough and are willing to accept the increased cost and (especially) delivery time that special order implies.

The common swing check has a seat inclined at about the same angle as that of a wedge gate valve, and usually has a stop on the cap or the

**Fig. 2-163.** Installation of swing check valves equipped with outside lever and weight to assist in closure. (*Courtesy of American-Darling Valve, division of American Cast Iron Pipe Company.*)

inside of the body, such that the disc travels about 60 degrees or a little more from full closed to full open. One of the reasons for having the inclined seat is that, when the valve is installed in the horizontal with the cap upward, the weight of the disc still exerts some force tending to hold it against the seat. In a vertical line, with the disc wide open, the disc has

**Fig. 2-164.** Internal view of butt-weld-end bolted-bonnet check valve with internal swing arm hanger. (*Courtesy of Pacific Valves, a unit of Mark Controls.*)

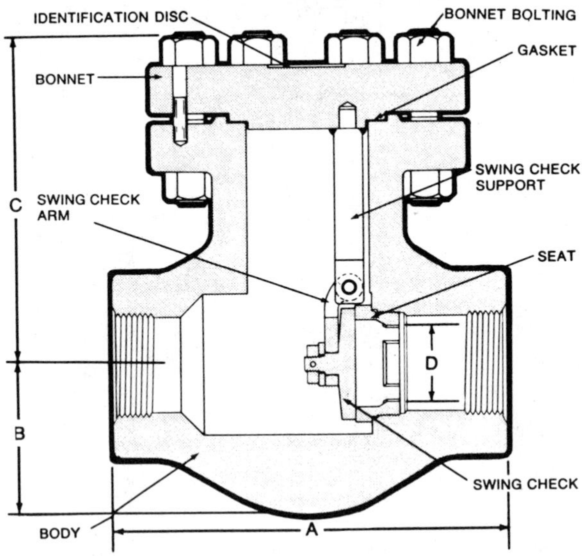


**Fig. 2-165.** Swing check valve with swing arm hanger attached to bonnet. (*Courtesy of Henry Vogt Machine Co.*)

to remain unbalanced. That is, there still has to be some gravity force acting on the disc to tend to close it when the fluid flow stops. (Note that virtually no check valves, except a few very specialized ones, are capable of functioning in a vertical line with the flow downward.)

Like its gate and globe counterparts, the swing check valve is available in almost any rating, material, and type of end connection and cap design. Physically, the swing check is about the same size as a globe valve from the bonnet flange on down. In the lower pressure classes, then, it is somewhat longer face-to-face than a gate valve is. In higher pressure classes, the potential exists for rather high differential pressures. This can make the disc get heavier and harder to swing, and can slam the disc shut rather forcefully. Therefore high-pressure check valves such as are

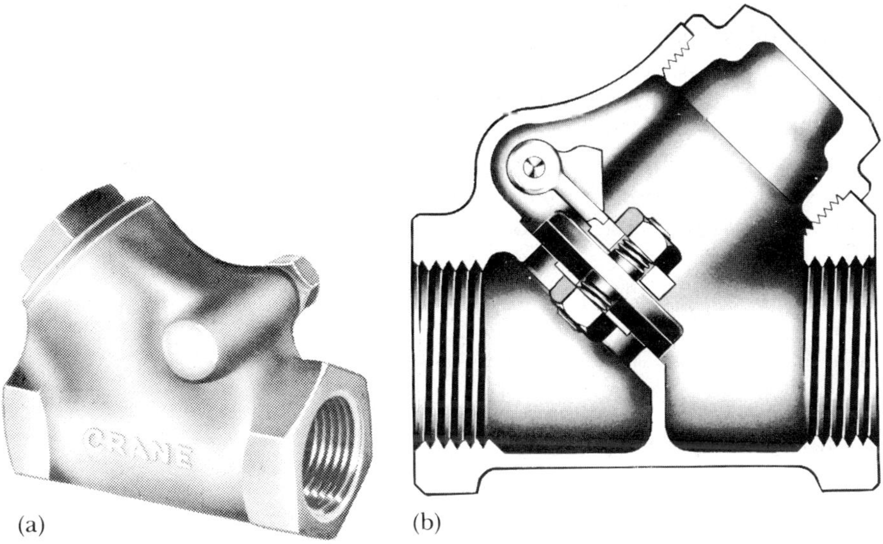

(a) (b)

**Fig. 2-166.** Y-pattern swing check with soft-seated disc and threaded cap. (*Courtesy of Nibco, Inc.*)

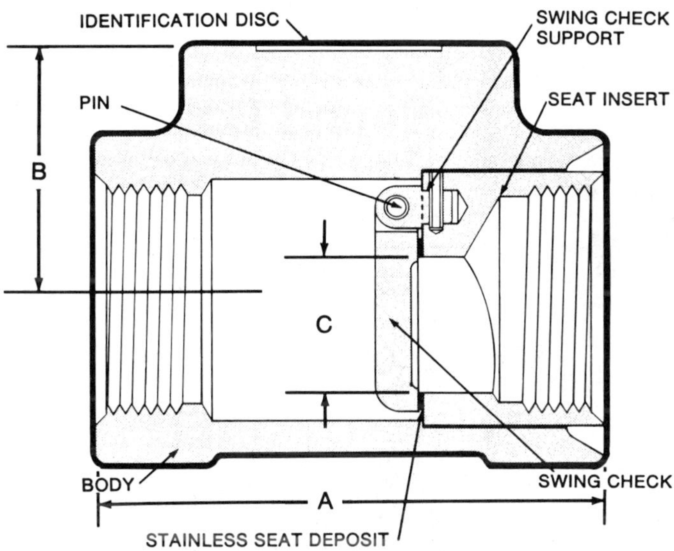


**Fig. 2-167.** Bonnetless swing check valve. (*Courtesy of Henry Vogt Machine Co.*)

encountered in steam and high-pressure process services tend to be other designs than the swing type.

In small sizes, the opposite problem exists. Although the fluid flow provides most of the force to close a check valve, gravity is also required most of the time. Small check valves, say under 2-in. pipe size, do not have much mass in the disc and are more prone to some mechanical problem preventing the disc from closing. In stainless steel and bronze bodies (Fig. 2-166a,b), the swing check is still fairly common since corrosion of the swing arm and pivot is not likely, but small carbon steel valves are more likely to be some other type of check. An exception is an interesting little swing check made from the same forging that is used to make a tee (Fig. 2-167), where the entire swing mechanism including pivot is inserted from the downstream end and welded in.

### *Tilting-Disc Check Valves*

The closest relative of the swing check is the tilting-disc check valve (Fig. 2-168). The tilting-disc moves the same way as the swing, except that the pivot arm runs right through the disc instead of being above the disc. The overall motion of the disc is thus less, and the motion is more rotational than translational. The pivot must still be offset a certain distance from the center of forces acting on the disc for the disc to open, but the distance is much less in the tilting-disc. The center of mass of the disc is much closer to the pivot point, and in fact is normally located even

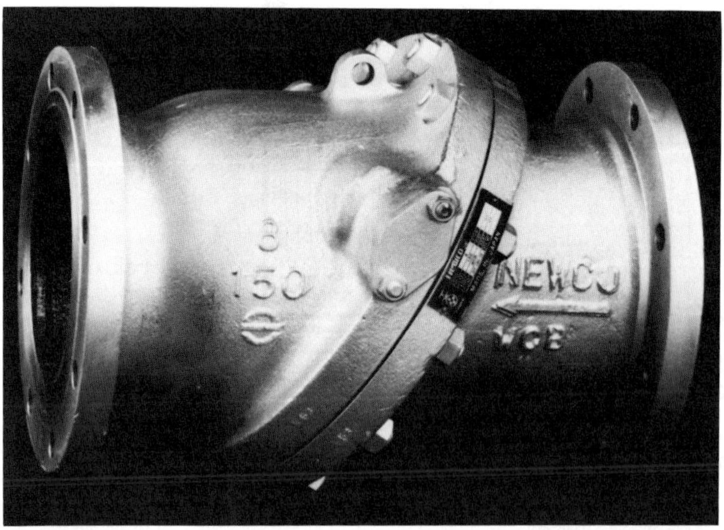


**Fig. 2-168.** Steel tilting-disc check valve. (*Courtesy of Newmans Inc.*)

closer to the pivot point than the center of opening forces. This means that the tilting-disc opens more easily (with less resistance to flow and consequent pressure loss) than the swing check. Of course, all these statements apply equally well to closing forces.

The tilting-disc check was thus the first "nonslam" check, and to this day its primary attractiveness is that it can close more gently than the swing check can. It can also be designed to close very fast, since the relative positions of the center of mass and the center of force can be varied in ways that are very impractical to do with a swing check.

It has certain disadvantages, mostly that it is not repairable in-line since it is not built with a cap. The body joint is along the seat line, at approximately a 45-degree angle to the ends and the pipe centerline. Otherwise, the mass and dimensions of a tilting-disc are comparable to those of the swing check. The angled seat is also one of the reasons for fast closing time: In the open position the disc turns in the direction of fluid flow like a butterfly valve, but in the closed position the disc is 45 to 60 degrees away from the pipe centerline, not 80 to 85 degrees as in a swing check.

The tilting-disc check is a rather specialized valve, since to most industries it is used only in applications where its closure characteristics are needed. Probably its most common use is in water lines, such as industrial cooling water or municipal water, as a pump discharge check valve. Water lines tend to be among the largest lines in a plant, and the mass of water being moved can be quite large. When water hammer and other fluid flow disruptions occur in lines this size, the forces imposed on the piping can be enough to knock pumps out of alignment or knock piping off its supports. The tilting-disc check is designed to assist in combating this problem by closing fast, almost as soon as the moving column of water reaches a standstill and before it has a chance to begin moving backward. One of the contributions to the piping loads is the inertia of the water column as it strikes an immovable object, the check valve disc that has closed in response to the water beginning to move backward through it. If the valve can close at the precise time that the moving water has reached its lowest average velocity, the least amount of force will result.

There is a second, related niche that the tilting-disc check occupies. The high closing forces that are a problem with large masses of water are also a problem with high-velocity vapor flows subject to sudden fluctuations, such as in compressor discharges. Here the valve is normally configured to respond fairly slowly to the changes in flow rate, to reduce the tendency to slamming, which is highly destructive not only to the valve but to the rest of the piping system subject to such loads.

The tilting-disc check valve is available down to about 4-in. pipe size (DN 100), but it is most common in sizes on the order of 24 in. to 48 in. (DN 600 to DN 1200) for municipal water and pump discharge check

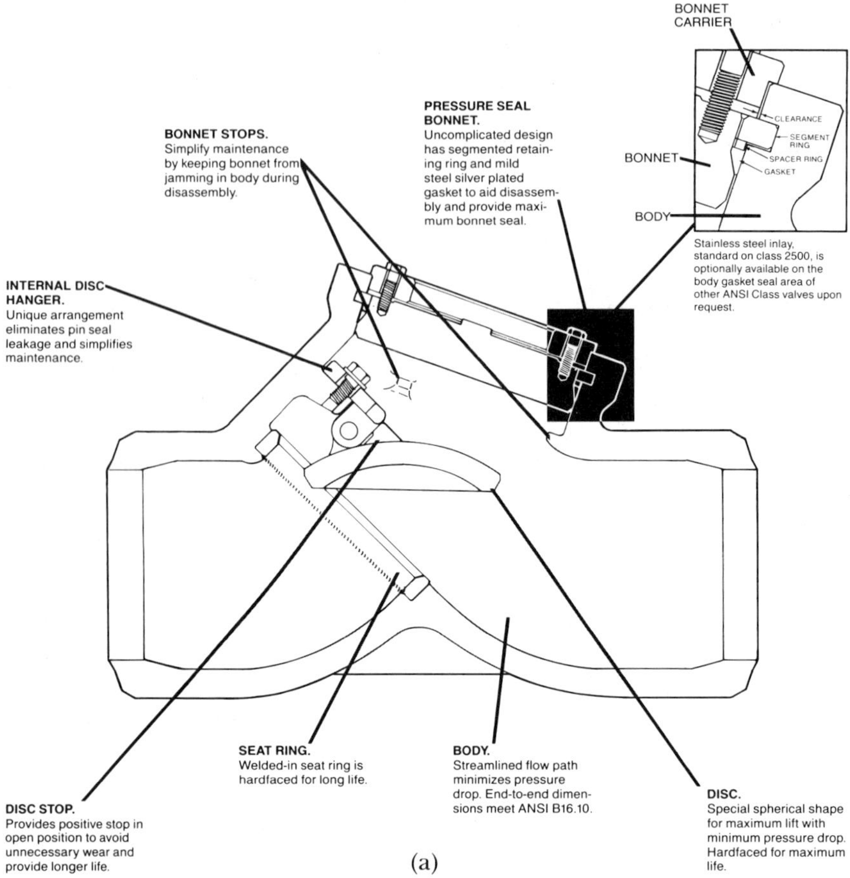


(a)

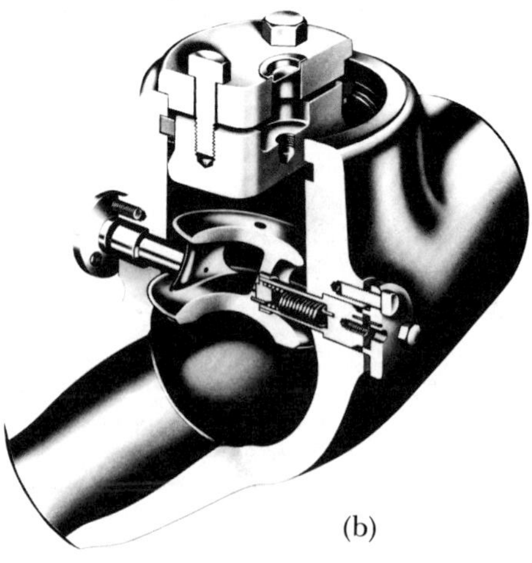

(b)

**Fig. 2-169.** (a) Cross-sectional view of high-pressure tilting-disc check valve with pressure seal closure. (*Courtesy of Pacific Valves, a unit of Mark Controls.*) (b) Cutaway of high-pressure tilting-disc check valve with pressure seal closure. (*Courtesy of Edward Valves, Inc., formerly Rockwell Edward.*)

service. In these services the valve is generally cast iron or steel, normally rated for class 125 (PN 10) or class 150 (PN 16). The other situation that the tilting-disc appears in calls for a steel body, high-pressure-rated valve (Fig. 2-169a,b) that is often butt-weld with a pressure-seal cover. These valves, by the way, are built more like a conventional swing-check in that the pressure rating makes the split body impractical so they have a bonnet or cover and a pivot arm that is accessible from the top. These two sets of service conditions are quite far apart in their requirements, and it can be very difficult to obtain tilting-disc check valves that fall in between these two extremes, for instance in a class 300 (PN 25) steel body.

### *Lift Check Valves*

The lift check valve operates on an entirely different principle from the swing and tilting-disc checks (Fig. 2-170). As its name implies, the lift check lifts when fluid pushes on it from below. It is also known as the piston check since the disc moves up and down like a piston.

While the swing check shares the door analogy with the gate valve, the lift check is more closely related to the globe valve. In fact, in the small sizes where the lift check is most common, the body is the same as that of a globe valve, just with a cap instead of a bonnet, and often the disc, or piston, is the same.

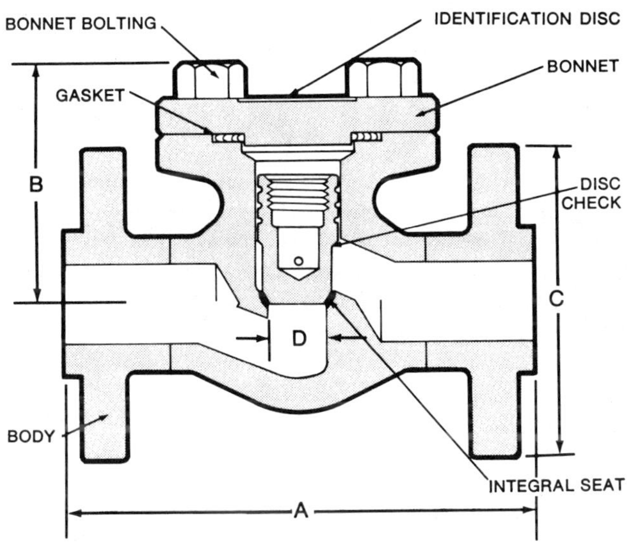


**Fig. 2-170.** Piston lift check valve. (*Courtesy of Henry Vogt Machine Co.*)

The piston moves up and down freely, but some sort of guide is required. The guides consist of an internal guide attached to the cap, inside the piston and occupying the space where the stem attachment would go on a globe valve, or either a cylindrical surface or three or four guide ribs external to the piston and bearing on the cylindrical area outside of the piston above the seating area. Smaller lift checks, 2-in. pipe size (DN 50) and smaller, are generally of the former type, while piston checks larger than 2-in. size (DN 50) are often of the latter type. Larger piston checks are special-purpose designs, often designed with damping devices and used like other special checks for pulsating flows or cases of water hammer.

In sizes 2 in. (DN 50) and smaller, the piston check is by far the most common type. It is available in any material, from brass to high alloy, and in almost any rating and type of end connection. Check valves are not a high-volume item in comparison to block valves of any description, so the ability of the lift check to share the same body as the same-size globe valve brings with it an economy that makes the lift check cheaper than some of its competitors. It does share the double 90-degree change in flow direction with the globe valve, and the consequent pressure loss.

The garden-variety lift check operates by a combination of fluid forces and gravity—gravity being probably the stronger force most of the time. The piston must move in a true up and down orientation, so the cap of the valve must be reasonably close to upright. In a vertical line the check will not work—the force of gravity holding the piston to one side is too much for the fluid flow to overcome. The piston is normally furnished with a pressure-equalizing hole through its side, above the seating area. This is required for the upstream line pressure to get behind the piston and help push it down.

In order to help the valve operate better, a spring is sometimes installed above the piston, pushing on the underside of the cap. This improves the operability of the valve, since it helps overcome any tendency toward sticking in the open position and closes a little more tightly. It does generally mean that a little higher differential pressure is required to open the valve (this is called cracking pressure). Normally this pressure is on the order of a fraction of a psi (0.1 to 0.01 bar) and is not worth considering. A spring-loaded valve might close slightly better in an off-of-vertical position, but it still cannot be expected to close reliably in a vertical line.

There is a piston check that will operate in a vertical line just as well as it would in the horizontal. This is the Y-pattern piston check (Fig. 2-171), and of course it works equally well in vertical and horizontal lines because the piston and spring area is in exactly the same orientation regardless of which orientation the end connections are in. As a general rule, most small Y-pattern piston checks are spring-loaded. They share the same advantage that Y-pattern globe valves do, that is, they have a straighter flow path and a lower pressure drop than does a comparable

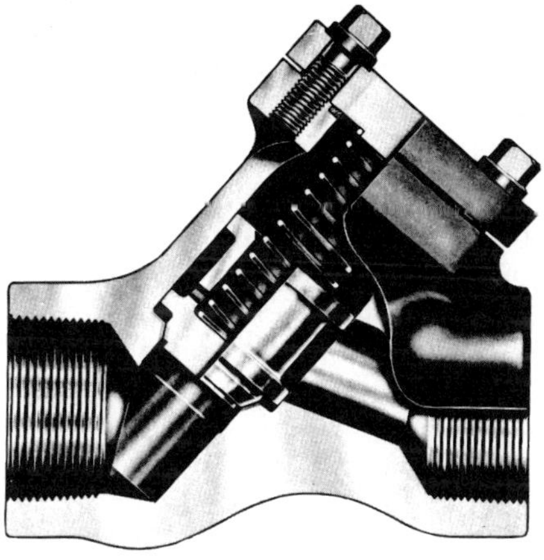

**Fig. 2-171.** Y-pattern spring-loaded piston check valve. (*Courtesy of Edward Valve, Inc., formerly Rockwell Edward.*)

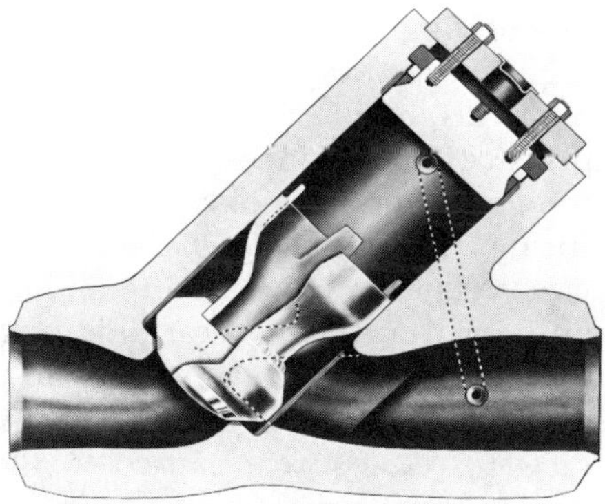

**Fig. 2-172.** Large Y-pattern piston check showing equalization line from downstream side to top of valve. (*Courtesy of Edward Valve, Inc., formerly Rockwell Edward.*)

straight or T-pattern check. They are, of necessity, larger than the straight pattern. A side benefit of the Y-pattern is that it is a little easier to tell from some distance away which direction the flow in the line is supposed to be, since the Y-pattern body "points" downstream. And the Y-pattern check also has the same body and disc as the Y-pattern globe.

Specification engineers would most commonly select the lift check in smaller sizes, up to about 1½-in. or 2-in. pipe size (DN 40 or DN 50) and swing checks or double-door wafer checks in sizes above that. But lift or piston checks are surprisingly common in larger sizes (Fig. 2-172) for several somewhat specialized purposes. One is for pulsating flow, where pressure and volume surges in a line, as from a piston or diaphragm pump, would cause some other types of check valves to bang open and closed with each cycle. Such high-cycle service is almost a guarantee of short life from check valves that depend on nothing but gravity to oppose the fluid flow forces. Since the piston check has to have a connection from the volume space above the disc to the downstream portion of the line anyway, valve designers can make use of the fluid flowing back and forth to dampen the piston's movement. There are a couple of ways to do this; one is to put an orifice into the piston to control the amount of fluid that flows in and out of the area above it and thus control its speed, and another, more adjustable method, is to run the line that connects the two volumes outside the body of the valve and put a small throttling valve into the line to allow adjustment of the piston's response time. Many also have a spring above the disc to help in dampening the opening forces.

Piston checks of this type are used quite often in higher pressures and smaller sizes where the swing check with the outside lever and weight might otherwise be used. This is because the packing gland required in the swing check is not needed here, so there is nothing to bind or obstruct the valve's movement or provide a leak path. High-pressure valves even without the consideration of fluctuating flow rates can make good use of the piston design, since a heavy disc, often with a spring, is a better match for a high opening force often found in such valves. These valves are often built as Y-pattern, which is not often true of the valves designed for pulsating flow. Valves built for high-pressure service also often have special design considerations, such as contours designed with aerodynamic effects in mind and hardfaced guide surfaces to reduce wear caused by the high forces involved.

Another variation on the lift check that is very popular in certain applications is the ball check (Fig. 2-173). The ball check is still a lift check, often with exactly the same body. The difference is that the piston has been replaced with a spherical ball, basically a ball bearing. It takes a slightly different guiding system than the piston, and does not need to be vented because it never totally blocks in the volume in the top of the valve. The main reasons for using the ball check are that since it has more internal clearance than the piston, in viscous services it will not gum up as easily, and that since the ball falls back on the seat in a different orientation each time, it generally wears better in high-cycling services. It is still possible to use a spring to aid in seating.

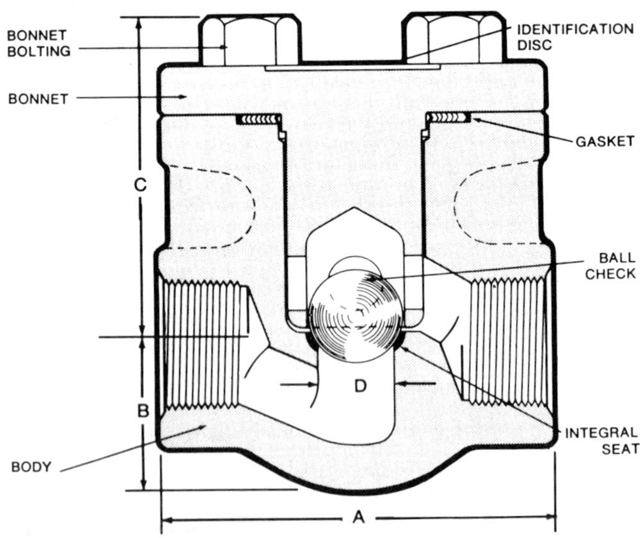


**Fig. 2-173.** Ball-type lift check valve for installation in horizontal line. (*Courtesy of Henry Vogt Machine Co.*)

Valves specially designed to be used in vertical lines are also ball checks, although with a different body construction (Fig. 2-174) because the flow around the ball and out through the top of the valve is easy to accomplish. These valves require a union or other joint in the middle for disassembly, since there cannot be a cap. They can be installed only where the flow is vertical up, since gravity is the closing mechanism. There are some ball checks suitable for either horizontal or vertical up installation (Fig. 2-175).

The same valve that is used as a vertical flow check valve can be placed in the horizontal, where it takes on an entirely different function. It can be used as an excess flow check valve, whose purpose is not to respond to fluid flows when they are of low magnitude. The check becomes active only when a high-velocity flow occurs. Basically, the fluid flow sweeps the ball into its seat and holds it there, thus shutting off the flow. This valve is useful for places where line ruptures or leaks might occur. Of course, since the ball must seat on the downstream side, it must be installed backward from any check valve.

### *Split-Disc Check Valves*

Yet another variation on the same theme is the split-disc or double-door or wafer check valve. All these names are commonly used to describe one product, and although the term "wafer check" is probably the most

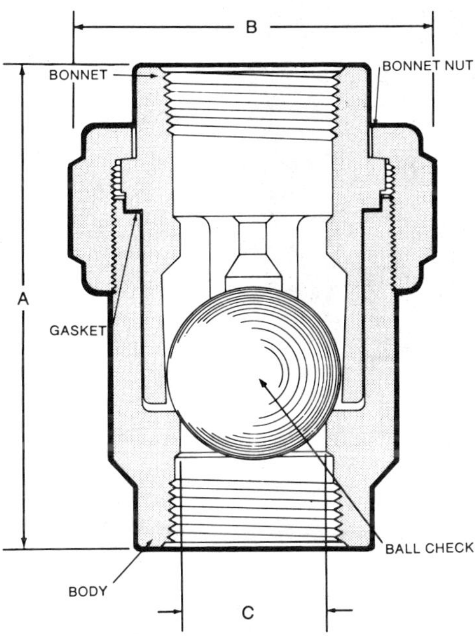


**Fig. 2-174.** Ball check valve for installation in vertical up-flow line. (*Courtesy of Henry Vogt Machine Co.*)

**Fig. 2-175.** Ball-type plastic-lined check valve for installation in horizontal or vertical up-flow line. (*Courtesy of Atomac, Durco GmbH.*)

common, it is also misleading in that other types of check can be wafer body. What is discussed here is a design (Fig. 2-176a,b) in which there are two discs, each in the shape of a half-circle and hinged on the straight side, which is also the vertical centerline of the valve, such that the two discs open toward each other. These discs are also spring-loaded, so that gravity is not required for the valve to operate.

Normally the two discs are mounted on the same hinge pin, which is inserted through the body wall in the same way that the pin of a swing check is. The springs are wrapped around this pin also. Normally there is a second pin, called a stop pin, just downstream of the hinge pin. The springs usually work against this pin to exert closing force in the discs, or doors, as they are more often called. Sometimes one spring pushes against both discs, but more often there are two springs, one for each door. In the open position the two doors normally stop against each other, but some designs use the stop pin for this purpose also. There are small plugs threaded into the outside of the body to cover the points where the pins are inserted. These are made of the same material as the body or of the pins, which tend to be of the same material as the rest of the valve trim.

With reference to the door analogy again (Fig. 2-177a,b), the fluid force required to open the valve is a great deal less than with a swing check since each door has less than half as far to travel as in a swing check, and not by lifting against gravity (unless the valve happens to be in a vertical line). Pressure drop is thus less than in a swing check. Since the parts are physically smaller, it is practical to install them in a wafer-

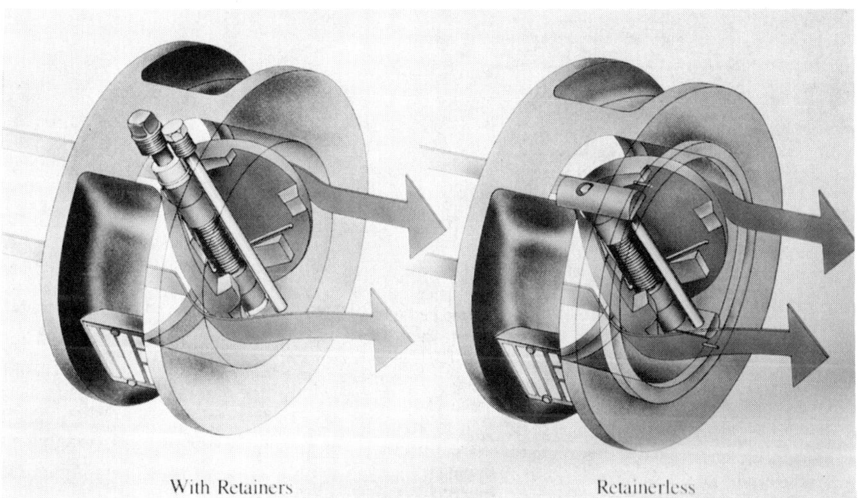


(a) (b)

**Fig. 2-176.** See-through views of standard double-door check and retainerless double door check. (*Courtesy of Marlin Valve, Inc.*)

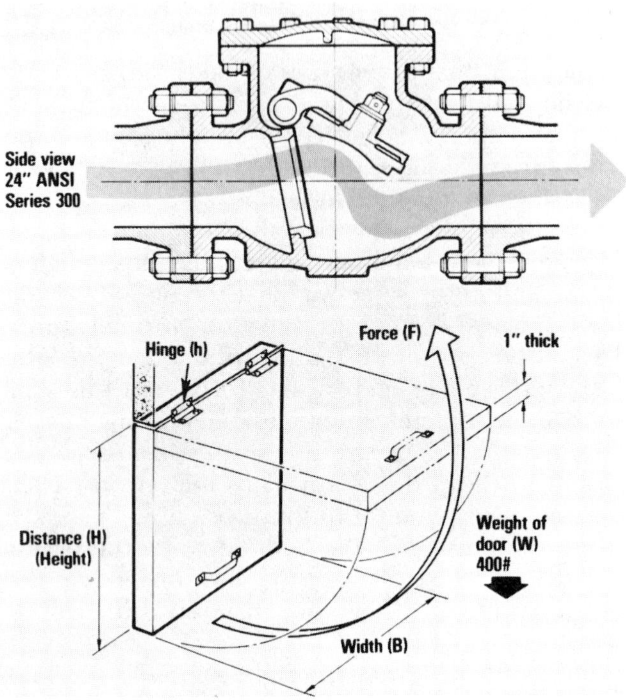


(a)

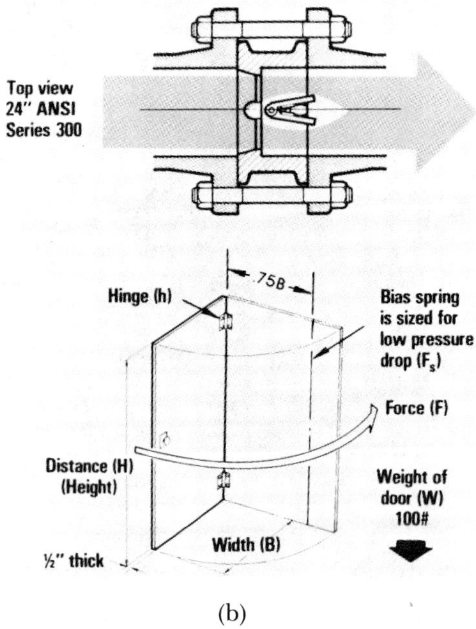


(b)

**Fig. 2-177.** Force diagrams showing check valve behavior. (*Courtesy of Marlin Valve, Inc.*)

type body, thus significantly decreasing the mass of the entire valve and thus its cost. All of these factors combine to make the double-door check one of the most efficient check valve designs.

A few valves, mostly in oil field applications but sometimes in high-alloy chemical service piping, are built to a swing design but in a wafer pattern. Such valves require the disc to swing beyond the downstream end of the valve. These valves share the weight advantage of the double-door check, but otherwise they are standard swing check valves in operation. Because of the limited space within the body of the valve, many are built with the seat at 90 degrees to the axis of the valve, rather than the 80- to 85-degree inclined seat in a flanged swing check. To compensate for this arrangement, which makes the disc hang loosely next to the seat instead of applying any gravity force to it, an elastomer O-ring is sometimes installed. Basically the only force available to seat the valve is positive back pressure. These valves tend to be very low in cost.

The double-door check is available in almost all materials, and because of its small body mass, in high-alloy construction its cost can be significantly less than that of the equivalent swing check. It has an economic advantage in large sizes, though, in just about any material, even cast iron. The springs are available in a number of materials, some of which make better springs than others. Inconel, Inconel-X, or some other variation of nickel-chrome-iron alloy is the most common since it has both good springiness and high corrosion resistance. Springs are available out of other corrosion-resistant material and high-temperature alloys also, some of which are good to much higher temperatures than the valve bodies are.

The springs are the source of one of the major disadvantages to the double-door valve—that they can break and go downstream. This unfortunate incident both renders the valve somewhat nonfunctional and potentially can cause damage to some downstream piece of machinery. This problem used to occur more often than it does now, although it is always a danger. When the two springs each push against the stop pin, then if one breaks the other half of the valve is still functional. Having only one-half of the valve working does not make it 50 percent as good as an intact valve, though, but perhaps 5 percent if the door with the broken spring does not close. Fluid backflow usually will still close the valve, even without the spring.

The springs also can be varied in springiness to close at greater or slower speeds, which helps for very large valves and for resistance to flutter or slamming. Generally, stiffer springs are used to close the valve faster to counteract slamming (the valve closes before the flow can build up any momentum in the backward direction), and softer springs are used if the valve shuts too forcefully against low flows.

In the double-door check, as in any other valve, a different material is used on seating surfaces for greater resistance to wear and impact loading. However, since the doors have less mass and travel less far than in a swing check (Fig. 2-178), there is less loading in the first place so less protection is often required. Hardfacing is rarely used. Often only an increased corrosion resistance is required, to maintain the integrity of the seating surfaces. An 11 to 13 percent chrome is often sufficient for services that require only carbon steel bodies. Since the metal mass in a double-door check is fairly small to begin with, in smaller sizes it is actually cheaper to make the entire body from the chrome steel rather than making a steel body with weld-overlaid seating areas. Conversely, in some very large valves, it may be cheaper to make the doors from carbon steel and overlay the seating surfaces rather than cast the entire piece from the alloy, just as is often done on gate valve wedges.

The wafer check, of course, is a through-bolting type. Extralong stud bolts are used, and the outside diameter of the valve body is the same as the inside of the bolting pattern, which is normally the same as the outside diameter of the gasket seating surface or very close to it. This means that the same casting can be used for a number of different ratings, each one being machined to a different outside diameter. It also means that the bolting lies totally exposed in the region between the flanges and outside the check valve wall, making the bolting more vulnerable to thermal expansion in fires or in high-temperature service. Some users do not permit wafer checks above a certain temperature, but others simply prefer a lug type. In check valves, the lug type serves the same purposes as it would in butterflies, that is to allow the valve itself to be actually bolted to the flange on either side, not just held in between. For purposes of avoiding thermal expansion from external heat sources, it is perfectly adequate to drill the lugs out and pass the studs through them rather than using cap screws; here the purpose of the lugs is merely to shield the bolting.

Where the four plugs used to retain the hinge and stop pins are considered as undesirable leak paths, they can be seal-welded. It is also possible to build a double-door check in which these pins are retained by an inner retainer (Fig. 2-179) in which the joint does not lead to the outside atmosphere.

In many piping configurations, a check valve is most likely to be found on a pump discharge and its purpose is to keep the fluid flow from spinning the pump impeller backward. Often the piping is vertical at this point, so the check valve becomes the lowest point of a volume to be drained. In standard swing checks the body is simply drilled and tapped. The wafer check is perfectly capable of being drilled and tapped also, but by its nature it is somewhat harder to do. The space available is very small, so in valves below 6-in. pipe size (DN 150) the standard drain size available may be oddball. In addition, the drain connection has to pass between two of the flange bolts. In many cases, a longer than

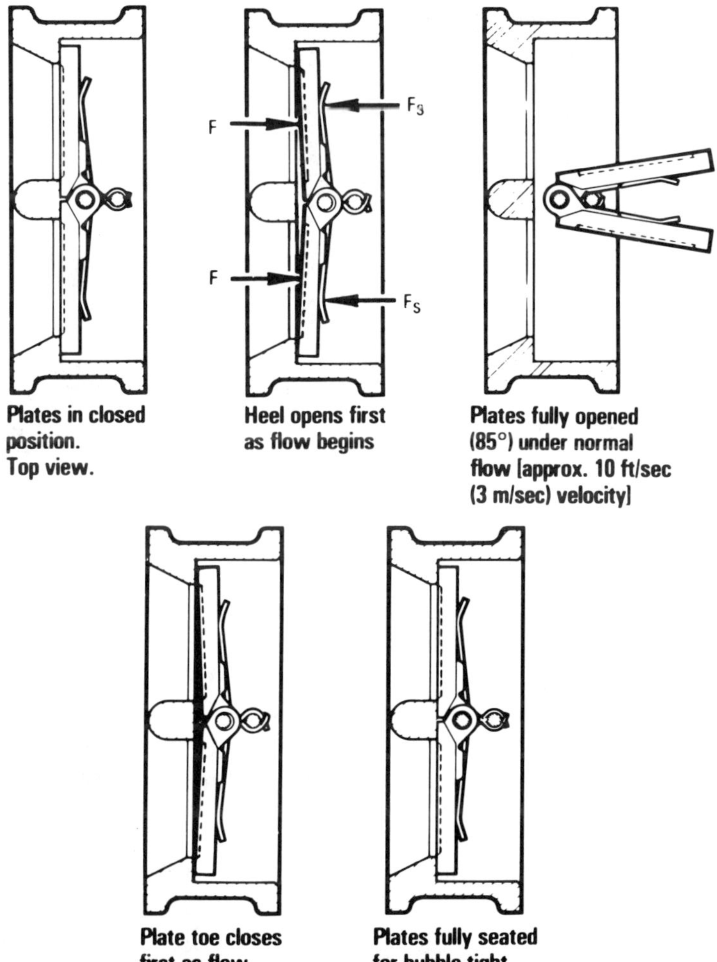


**The fluid develops a Resultant Force (F) which acts at the center of the area of sealed surface [0.424 in. (11 mm) radius]. The point of Spring Force ($F_s$) acts beyond the center of area of each plate [½″ to ¾″ (13 mm - 19 mm) radius]. This fulcrum causes the heel to open first, preventing rubbing of the seal surface prior to normal plate opening. At closing, when the fluid flow decreases, the point of Spring Force causes the toe of the plate to close first.**

**Fig. 2-178.** Opening and closing behavior of double-door check valves. (*Courtesy of Marlin Valve, Inc.*)

**Fig. 2-179.** Retainerless double-door check valve, partially disassembled. (*Courtesy of Marlin Valve, Inc.*)

normal pipe nipple is thus required to provide a valved drain. The drain actually comes from the area immediately downstream of the seat plane, and in some cases actually notches into the seat slightly at a point between the doors so it does not interfere with seating. This, of course, is the lowest possible drain point.

## *Centered Spring Check Valves*

A common type of specialized check valve is the centered spring check (Fig. 2-180). This valve is widely manufactured and is used as a standard type of check valve in building service piping and a number of other industrial applications. This type is also the simplest check valve for very small sizes (Fig. 2-181). But its most significant application is in the prevention of water hammer.

The centered spring check works somewhat like a globe valve or lift check, in that the disc moves directly into the flow. Unlike the globe or lift check, no right-angle turns are required. This movement is accomplished by having a centering device suspended from the sides of the valve, downstream of the seat, holding a short stem or guide that is attached to or part of the disc. The force that moves the disc into the seat is supplied by a spring that wraps around this guide and pushes against it.

The significance of this design is that the spring force directly opposes the force of flowing liquid. As the flow of liquid slows, the spring begins to push the disc closed. When the fluid flow stops completely, the disc seats and stays seated until the differential pressure of fluid is enough to lift the disc back off the seat. In many other check valve designs, the movements of the disc or discs are spring-loaded or otherwise guided by a force other than fluid flow, but the two forces are not in line. A swing check, for instance, moves by gravity, but the resultant of its downward movement and the influence of the pivot arm is the actual

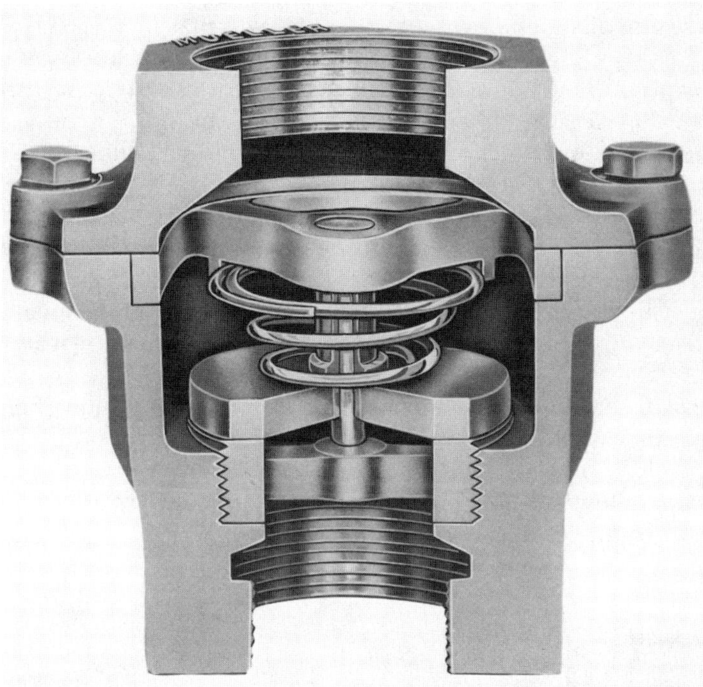


**FIG. 2-180.** Center spring check valve cutaway. (*Courtesy of Mueller Steam Specialty Co.*)

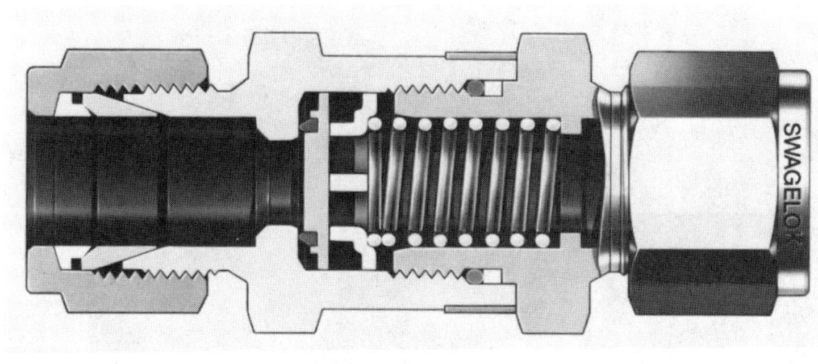


**FIG. 2-181.** Poppet-type spring check valve with flareless tube ends. (*Courtesy of Swagelok Co.*)

force that opposes the fluid force. Although these may seem to be insignificant differences, the differences in behavior are really enough to justify a special check valve where the water hammer problem exists.

It must be noted that in severe water hammer situations, often no one single change will effect a cure. Often, a combination of solutions including special check valves, reorientation of piping to reduce gravitational influence, better anchoring of pumps, adding chambers or piping to allow for diffusing of shock waves, or other changes may be necessary.

An additional advantage claimed for the centered spring check valve is its lack of closure noise. The valve indeed closes silently, or at least with much less noise than a swing check, but in general the noise is only a symptom of a problem and not a problem itself. In building water piping in a basement, the lack of check valve banging is probably significant. But what really makes this important is the fact that the noise is caused by impact of the disc onto the seat. When the valve seats gently, as do a number of other check valves under the right conditions, less wear occurs and the valve lasts longer. In many cases the centered spring check is built as a wafer type, often in iron with bronze trim for water service.

Other types of centered spring check valve are more exotic (Fig. 2-182), available in a wide variety of materials and used in pulsating or widely varying flows. Some of them look like a wide spot in the piping or like a snake that swallowed a rat. This bulging design is very common in Europe, but is also imported to the United States. It is also possible to

**Fig. 2-182.** Center-spring piston check valve designed for pulsating liquid service. (*Courtesy of Daniel Industries, Inc.*)

configure this valve as a control valve by bringing a rotary stem in from one side and adding a rack gear to the sliding stem. With a pinion attached to the rotary stem, a standard rotary actuator can be used to open and close the valve. Because of its streamlined flow path, a very low pressure drop can be obtained.

These valves are normally built in welded or cast steel, sometimes with a soft-seat component but also with hardfaced seat. The spring also has to be resistant to temperature and corrosion. Because of their streamlined design, they generally have fairly low pressure drop and high flow capacity. However, they are also somewhat longer than other valves.

### *Backflow Preventer Valves*

Another specialized type of check, known as the backflow preventer, is actually two check valves plus one or two block valves in one (Fig. 2-183a,b). The backflow preventer is a legally mandated device to separate potable water systems from nonpotable systems. Generally, the backflow preventer is used whenever there is a hard-piped connection from municipal water systems to an industrial water user such as a building sprinkler system, or in residential water systems (Fig. 2-184) to guard against lawn chemicals being sucked back into the household water supply, or in any open water system that has the possibility of being infiltrated with bacteria.

Numerous incidents of sickness and death resulting from contaminated water are behind the requirements for backflow preventers. They are made by a number of manufacturers that specialize in the municipal business and are generally flanged or threaded end, in cast iron or bronze. Health codes generally prescribe where they are required, with some states requiring them more frequently or in more places than others.

The reason for the interconnects in the first place is often that the drinking water supply can be made available for short periods of time for fire fighting or water makeup. Often a backflow preventer is not required if no permanent connection exists and a backflow is not possible, for instance, if a source of potable water can flow by gravity into a tank.

The design of the backflow preventer consists of two check valves in series, usually with soft inserts in the disc to seat more tightly, with a drain between them such that when the valves close, the drain opens to prevent any seepage from going back upstream. The block valves, upstream and downstream of the check valves, are there to act as reinforcement to make extra sure that there is no possibility of upstream leakage. In smaller sizes, this entire assembly is made as one solid body casting.

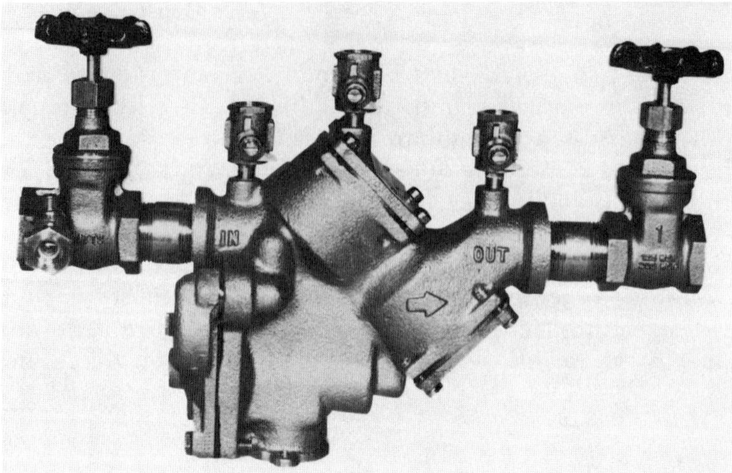


(a)

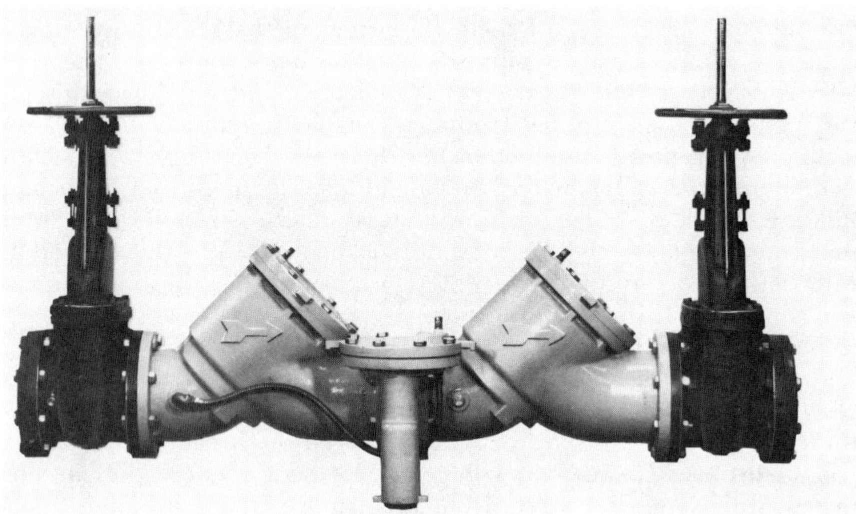

(b)

**Fig. 2-183.** Backflow preventers, in bronze with threaded ends and in cast iron with flanged ends, each provided with gate valves at upstream and downstream ends. (*Courtesy of Watts Regulator Company.*)

**Fig. 2-184.** Small backflow preventer designed for household plumbing installation. (*Courtesy of Watts Regulator Company.*)

Bodies are usually cast bronze up to 2-in. size (DN 50) and cast iron with bronze trim in larger sizes, with available sizes ranging from ½-in. (DN 12) to about 10-in. pipe size (DN 250).

There are a number of variants on the backflow preventer, to account for various severities of service. Many household models are designed to fit into the space next to the water meter, or on lines leading to the outdoor faucets. Most household models are designed for low noise during operation. Other, larger ones for industrial use have positive means to expel the water trapped between the two checks to ensure that there is an air gap between the different water services at all times that positive flow (in the correct direction) is not occurring.

### *Other Check Valves*

Among the miscellaneous types of check valves in service are the ones used for highly variable flows. These valves must open at relatively low flow rates, yet present low-pressure drops at high flow rates. The most common occurrence of these conditions is in compressor discharge lines, where often there is also a significant pulsation problem.

There are a number of designs of compressor check valve, many made by compressor manufacturers themselves for use between stages of the compressors but also needed on the piping outside of the compressor. There are a few manufacturers that specialize in this valve. Typically these valves have many, very small passageways, which not only help to break up the pulsations but also provide very small flow passages for the check mechanisms to operate across. Some designs use a metal reed of spring steel that snaps shut across each of the flow passages when the flow stops. Others use a plate that contains a number of reeds or fingers, sometimes provided with tiny coil springs in addition, which both break up the flow into many small passages and provide quick closure (Fig. 2-185). This whole assembly generally is built like a wafer, which is then mounted into a housing to be bolted into the pipe line.

All of these valves, and especially the spring-loaded ones, need to be designed for the particular location they are to be installed in. The flow rate, gas density, pressure, and amount of pulsation, in terms of the speed and volume of the cylinder discharges, must be known. The orientation of the valve in the piping is helpful information also. The manufacturers can size these valves, and can determine the size of the pulsations from the size and speed of the compressor. This is generally a very complex calculation, almost always computerized.

Standard check valves are not suitable for most compressor discharge lines, and if installed there are almost always doomed to a short life. Reciprocating compressors produce a pressure pulse with every dis-

**Fig. 2-185.** Special check valve for discharge of reciprocating compressor. (*Courtesy of Hoerbiger Corporation of America.*)

charge, and the resultant pressure waves produce noise, sometimes even amplified by the piping through resonance, and vibration. The plate-type compressor discharge check valves are specifically designed to dampen these pressure pulses without imposing significant pressure drops. They are also an effective solution for discharge lines on centrifugal compressors with widely varying duties, although the pressure pulsation is essentially absent in a centrifugal compressor.

Another useful design is the recirculation check, which is a check valve and flow-splitting valve in one (Fig. 2-186a,b). Its use is on the discharge side of a process pump, where a certain minimum flow through the pump is desirable or necessary. The recirculation check acts like a centered-spring check in reverse, with a spring-loaded disc that is connected to a cage and plug in the center of the valve, with an additional side outlet (Fig. 2-187). Under full flow conditions, the check is wide open and the entire flow goes down the pump discharge line. As the flow through the valve decreases, the flow passageway leading to the recirculation line starts to open and some of the fluid is diverted to the pump bypass. This allows at least some fluid to be flowing through the pump at all times, even if the discharge line is completely blocked in downstream of the recirculation check.

**Fig. 2-186.** Exterior and cutaway of automatic recirculation check valve with recirculation outlet on the side. (*Courtesy of Yarway Corp., a subsidiary of Keystone International, Inc.*)

**Fig. 2-187.** Diagrams of flow through automatic recirculation check valve. (*Courtesy of Yarway Corp., a subsidiary of Keystone International, Inc.*)

**Fig. 2-188.** Resilient-body check valve used on sewer line discharge. (*Courtesy of Red Valve Company.*)

Yet another type of valve that operates similarly to the pinch valve is the resilient body check (Fig. 2-188). This check operates in a manner similar to the check valves in blood vessels. It is capable of shutting off around solid objects and opens under very small heads of water.

# 3

# Operators and Actuators

## Displacement Actuators

### *Cylinder Actuators*

When a human being is not available to operate a valve, or it is not convenient to send a human being to operate one, it then becomes necessary to provide some mechanical means to actuate the valve, and to send a signal from some other location to cause the actuation to occur. There are many methods of mechanical actuation of valves. The simplest ones, and the oldest ones, are those in which power is transmitted directly from a distant location. More commonly, there is an operating device on the valve that takes its power from a source of energy available right at the valve, in response to a signal that is sent from some distant location.

The choice of type of actuator also depends on the type of valve. Valves that open by means of a short-movement, high-force stroke are naturally different from those that open by means of a long-duration, lower force stroke, and the best choice of actuator for each is one that matches the characteristics of the valve. Valves that are only full-open or full-closed are different from ones that spend even a portion of their service life in a partly open throttling or flow-restricting mode. The power sources that can be made available are also a consideration, since many valves can be actuated quite well with several different types of actuator. Sometimes the choice of signal transmission is a consideration, although that factor is less important as long as there are reliable interface devices to convert signal types to actuator control behavior.

The earliest power source used to actuate valves, other than human muscle, was air. The use of cylinders powered by expanding vapor goes back to the eighteenth century, first using steam. Eventually compressed air came to be used to transmit power over short distances, such as for railroad braking in the 1870s. Even with crude cast-iron cylinders and pistons, it was possible to have air pressures of 100 psi (7 bar). Cylinders were used for direct actuation of a number of devices. At the time, the only air source was direct-action, steam-driven-cylinder air pumps. The invention of the rotary compressor eventually provided an unlimited source of compressed air.

Today, pneumatic-cylinder-actuated valves are found in a number of common applications. As usual with the two major categories of valve motion, there are cylinders for rising-stem valves and cylinders for rotary-motion or quarter-turn valves. The rotary-motion applications are more common. Air cylinders can be used for both on-off applications and modulating applications. On-off actuation is relatively easy—it only requires sending air to one cylinder port until the piston moves all the way to the end of the cylinder, where it will stop because of mechanical interference or because a control device turns off the air flow (Fig. 3-1). Getting the piston back to its starting point is done either by

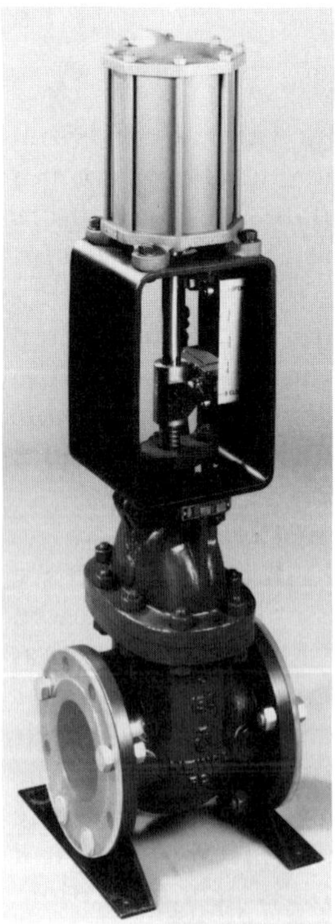

**Fig. 3-1.** Linear air cylinder actuator mounted on gate valve. (*Courtesy of Newmans Inc.*.)

pushing it in the opposite direction with air, or by releasing the air pressure and letting a spring push it back. Modulating or control applications are somewhat different. Here, the degree of valve opening is varied based on a control signal. Often, but not always, a feedback device is required to determine what the actuator's position is and to adjust it until it matches the signal being input.

All pneumatic cylinders, regardless of what type of valve they are mounted on, have several common elements (Fig. 3-2). The first is the cylinder itself, which is most often a steel tube honed to very tight tolerances on internal diameter, concentricity, and straightness. In place of steel, it is also possible to use a filament-wound fiberglass cylinder similar to fiberglass pipe. The cylinder is formed on a mandrel that is very highly polished and imparts very good internal surface characteristics to the finished cylinder, which is then normally usable as is, without any further internal cleanup. These cylinders are most often found in low-pressure applications or ones where weight or corrosion resistance is a factor.

In the last 10 or 15 years a new actuator design made of extruded aluminum has become common (Fig. 3-3). These are normally designed as a modular or prepackaged actuator, with only a few choices of cylinder size. The primary advantages of the aluminum extrusion are that the outside can be shaped with flat bottom and special contours to mount directly onto valves without complicated brackets, and the air

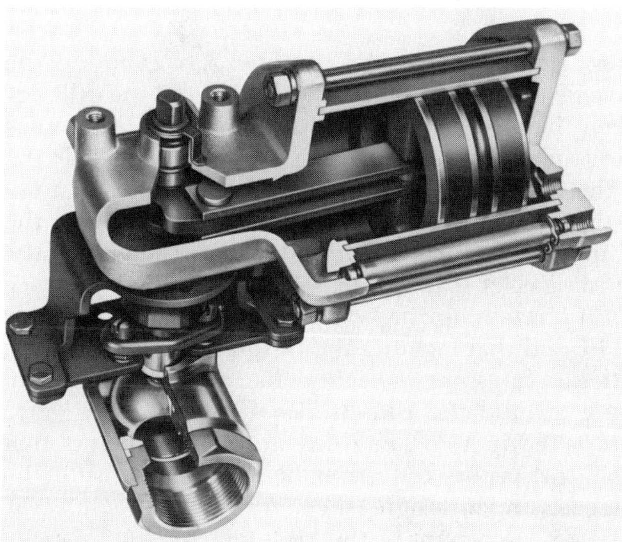

**Fig. 3-2.** Air cylinder actuator mounted on ball valve. (*Courtesy of Neles-Jamesbury, Inc.*)

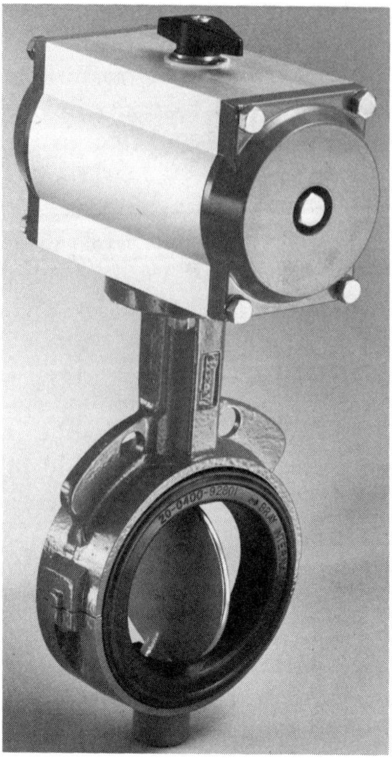

**Fig. 3-3.** Extruded aluminum one-piece actuator with position indicator, mounted on rubber-lined butterfly valve. (*Courtesy of Bray Valve and Controls.*)

passages and linkage openings can be easily machined directly into the one-piece cylinder. The interior of the cylinder requires only a slippery coating.

The next component is the end. In steel and fiberglass cylinders the two ends normally have tie rods that connect them and hold the cylinder locked in place between them. Usually one end is blank, and the other end has the opening for the piston rod. The air ports are normally in the heads, sometimes in the center of the end, sometimes in the area between the center and the outside edge of the end, and sometimes on the side with an internal air passage through the side below the end of the cylinder itself. There is a gasket seating surface and a centering device or, more commonly, a full retaining ring to accept the outside diameter of the cylinder end. The end that includes the piston rod opening has a seal, often an O-ring design but sometimes more like a packing gland, to keep air from escaping around the rod. It also functions, as a valve packing gland would, as a wiper to keep dirt and contaminants that might have settled on the piston rod from being pulled inside the cylinder. In prepackaged cylinder designs, the end is usually retained by cap screws threaded into the corners of the body extrusion.

The last necessary component is the piston, which together with the piston rod transmits its motion to the outside of the cylinder. The outside rim of the piston has one, or sometimes two or three, rubber O-rings, formerly leather cup seals, to act as a seal and a wiper to keep pressure from bypassing the piston. The cross section is quite thick at the rim, then quite thin through the web, and thick again in the center. The piston rod is normally slightly smaller through the piston area than through the rest of its diameter, providing a shoulder for the piston to push against. The piston is retained by a nut or a retaining ring on the end of the piston rod, which also sees the piston force when the air is applied from the side with the piston rod. Extruded aluminum cylinder bodies normally have corresponding die-cast pistons that include the rod, rack, or yoke mounting as one piece (Figs. 3-4 and 3-5).

A third type of cylinder actuator uses a one-piece cast dome or can-shaped cylinder (Fig. 3-6). The remainder of the components are the same as described above. These actuators are sometimes used for true control valves and can be built with air actuation in both directions or with a spring return (Fig. 3-7). The cylinders are specially designed to move easily under slight changes in air pressure, which is a required characteristic for actuators on modulating valves.

Direct-actuated cylinders are used on gate valves and sometimes on globe valves (Fig. 3-8), generally for actuation from some remote location. The most common use of such valves is for emergency shutdown

**Fig. 3-4.** Cutaway of one-piece actuator showing integral piston and rack design. This actuator is configured as air-to-open, air-to-close. (*Courtesy of Bray Valve and Controls.*)

systems. Since gate valves are not intended for throttling anyway, no use is made of the cylinder for controlling the valve position. There are some applications where an air cylinder is used just to open and close very large valves that it is not practical to operate manually, or where the operation occurs regularly (say, every 5 minutes or so) and it is just not practical to have a human being operate such a valve. Such air applications were more common in the past than they are now; now it is more likely for these valves to be equipped with an electric motor operator.

Generally, air cylinders for gate valves are rather large and are more likely to be custom fabrications (Fig. 3-9) than the more usual standard-sized drop-in air actuators for rotary valves. Not many gate valves are equipped with mountings specifically to handle air cylinders. Generally, the cylinder is mounted to a bracket that is bolted to the yoke in some manner, or to the bonnet, even sometimes taking the place of some of the bonnet bolting. This mounting has to sustain the force of the cylinder pushing or pulling against the valve to move the disc closed or open. Sometimes there are no separate stops, so the piston moves until it hits the end of the cylinder. Often, there will be one stop, near the top of travel, to keep the piston from trying to take the top off the cylinder. Generally, the wedging action of the gate seating into the valve is sufficient to adequately stop the travel in the downward direction.

**Fig. 3-5.** Cutaway of same actuator configured as spring return. (*Courtesy of Bray Valve and Controls.*)

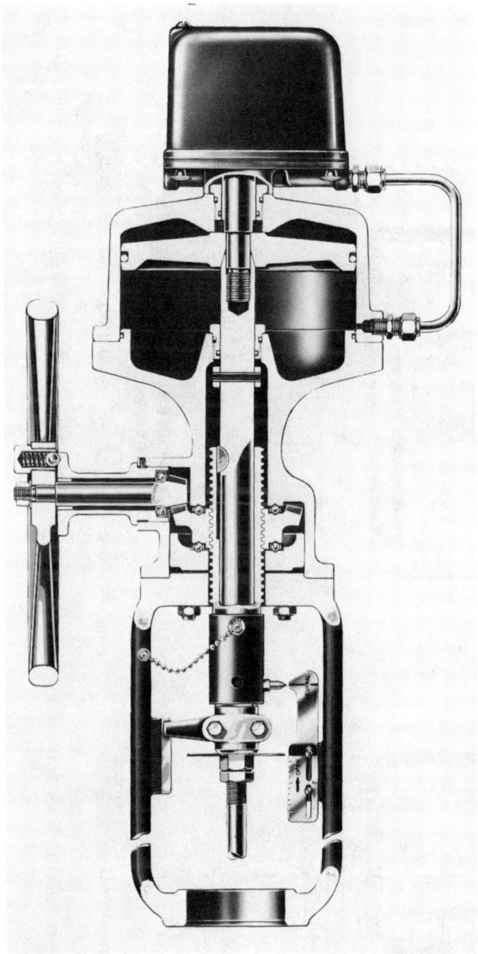

**FIG. 3-6.** Cylinder actuator with one-piece housing, suitable for either on-off or control application. Positioner (top) is required for control. Also has manual gear override. (*Courtesy of Fisher Controls Intl.*)

Since the travel is limited to the stroke of the valve itself, if more power is required, the cylinder diameter must increase. The extruded aluminum one-piece prepackaged actuators are not used on rising-stem valves.

In a cylinder-actuated gate valve, the piston rod generally does not have to be the same diameter as the valve stem. The valve stem has to be sized to include the thread depth and a few other factors that don't really apply to the piston rod. For the sake of convenience, though, sometimes the two are the same or almost the same size, especially on smaller valves. Since very few gate valves are built purely for cylinder actuation, the existing valve stem with Acme threads for the operating nut normally is retained. The coupling to connect it to the piston rod is also machined with Acme threads on the bottom end and threaded onto the valve stem

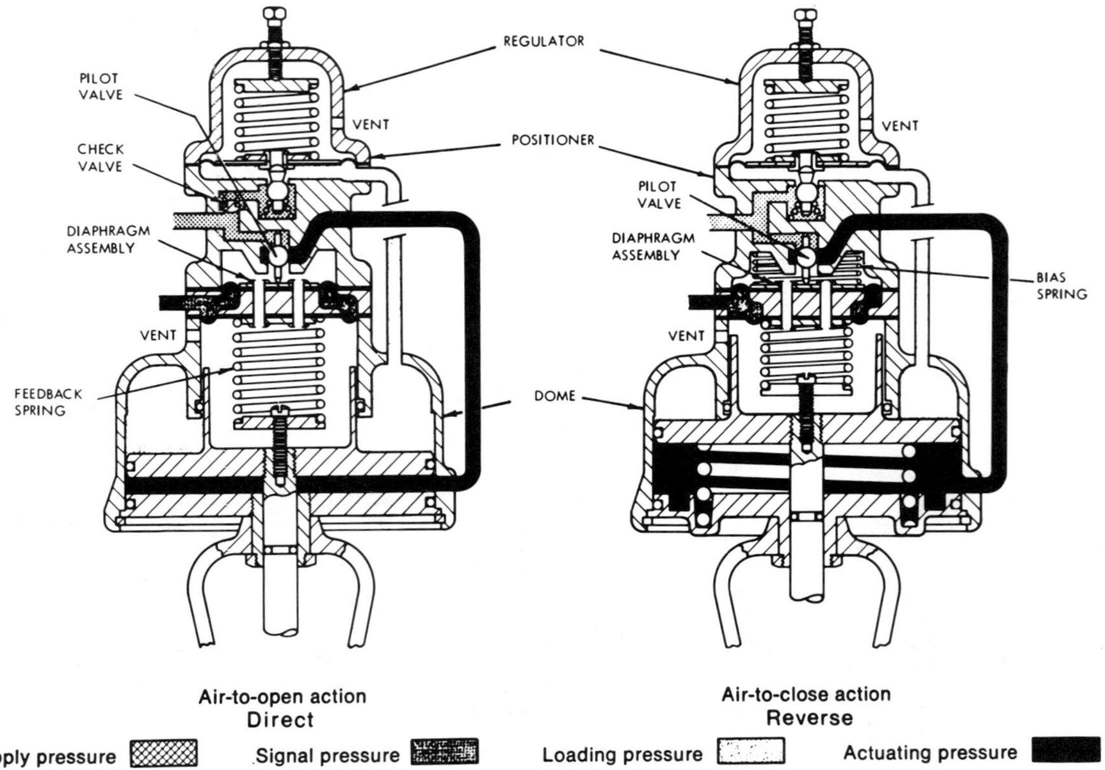


**Fig. 3-7.** Diagram of operating sequences for air actuator used for control. (*Courtesy of Masonielan/ Dresser Valve and Controls Div.*)

**Fig. 3-8.** Cylinder-actuated globe block (isolation) valve undergoing seismic testing. (*Courtesy of Edward Valves, Inc., formerly Rockwell Edward.*)

**FIG. 3-9.** Multiple-cylinder hydraulic actuator for gate valve. (*Courtesy of Pacific Valves, a unit of Mark Controls.*)

to provide coarse adjustment (Fig. 3-10). The upper end that connects to the piston rod is either threaded or pinned. Fine adjustment of travel should be done by the stops in the cylinder, since there is and needs to be a fair amount of slop in the disc-to-stem and stem-to-piston rod connections.

Control of the air supply to the direct-coupled cylinder is through the conventional solenoid valve or pilot valve. Actuating speed is not normally a concern, especially in valves that are air actuated only because of their size. However, valves for emergency shutdown may be required to close in a specified time, which may be on the order of 5 seconds or less. Here, since the volume of air required may be very large, it is necessary to check that the size of the inlet or exhaust tubing does not restrict opening and closing speed by limiting the volume of air that the tubing can pass. If this occurs, it may be necessary to increase the supply line size, use a quick exhaust valve, and/or supply an auxiliary air bottle located immediately next to the valve to supply all the air required, rather than depend on air coming from the air header. The air

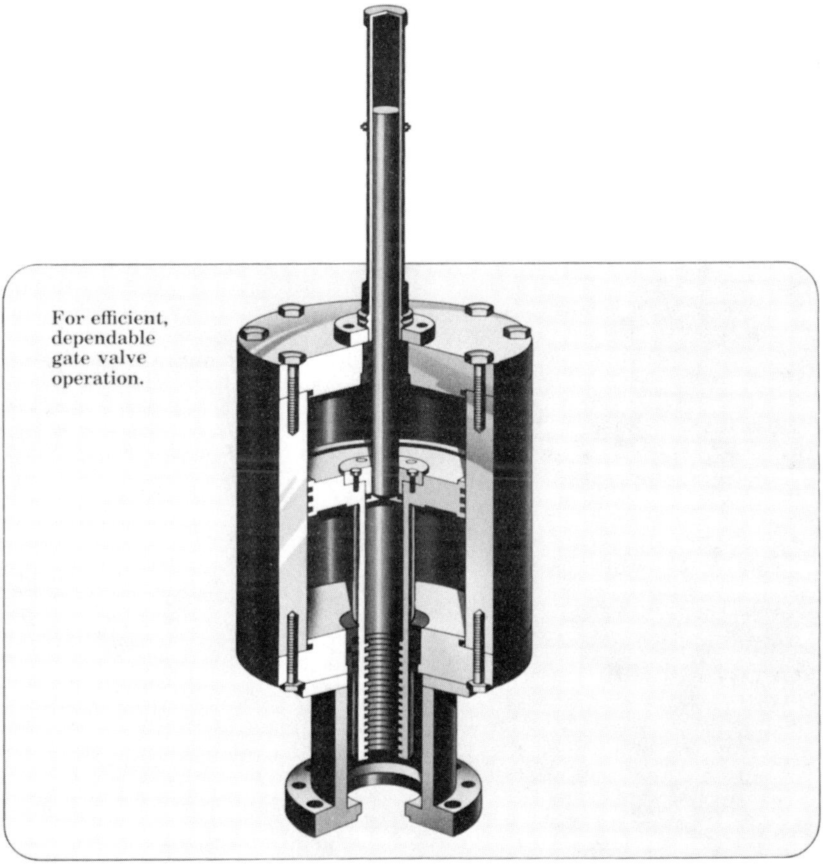


**Fig. 3-10.** High-pressure linear air actuator with coupling for rising-stem valve. (*Courtesy of Shafer Valve Company.*)

supply tanks are often required to contain enough air to actuate the valve through one or two complete open-and-close cycles.

Air cylinders for quarter-turn valves are considerably different from cylinders on rising stem valves, since the motion involves a change of direction. In many ways, the design of quarter-turn actuators can be more flexible since the stroke is not limited to that of the valve stroke and it is easy to apply two or more cylinders to the same valve (Fig. 3-11). The extruded aluminum cylinder described above has made mounting much easier than with steel cylinders since a bracket is generally not required and often a separate coupling is not required either.

The cylinders themselves are pretty much the same as any other air

**Fig. 3-11.** Mounting of multiple air cylinders on quarter-turn high performance lug type (single flange) butterfly valve. (*Courtesy of Neles-Jamesbury, Inc.*)

cylinders, except that the stroke is variable by the length of the linkage arm available. For the same torque applied to the valve stem, a longer linkage arm means a longer end travel, thus a longer stroke, thus a smaller cylinder diameter since the power produced by the cylinder is proportional to its piston area times the stroke. It is more common to have longer, smaller diameter cylinders. In fact, when increased torque is required, it is more common to add a second cylinder than to increase the cylinder diameter unduly. The use of two cylinders is advantageous because the force on the valve stem bearings becomes more of a balanced load. In the aluminum-body actuator designs, the double-cylinder design is used most of the time.

There are several types of linkages in common use. The simplest uses a lever arm mounted on the top of the valve stem and pinned to the end of the piston rod (Fig. 3-12). Another, called the Scotch yoke, consists of a clevis or yoke on the end of the piston rod which holds a pin that passes

**FIG. 3-12.** Air actuator with spring return and manual control block, for quarter-turn valve. (*Courtesy of Shafer Valve Company.*)

through a slot in the lever arm (Fig. 3-13). The piston rod can move linearly without side thrust while the lever arm moves in a true arc, but not at a constant rotational velocity. Viewed from the perspective of the lever, the yoke (piston rod) is farther out, moving more slowly and providing more leverage at the ends of the stroke than during the center of the stroke, where the lever arm is rotating more rapidly but with lower torque being applied (Fig. 3-14). This configuration is a good match for the behavior of most quarter-turn valves, where the higher torques are needed at the ends of the stroke and higher rotational velocities are acceptable in between. The Scotch yoke is quite common in single-cylinder applications but even more so in double-cylinder designs, where the lever arm is normally in one piece.

**FIG. 3-13.** As the Scotch yoke mechanism rotates, the torque exerted is the greatest at each end of the stroke and the angular velocity is greatest in the center of the stroke. (*Courtesy of GH Bettis.*)

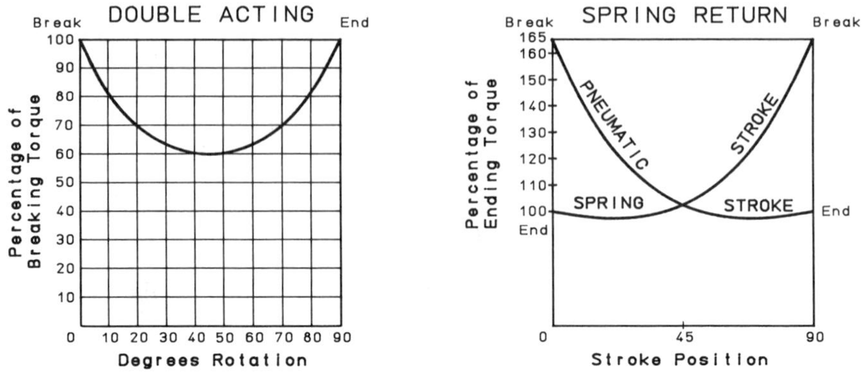


**FIG. 3-14.** Torque characteristic curves for typical quarter-turn actuator with spring return. (*Courtesy of Shafer Valve Company.*)

There are also a couple of toothed gear arrangements. One uses a rack section that engages a toothed lever arm or gear segment mounted to the valve stem, while a similar design used in one-piece actuators has a small pinion attached to the actuator output shaft. Both of these designs can be built as single- or double-sided for one or two cylinders. The lever or pinion is often equipped with a stub shaft to be attached to a coupling that mates to the top end of the valve stem. In the aluminum one-piece actuators, the pinion and shaft are often one integral piece. There is also an actuator with four cylinders and pistons (Fig. 3-15) each piston oriented at 90 degrees from the next around a central pinion. This design is even more compact than the conventional one-piece actuator since the cylinders can be smaller in diameter.

Since the purpose of an actuator for rotary-shaft valves is to provide torque, output ratings are described in torque units (most commonly inch-pounds in the imperial system) rather than in terms of force output like direct-acting cylinders. Torque ratings are for breakaway, or the torque required to overcome inertia and seating forces to start the valve moving. Rated torques are normally different for air-powered stroking and spring-powered stroking (single-acting, spring return) because the spring force is often not exactly the same as the force available from the air and also because the spring exerts less force in the most-extended position than it does in the most-compressed position. Available torques for aluminum body actuators range from 1000 to over 50,000 inch-pounds (125 to over 6000 Nm) at 80 psi (5.5 bar), which is the reference actuating pressure. Steel actuator assemblies tend to be larger, although many small ones are available. Torques available from steel actuators can exceed 1,000,000 inch-pounds (125,000 Nm). Where higher torques are required or where hydraulic power is more suitable than pneumatic, a

**Fig. 3-15.** One-piece four-piston actuator mounted on a three-piece bolted-body ball valve. (*Courtesy of SVF Inc.*)

hydraulic cylinder can be used instead of the air cylinder (Fig. 3-16). Although the hydraulic cylinders are smaller in diameter, they are designed to be interchangeable with the air cylinders.

In the standard steel actuator assembly (Fig. 3-17), the linkage frame includes a bottom mounting pad that is designed to be bolted to a bracket, normally fabricated from bent steel plate or rectangular steel tubing, which is then bolted to the mounting pads on the valve. The linkage is sometimes open but is normally fully enclosed like a manual gear operator, often in a cast housing with a cover plate that helps to hold all the components in alignment. Normally there is an indicator attached to the top of the linkage or pinion that shows the position of the valve. The actuator can be rotated 180 degrees on the mounting bracket and the indicator will be correct, but if the actuator is rotated at 90 degrees from its normal position (which is unusual), the indicator will have to be detached and itself turned 90 degrees in order to correctly indicate the valve position. For the best conservation of usable space, the cylinder orientation is normally parallel to that of the pipe.

If it is necessary to determine the valve's position by viewing it from some distance away, there are some special indicators that can be added to enhance visibility. The simplest varieties are just sheet metal arrows or targets attached to the outside of the existing indicator plate so as to

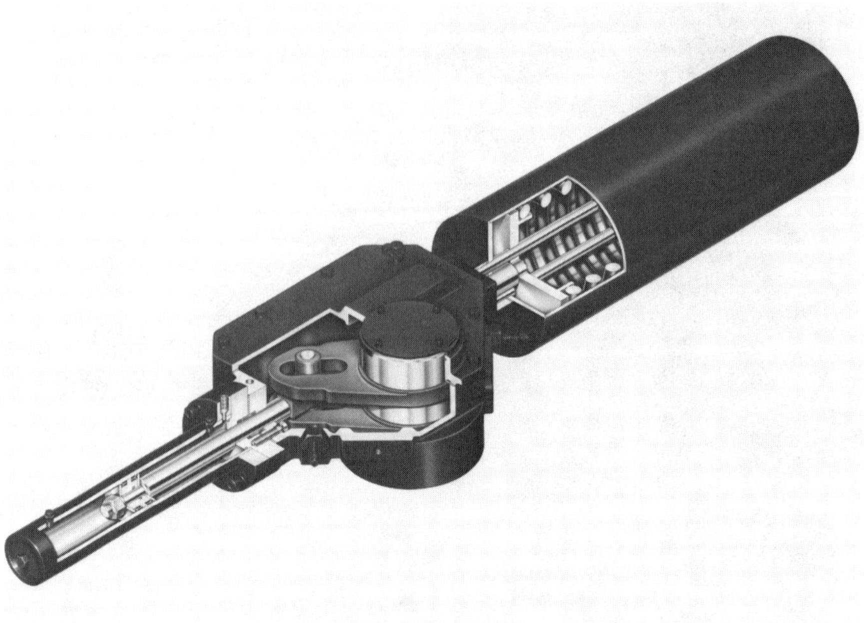

**Fig. 3-16.** Hydraulic cylinder actuator with spring return and Scotch yoke mechanism. (*Courtesy of GH Bettis.*)

**Fig. 3-17.** Double-acting air actuator showing filter and regulator, and limit switches on inside of bracket, mounted on top-entry butt-weld-end ball valve. (*Courtesy of McCanna, Inc.*)

stand up higher. Other varieties are like rotating beacons that have a transparent shell with segments painted on the inside and a core that rotates with the valve stem which has contrasting colors painted on it (Fig. 3-18). When the valve is in one position, the indicator displays solid yellow or another bright color. The painted segments on the core line up behind the segments of the shell so that the indicator appears as one solid color. In the other position, 90 degrees rotated from the first position, the contrasting colors on the core are visible and the indicator appears to be displaying two colors, such as yellow and black. This device can also be designed to show the position of a three-way valve (Fig. 3-19).

Like cylinders used to actuate rising-stem valves, the cylinders used on quarter-turn valves are sometimes used merely to power the valve from open to closed or vice versa. However, they can also easily be used to position the valve to some increment between open and closed. This is for use in modulating applications. Such an application requires a spring for the air to push against so that the more air pressure is supplied, the farther open the valve goes. The use of springs also permits a fail

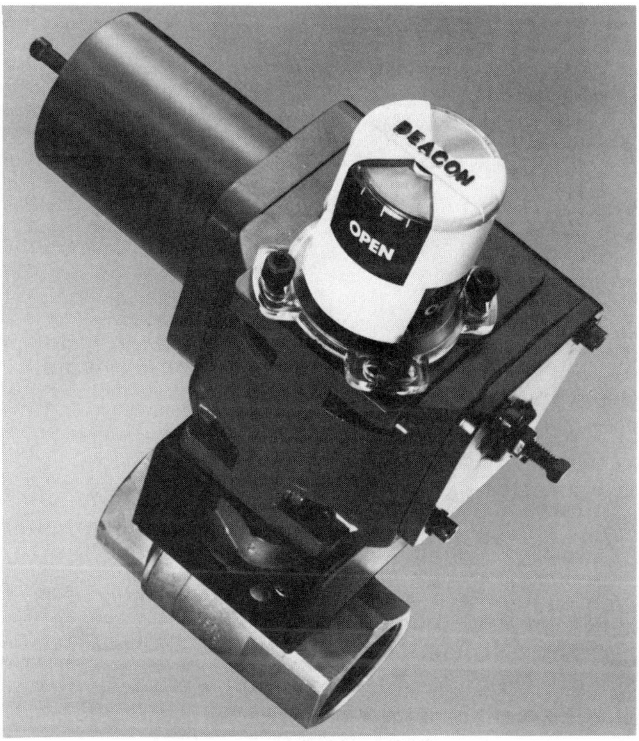


**FIG. 3-18.** Indicator beacon to display valve position. (*Courtesy of Westlock Controls Corp.*)

**Fig. 3-19.** Indicator beacon for three-way valve, displaying valve position. (*Courtesy of Westlock Controls Corp.*)

position. When the air pressure is released, deliberately or accidentally, the springs move the piston rods to the end of their travel. Depending on how the actuator is assembled, the valve then either goes fully closed or fully open. Some actuator brackets are symmetrical about a horizontal plane, with mounting bolt patterns and coupling openings on both top and bottom so that if you want to reverse the direction of the springing from fail open to fail closed, all you have to do is take the indicator plate off, remove the actuator, and flip it over.

Manual overrides can be furnished on most cylinder actuators for rotary-shaft valves (Fig. 3-20), depending on the torque requirements of the valve. On smaller valves with one-piece aluminum actuators, the top end of the actuator shaft is identical to the bottom end. A valve wrench can be used, although the indicating device may have to be removed. A worm gear operator can also be provided, with a clutching device (Fig. 3-21). Many larger actuators use a jack screw or hand jack that is part of the spring return assembly. This is basically a threaded stem that can be manually operated from the end with a crank or a handwheel. The jack

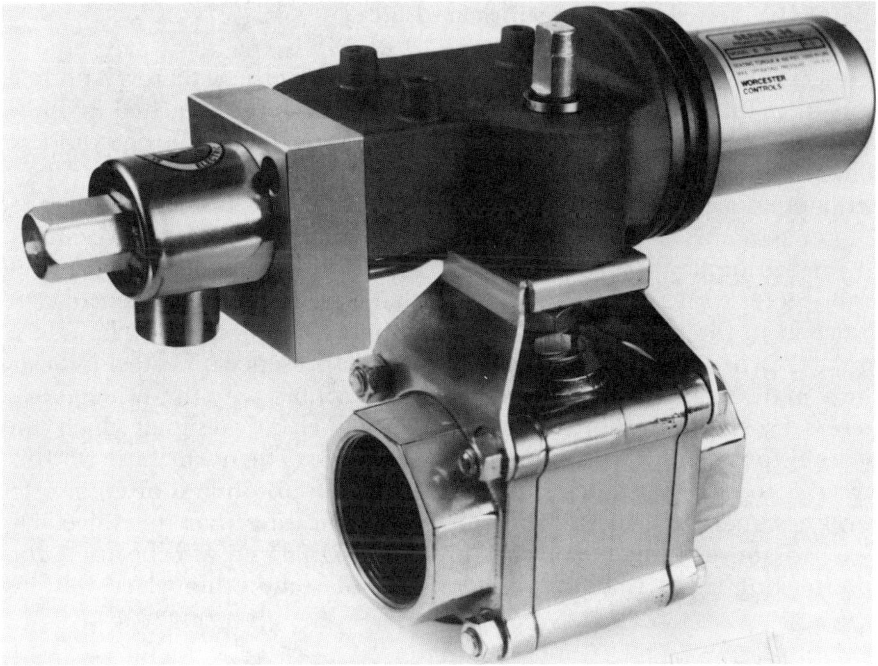

**Fig. 3-20.** Small air actuator for quarter-turn valve, with solenoid valve attached. Provision for indicator mounting or manual override on top of the actuator. (*Courtesy of Worcester Controls Corp.*)

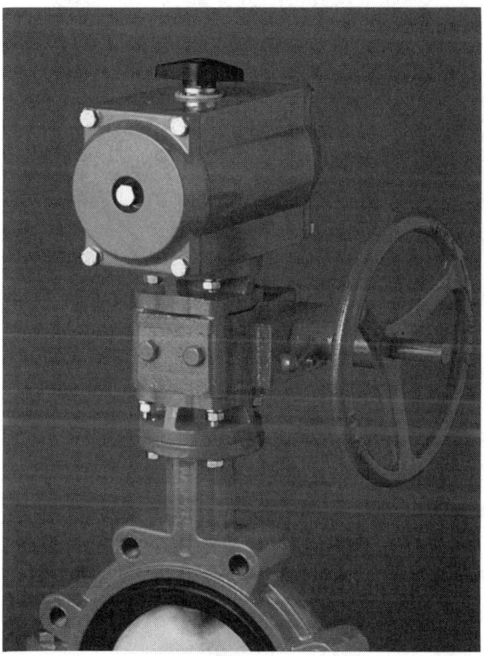

**Fig. 3-21.** Worm gear override on actuator. (*Courtesy of Bray Valve & Controls.*)

screw usually travels in and out with the motion of the actuator. Another type of manual override is a small hydraulic cylinder, attached in addition to the spring return, with a hand pump for operating the cylinder. These hydraulic cylinders can also double as a device for control of actuating speed.

On actuators used just for full open or full closed, it is also common to furnish limit switches (Fig. 3-22). These are mechanical to electrical transducers with a mechanical contact that is pushed closed when an arm attached to the valve shaft or actuator shaft contacts the switch. There are also proximity switches which detect the presence of a steel indicating arm by magnetic means so that physical contact with the switch is not necessary. Either type of switch then sends an electrical signal which can be used for any number of purposes (Fig. 3-23). The normal use for this signal is to close the valve supplying air to the piston and often also to send a signal back to the control room indicating that the valve has reached the end of its travel. This same signal can also be used for interlocking, to permit the start or stop of some other electronically

**Fig. 3-22.** Mechanical action limit switches mounted on air actuator for quarter-turn valve. (*Courtesy of Worcester Controls Corp.*)

**Fig. 3-23.** Position indicator transmitter mounted on manual-operated ball valve. (*Courtesy of Swagelok Co.*)

controlled piece of equipment, such as a pump, whose successful operation depends on whether the valve has or has not opened or closed.

The tubing on the valve itself, which transmits the actuating air to and from the cylinders, and the tubing used for the field connections to the actuating air supply and the signal lines (for pneumatic controls), is often copper tubing with some plastic coating for corrosion protection if no fire resistance is needed, or stainless steel if fire resistance or a higher degree of corrosion protection is required. The size of the tubing depends on the amount of air required. Tubing of ⅛ in. (3 mm) may be used on very small actuators, ¼ or ⅜ in. (6, 8, or 10 mm) is common, and some large-volume air cylinders may take up to ¾-in. (19-mm) tubing. When fairly large volumes of air are needed, sometimes a small air bottle is supplied. This may look like a sample bottle, maybe 1½ in. (38 mm) in diameter and 18 to 24 in. (450 to 600 mm) long, or perhaps like a standard propane bottle, or even larger, to supply a volume of air located immediately adjacent to the cylinder.

The speed of actuation can also depend on the size of the ports in the solenoid valve, as well as the port in the exhaust valve (if separate from the inlet valve) which releases air from the cylinder to the atmosphere as the spring or air pushes on the other side of the piston. The opening and closing speeds can normally be regulated independently by means of tiny needle valves that can be adjusted with a screwdriver or small wrench to regulate the flow of air in and out of the cylinders. Speeds normally can be regulated over a 2 : 1 to 3 : 1 ratio, such as a maximum of 6 seconds and a minimum of 2 or 3 seconds.

In addition to the components described so far, other parts normally present on an air actuator include a small strainer in the air line to keep rust particles out, a bug screen to cover the end of the exhaust valve, and pressure gauges to show the pressure in the supply lines. Sometimes these parts are mounted on a separate bracket with the associated controls, but on the modular or prepackaged actuators they are sometimes an integral design mounted right on the housing. The most important other components, the pilot or solenoid valves and the feedback devices, are discussed at the end of this chapter since their design and functions are common to just about all types of actuated valves.

Signals to actuate the valve come directly from the distributed control system or from some nearby device that senses a pressure or flow and converts that information to a control signal. Control signals can be either electric or pneumatic, although pneumatic controls are becoming less popular. With an air signal, a pilot valve is used; a solenoid valve is used with electrical signals, but the valve itself is basically the same. Control signals are normally 3 to 15 psig (0.2 to 1 barg) for pneumatic systems. Electrical signals may be 120 V alternating current (VAC), 12 or 24 V direct current (VDC), or any of several other levels.

Air for actuator power is normally supplied from an instrument air compressor, which uses a dryer to remove moisture from the air in addition to boosting its pressure to a useful level, generally 80 to 150 psi (5.5 to 10 bar) gage pressure. The dryer is normally designed to lower the dew point of the air (the temperature at which water would start condensing out) to –40°F (–40°C) or thereabouts. Slugs of water in the system can be very damaging, both because of corrosion and because water is basically incompressible and could be mechanically damaging to the actuator during compression and decompression cycles. The air is then piped through a dedicated header to be used only by air-operated controls and instruments. Occasionally, a cross-connection to utility air systems exists for emergency purposes. Main headers are usually galvanized steel to prevent rust pickup or, at the very least, are flushed before being put into service. Smaller subheaders are often stainless steel since they can often be opened to the atmosphere. They could be copper or brass, but the stainless steel offers a better degree of mechanical and fire-resistant protection. Valves in the system are generally brass or bronze. The connections leading to individual air consumers, such as actuated valves, are generally ½-in. (DN 12) pipe up to a point, then the final connection is usually made with tubing to permit easy field adjustment.

Air is almost universally used as the actuating medium, but other gases are used for specialized applications. The use of nitrogen is quite common for higher pressure applications because the nitrogen is normally available in bottles at 2200 psi (150 bar) or in the form of vaporized liquid nitrogen at up to 4000 psi (280 bar). High-pressure air is also used, but it, like the nitrogen, must be piped separately from the standard instrument air or be supplied in bottles. High-pressure actuation is normally considered only where very large or very high-pressure valves are involved or in cases where very high actuation speeds are required. Sometimes nitrogen is used because the cleanliness and lack of moisture of the nitrogen is assured, while perhaps that of the available air is not. Nitrogen is also sometimes used just because it is available, often in cases where a great deal of nitrogen is consumed anyway for such things as inert gas blanketing so that it becomes easier to get nitrogen than air. As far as actuation is concerned, nitrogen behaves essentially the same as air. Sometimes, specialty gases may be used for actuation, such as pipeline gas in remote locations. The density of the gas may have to be considered if it is considerably different from air or nitrogen.

There is a specialty variant on the air cylinder actuator known as the self-contained actuator (Fig. 3-24). Here, in addition to the standard cylinder, linkage, and pilot valves, there is an integrally mounted tank or accumulator which contains enough gas or hydraulic fluid to operate the valve through one or more open and close cycles. The tubing connecting the tank to the cylinder is sometimes specially protected from vibration and mechanical damage. The purpose of this design is to make as sure as humanly possible that the valve will actuate when called on. Typical services for these valves are main shutoff valves in nuclear plants which are designed to withstand a certain amount of *g*-force during seismic loading, as required by nuclear regulation. There are a number of similar applications for cross-country pipeline work, not necessarily with the reservoir mounted on the valve but with custom-designed cylinder actuators that directly mount onto their valves, often with a single stem that is both the valve stem and the actuator stem in one. These designs are inherently more secure than an actuated valve assembled from components, but are generally more expensive.

These valves typically operate at high actuation pressures, up to 2000 psi (140 bar) like the high-powered actuators, partly because of the increased force required due to fast operating speed or valve size, and partly to decrease the actual volume of gas required to be stored.

Any valve that operates at a higher actuation pressure than standard must be specially designed for this service. Typically, the tubing and tube fittings are the same since standard copper tubing can withstand up to 1000 psi (70 bar) and stainless tubing can take 2000 to 3000 psi (140 to 210 bar) or more. Almost all other components must be specially selected or designed for this service: the cylinders, for maximum work

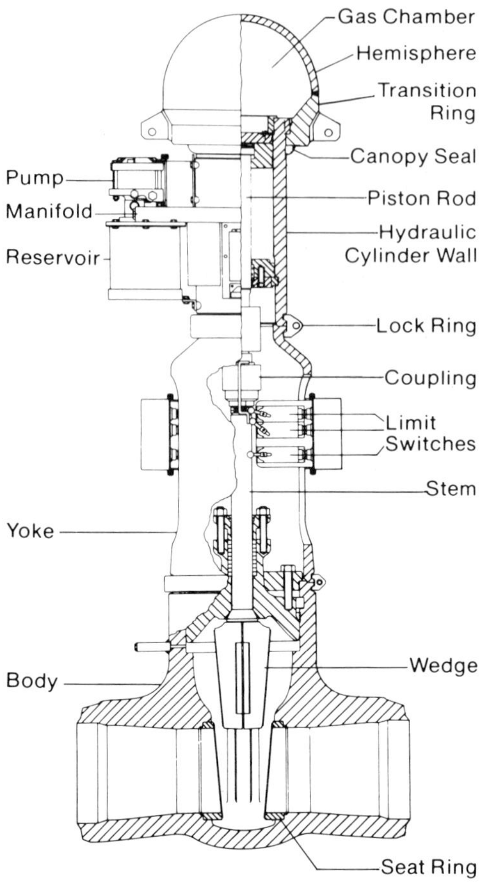


**Fig. 3-24.** Self-contained gas-hydraulic actuator for emergency shutdown valve. (*Courtesy of Edward Valves, Inc., formerly Rockwell Edward.*)

ing pressure, the gauges, for correct range, the pilot or solenoid valves, for maximum working pressure, and the sizing of tubing and valve ports to see that the required standard volume of gas can be passed through the ports in the time allowed for the valve to stroke.

Sometimes the power required to operate a valve is higher than that capable of being provided by an air actuator, either by increasing cylinder size and stroke or by increasing the actuating pressure, or the speed of actuation is unattainable with an air actuator. Then it becomes necessary to step up to a hydraulic actuator. The fluid used is normally a standard hydraulic fluid. The hydraulic fluid, like the air, really only transmits power from some other source or location. The power input to the hydraulic fluid is commonly either a hydraulic pump or a source of air pressure, in which case the combination is referred to as a hydraulic/pneumatic actuator.

Besides the greater power, there can be several other reasons for the choice of a hydraulic actuator. If space is at a premium, the hydraulic actuator can deliver the same power in a much smaller package. It is easy to provide a system that will function during a plant failure, where it is a little more difficult to ensure that an air system will continue to function. The operation of a hydraulic system tends to be smoother than a pneumatic system, and the positioning accuracy can be greater.

Sometimes the hydraulic fluid is supplied from a central source, like instrument air, but this is rare in process plant design. Aircraft controls are done this way, but standard actuated valves generally are not unless an entire installation is designed with hydraulic actuators and unless the total area involved is very small. The conventional arrangement is to provide a hydraulic reservoir adjacent to the valve with pressure supplied by an electric-powered hydraulic pump, an air motor driven by the 80 psi (5.5 bar) air supply, or by a high-pressure gas reservoir, often nitrogen. These gas reservoirs, known as accumulators, consist of a high-pressure bottle with nitrogen supplied to the top from bottled nitrogen or another high-pressure source. The hydraulic fluid is also in the accumulator. The solenoid valves and manual override valves, gauges, electrical connections, and tubing are generally installed in a skid near the valve. The same accumulators, with different actuation controls, can sometimes supply several valves in the near vicinity.

These hydraulic-pneumatic actuators exist because the speed of power transmission is the speed of pressure wave propagation (basically the speed of sound), which is much higher in liquids than it is in any vapor. The expanding gas in the reservoir only has to move a short distance, and the hydraulic fluid can more efficiently and rapidly transmit the work done by the expanding gas to the movement of the pistons.

The hydraulic actuator is pretty similar in appearance to the air actuator. The actuating pressures are much higher than for pneumatic actuators. Typical hydraulic systems will use actuating pressures on the order of 2000 psi (140 bar) with pressure supplied by bottled nitrogen at 2200 to 2600 psig (150 to 180 bar), and overpressure relief valves located near the power source set at about 2900 psig (200 bar). The cylinders are normally steel with sturdy tie rods, but the cylinder area will be smaller than an equivalent pneumatic cylinder. The pilot or solenoid valve is generally a solid machined-steel design to withstand the pressure, rather than the typical cast-brass valve suitable for air. Even the tubing is different: air systems use brass or stainless steel with compression fittings, but hydraulic tubing is heavier wall, often carbon steel, with heavy-duty hydraulic fittings. Incompressible hydraulic fluids do not pose the same kinetic energy release dangers if a rupture should occur.

### *Diaphragm Actuators*

For many years the most common actuator used on common control valves, the "classical" control valve with actuator, has been the diaphragm type. The diaphragm actuator is easy to manufacture, easy to understand, and it is directly coupled so that for every fraction of an inch that the actuator lifts, the valve disc moves by that much. (There are also diaphragm actuators for rotary valves, for which the preceding statement is not true.) It is usually easy to tell by looking at an actuator if it will fail open or fail closed. There are hundreds of thousands of these actuators in service all over the world.

The diaphragm actuator is almost like a cylinder actuator without the cylinder (Fig. 3-25). The diaphragm, for which the actuator is named, provides a seal to keep air from passing around the piston, or plate, as it is more commonly known. The air actually acts on the entire diaphragm,

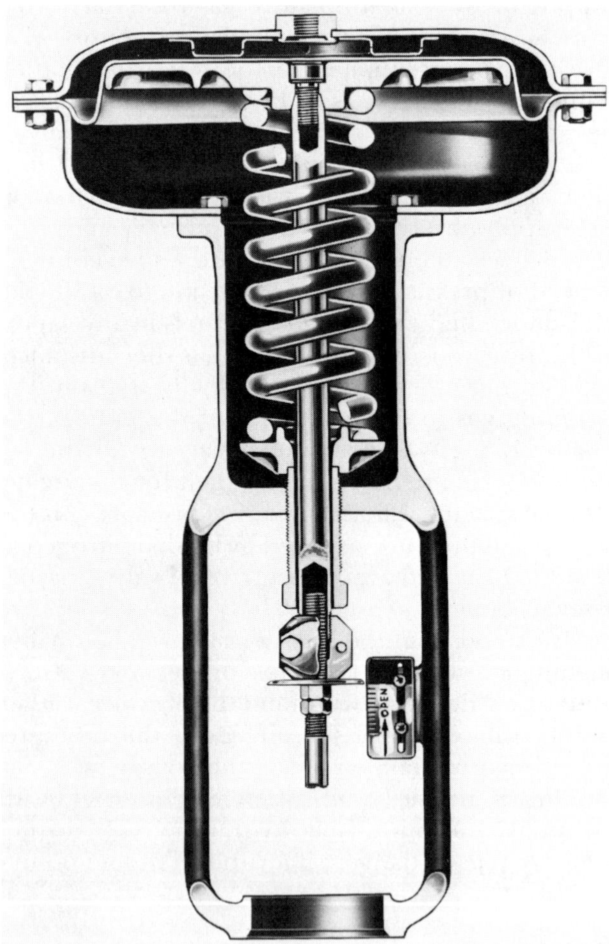

**Fig. 3-25.** Spring-return diaphragm actuator configured as direct acting, push down to close, fail open. (*Courtesy of Fisher Controls Intl.*)

and as the plate moves from the top to the bottom of its stroke, the diaphragm flexes and the effective area that the air pressure acts on changes slightly. For most cases, though, this effect can be neglected. The basic operational differences between a diaphragm actuator and a piston cylinder actuator are that the diaphragm actuator will have a much shorter stroke and somewhat less power output (thrust) than the cylinder but will be capable of providing more precise positioning. It is possible for the diaphragm actuator to be more precise and to move in smaller increments because the diaphragm introduces essentially no frictional resistance or breakaway force, while the O-ring seals in a cylinder often provide quite a bit of resistance to motion. This is especially true if the cylinder actuator has not stroked for a while.

The common types of diaphragm actuators consist of a two-piece housing, the plate and rod or stem, the diaphragm itself, and a mounting yoke or bracket assembly. Since almost all diaphragm actuators are on valves built especially for control, they are sometimes built almost as part of the valve. The specialized control valve known as a regulator is often built with actuator mounting integral with the valve bonnet and, in some cases, the actuator itself is integral with the body so that the valve stem and actuator shaft are the same and there is one single pressure boundary that includes the valve body and the actuator shell.

There are a fairly wide range of sizes of diaphragm actuator, but these vary from fairly small to medium-large in relation to other types of actuators. Almost all diaphragm actuators are powered by standard instrument air. The air is brought to the vicinity of the valve at 80 to 120 psi (5 to 8 bar), but the diaphragms usually operate in the 20 psi (1.4 bar) range, and most are only designed for a maximum of 40 psi (3 bar) or so. Larger thrusts can be obtained by making the diaphragm larger, but the physical size limit is about 50,000 lb (22,000 kgf) of thrust. Bear in mind that this is not as severe a restriction as it seems. By their very nature, diaphragms are not capable of long strokes. Many can stroke about ¾ to 2 in. (20 to 50 mm), and a 4-in. (100-mm) stroke is about the longest achievable for a standard actuator. They are therefore only capable of powering valves which move a short distance from full open to full closed, but which need very close regulation. This basically includes globe-type valves and quarter-turn valves. Both of these valve types can be constructed so that the forces involved in opening and closing can be balanced against each other or reduced to some degree. For these valves, the diaphragm provides the best actuating characteristics, and other actuators are more suitable for operating other types of valves.

The actuator housing is generally either cast ductile iron or pressed sheet steel. The housing is in two pieces, and the outer edge of the diaphragm usually serves as the pressure-retaining gasket between the two halves. Otherwise a separate gasket of soft material is provided. The two halves are bolted together around the edge with four to twenty or more bolts, depending on size. Some cast housings have the yoke cast as part of the lower half of the housing, but most yokes are separate. The yoke is generally cast ductile iron or cast steel and is often designed to mount directly onto a valve body since many true control valves are almost never installed without a diaphragm actuator. The yoke is also used for mounting all of the auxiliary devices that may be required.

The piston or plate is usually also cast ductile iron or pressed steel. The piston takes up between 75 and 90 percent of the inside diameter of the actuator. Its stroke is only a relatively short distance compared to its diameter. The actuator shell, because of this short stroke, can be curved into an ellipsoidal shape so it can be fairly thin and still adequately withstand the actuating pressures.

The shaft is normally steel, or brass in light-duty actuators, threaded at its top end to accept the piston and diaphragm retainer. Its bottom end passes through the actuator shell with a bushing or other guiding device. If the bottom space of the actuator is not pressurized, no O-ring or other pressure retaining device is required. The bottom of the actuator shaft has some mounting configuration to mate with the top of the valve stem. This is often a female thread in the bottom of the actuator shaft with a male thread and a locknut on the valve shaft. Often a position indicator is sandwiched into this connection.

The position indicator is generally a metal disk, sometimes a pointer, attached to the stem to give a clear visual indication of the position of the shaft (Fig. 3-26). There ordinarily is also a scale attached to the mounting bracket, graduated in divisions and slightly adjustable to line up with the actual fully open and fully closed positions of the shaft. If limit switches are installed, the arm of the limit switch is positioned to touch the position indicator. Usually, if both open and closed limit switches are furnished, they will be mounted on a plate which is then mounted to the actuator bracket. When limit switches or other devices requiring mechanical contact with the shaft are not required, some manufacturers will supply a clear plastic cover to protect the indicator from mechanical

**Fig. 3-26.** Position indicator for reciprocating stem control valve, with limit switches to detect full-open and full-closed positions. (*Courtesy of Valtek, Inc.*)

damage such as an object being trapped under the indicator while the valve is closing. It is also possible to mount a rotary transducer to a linear actuator by means of a linkage (Fig. 3-27).

The diaphragm itself is normally an elastomeric rubber material or a fabric coated with an elastomeric rubber. If the actuator itself must be exposed to higher than normal temperatures, beyond about 150°F (65°C), special elastomers are available to extend the ambient range up to as much as 300°F (150°C). Some diaphragms are premolded for less deformation in service. They can also be stainless steel or bronze for applications such as regulators where the process fluid is in contact with the bottom of the diaphragm. Metal diaphragms, of course, have much lower flexibility. The diaphragm spans the entire width of the actuator with only a hole in the center for the shaft and piston retainer. The air is, in effect, acting on the diaphragm, and the diaphragm basically just transmits the pressure force through its thickness to the piston. As mentioned above, because of the flexure of the diaphragm as it moves through its travel, it has a slightly greater effective area at the top of its travel or its most stretched position. It has a slightly smaller area at the

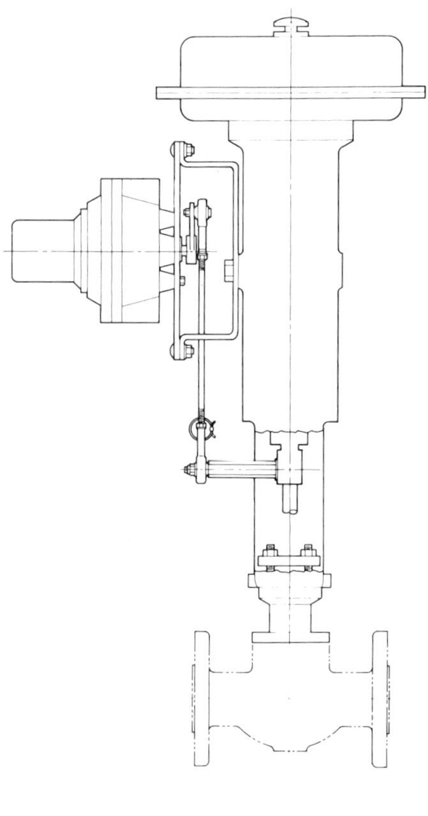

**Fig. 3-27.** Indicator beacon with position transmitter mounted on reciprocating-stem control valve with linear/rotary converter. (*Courtesy of Westlock Controls Corp.*)

end of its travel or in the middle, depending on its exact geometry. There are not any particular advantages stemming from this behavior, and this nonlinearity may have to be considered in cases where an actuator sizing is marginal.

Most diaphragm actuators are single-acting, that is, the actuating medium only pushes on one side of the diaphragm and a passive mechanism, a spring, pushes back from the other side. Some actuators are double-acting, where air is applied from one side to cause movement in one direction, and air is applied to the other side to cause movement in the reverse direction. Double-acting actuators are more expensive since they require more hardware and, normally, a plate on each side of the diaphragm. The spring is normally incorporated in the same housing with the actuator. Inexpensive designs sometimes have the spring exposed, outside the body of the actuator. A standard spring steel material is normally used, and standard spring sizes are often provided, rated in terms of the air pressure required to balance the spring force.

Another characteristic to be considered is whether the actuator should be direct-acting or reverse-acting (Fig. 3-28). For two actuators in which the spring is between the diaphragm and the valve, a direct-acting

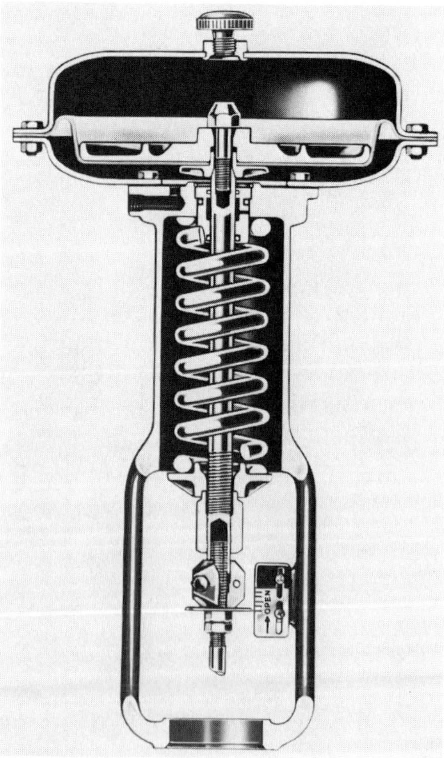


**Fig. 3-28.** Reverse-acting diaphragm actuator with adjustable travel stop. (*Courtesy of Fisher Controls Intl.*)

one requires air from above to push downward on the spring, thus the stem moves down or toward the valve when air is applied. A reverse-acting one has the air connection on the bottom, pushing the diaphragm upward to pull on the spring, thus the stem moves up or away from the valve body when air is applied.

The choice of direct- or reverse-acting will be determined first by the valve body design. Single-port globe valves open by lifting the stem. This is true whether the flow is over the plug or from under the plug. Double balanced-port valves often open by pushing down on the stem. (Identification of up and down assumes that the valve is in the standard position—horizontal pipe orientation with the stem vertical and emerging from the top of the valve body.) For rotary valves, it depends on which side of the pipe the actuator is mounted. Secondly, the fail mode must be considered. In cases where continued flow through the valve is important, if air pressure or power is lost, the valve must fail open. In other cases, where the valve is expected to close when an incident occurs, say to prevent flammable material from flowing into an area where fire or damage may be occurring, the valve must fail closed. A direct-acting actuator will fail open when the spring is below the diaphragm on the nonpressure side, which is the most basic combination. A reverse-acting actuator will fail open. If the opposite action is required, either the valve selection must be changed to one which opens in the opposite direction or the spring must be moved to the area above the diaphragm (which is the only possible solution for certain valves, such as diaphragm valves that only open one way and are not easily replaced by another type of valve).

Most of the mechanical components of direct-acting and reverse-acting actuators are completely different. On a direct-acting actuator with spring below, the air inlet is dead center on top. The spring adjustor, which controls the exact length of stroke, is on the bottom. A reverse-acting actuator has the air inlet on the lower side and has an adjustable travel stop on the top instead of a spring adjustor. The graduated indicator plate, which has an arrow to indicate direction to open, is turned in opposite directions on the two types of actuator. A manual override handwheel can be provided for both that can operate the valve regardless of whether air is being applied to the valve or not if the operator turning the handwheel is strong enough. All that the handwheel can do on the direct-acting type is to push down, since the spring force is always pushing up. On the reverse-acting type, some handwheels can only pull up on the stem, while other designs can both push and pull. On some of the largest diaphragm actuators this handwheel is side-mounted and geared to the override stem.

Another accessory that is sometimes required for unstable or rapidly changing flows is a hydraulic snubber. This component goes between the diaphragm and the spring, and uses a piston mounted on the actuator stem to work against a volume of hydraulic fluid to dampen out some of the short-lived forces imposed by the flow stream on the valve disc. This gives longer life to the parts of the diaphragm and the valve and reduces the fluctuations that would be caused by the feedback devices constantly reacting to changes in the valve position.

Many actuators that normally are used on globe-type control valves (sometimes referred to as reciprocating stem valves) can also be used on rotary-stem control valves (Fig. 3-29). However, some slight modifications can be made to make them more suitable for rotary operation. In most cases the yoke is replaced with a cast housing for the lever arm. The lever arm is mounted onto a shaft which then is coupled to the valve shaft. Since these valves are intended for control, the mechanical or hysteresis must be controlled as tightly as possible. Often this is done with a spline coupling rather than just a square stem top. The actuator rod is coupled to the lever arm, which usually requires some side play in

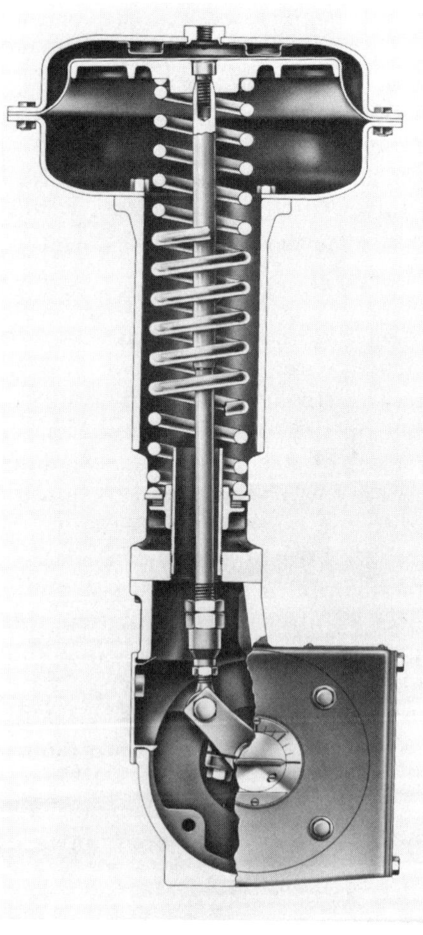

**Fig. 3-29.** Diaphragm actuator designed for rotary-stem control valve. (*Courtesy of Fisher Controls Intl.*)

the actuator rod. This can be done with direct-acting diaphragms just by placing the diaphragm higher to give a longer rod length since the diaphragm itself permits some degree of swiveling. Also, many rotary control valves do not rotate a full 90 degrees. A 60-degree rotation requires less play in the diaphragm rod.

The lever arm housing is generally reversible, with a bolted cover that also contains a position indicator. Since most rotary control valves are designed with a preferred orientation, the actuator is designed to be located in any position to suit the available space (Fig. 3-30). The optimum installation position for many rotary control valves is with a horizontal shaft, which allows the actuator to be in its preferred verticalorientation. The lever arm housing can simply be rearranged to suit the valve having its topworks on either the right or left side of the line.

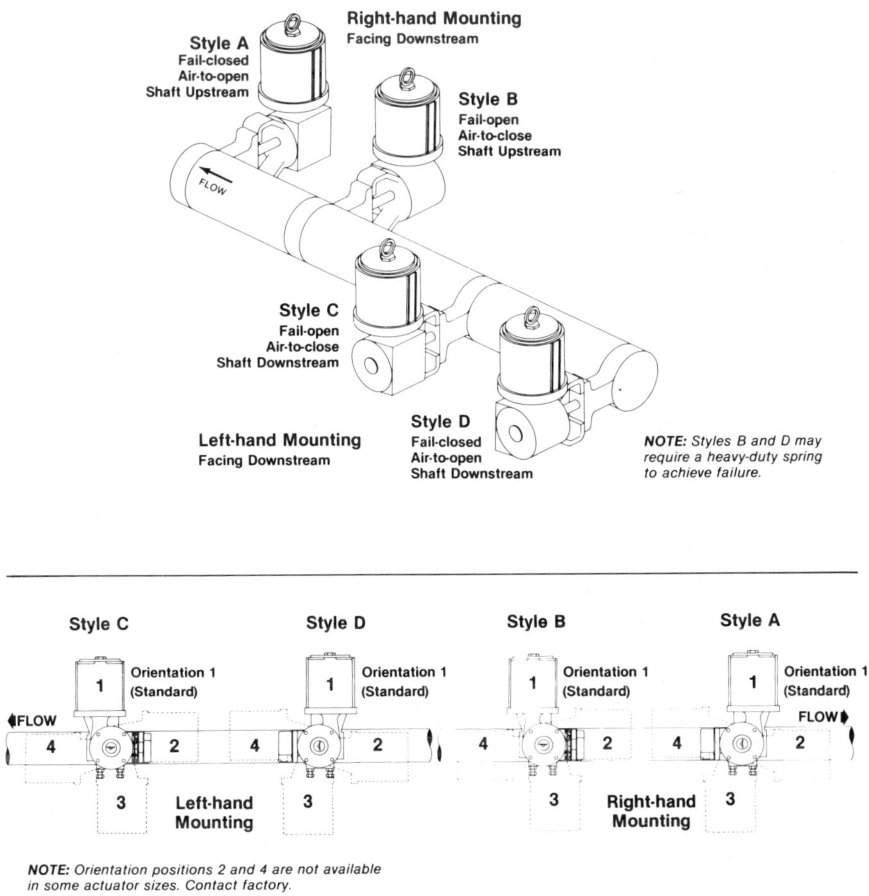


**Fig. 3-30.** Possible actuator orientations in relation to valve and line orientation. (*Courtesy of Valtek, Inc.*)

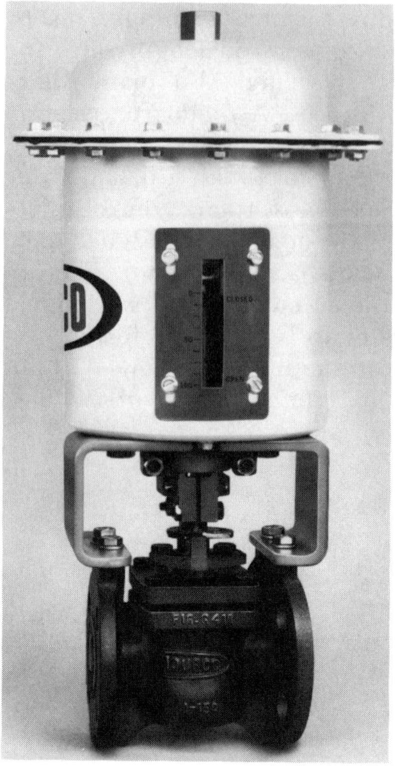

**Fig. 3-31.** Diaphragm actuator with cammed coupling for direct mounting on plug valve. (*Courtesy of Duriron Company, Inc., Cookeville Valve Division.*)

A totally different means of linking a diaphragm actuator with a quarter-turn valve is the design found on some plug valves and, rarely, other types of quarter-turn valves (Fig. 3-31). Here, the diaphragm shaft is in line with the valve stem, but the diaphragm shaft moves up and down while the valve stem rotates. The lower portion of the actuator contains a cammed bearing that the shaft pushes against and which, in turn, is coupled to the valve stem. These actuators work well with the relatively high torques required by the plug valve.

### Rotary Vane Actuators

The rotary or vane actuator is a seemingly close relative of the cylinder actuator. It does the same job; it is generally interchangeable; it takes the same signals and the same energy source. However, it is an entirely different creature, with its own strengths, because it totally does away with the mechanical components required to transform linear motion into rotary motion. It applies air pressure to directly produce rotary motion.

The principle of the rotary actuator is simple. Air that is introduced into a chamber expands and pushes on a rectangular vane which responds like a piston by moving in the direction that the air is pushing. While the piston of an air cylinder is attached in the center and moves linearly, the vane is attached all along one edge and moves rotationally, like a door. The rotational movement is translated directly to the shaft that the vane is attached to. This shaft is the actuator shaft which transfers its movement to the valve shaft.

The mechanics of this system are not so simple. The sealing surfaces are not round, and there is no easily located cylinder end to stop the motion. But some rotary actuators can provide significantly higher torques than a cylinder actuator can, and more importantly, they can be very much smaller for the same amount of output torque. The rotary actuator really competes with the hydraulic actuators for the same applications. And for really high torques, it is possible to build a hydraulic rotary actuator.

There are really two totally different types of rotary actuators. The first, which is more commonly called a vane actuator (Fig. 3-32), is used on fairly conventional quarter-turn valves with relatively low torque

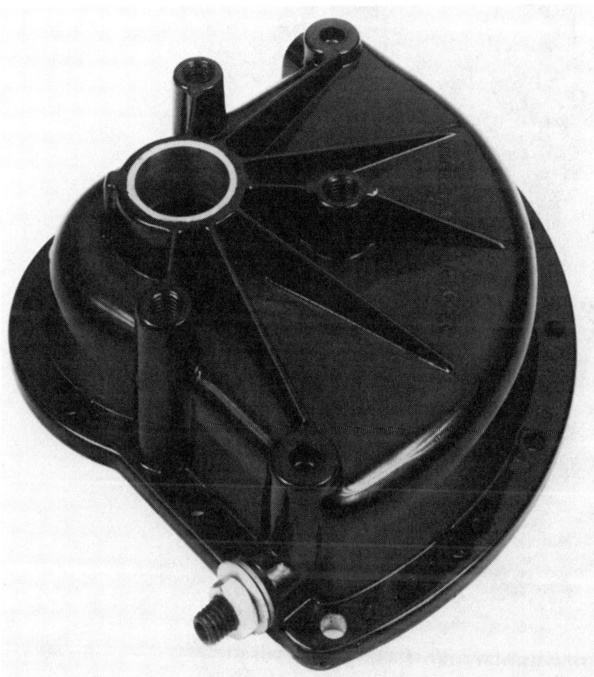

**Fig. 3-32.** Rotary vane actuator. (*Courtesy of Xomox Corporation.*)

requirements. Its primary attraction, besides eliminating the external linkage associated with cylinder actuators, is its small size and low cost. The other variety is a high-muscle actuator used in circumstances where a large amount of torque is required and not a lot of space or weight is available. The primary difference between the two is that the vane actuator is one-sided; that is, there is only one "cylinder" with one vane being moved back and forth, whereas the true rotary actuator is radially symmetrical, with two identical vanes.

The physical appearance of the vane actuator is a lot like that of a gearbox or the linkage of an enclosed cylinder actuator. It consists of a die-cast metal box with a cover, and a shaft and vane that is retained in the cover and the bottom of the box (Fig. 3-33). The difference is that the box must be pressure-tight, so there is some type of static seal on the cover and O-rings on the shaft. The active member is a rectangular vane mounted on the shaft or cast in one integral piece with the shaft. The seal must pass around the shaft, that is, the air pressure cannot push around the back side of the shaft to the space on the opposite side of the

**Fig. 3-33.** Interior view of vane actuator showing travel stops and sealing mechanism. (*Courtesy of Xomox Corporation.*)

vane. The edges of the vane also must seal against the front and back of the enclosure. These seals normally are a one-piece rubber seal that wraps around the pressure side of the shaft on top and bottom just away from the edge of the shaft. In effect, the seal and the surface of the vane are folded around the shaft.

Externally, the actuator has a bolt pattern drilled into the bottom and top for bracket mounting. Turning it upside down reverses its direction of rotation, although this is seldom done since almost all quarter-turn valves by convention open counterclockwise, and the actuator is double-acting only. It has air inlet and exhaust connections on each side, and the shaft has the usual coupling on the bottom and top. For manual override, it is possible to use a lever on the end of the shaft. The connections and controls are the same as on a standard cylinder actuator. When stops are provided, they are in the edges of the enclosure, under the end of travel of the vane, the same place that they would be in the ends of a traditional cylinder. Available torques are in the range of 400 to 30,000 inch-pounds (50 to 3750 Nm).

The vane actuator is almost always furnished as a prepackaged unit for butterfly and plug valves with relatively low torque requirements. It is physically very compact compared with a cylinder, and it can be fairly inexpensive because of its size and design. In addition, because the air directly turns the shaft, there is only one mechanical connection, the one between the shaft and the valve stem, rather than the two or three in a cylinder for rotary-shaft valves. This significantly reduces the slop, or hysteresis, associated with mechanical joints. However, there is no really easy way to provide a spring return. Most spring-return devices that are provided are separate, and are like a manual spring closure device added to the top of the actuator. It is possible to make the actuator fail-safe without a spring, by providing a small air accumulator next to the housing with a pilot or solenoid valve (Fig. 3-34), such that when the air supply to the actuator fails, the pilot valve admits the air from the accumulator into whichever side of the actuator has been tubed up to be the fail-safe side, either fail-open or fail-closed.

For actuating a three-way valve that rotates through 180 degrees, two vane actuators are required, stacked on top of one another (Fig. 3-35 a–c). The first 90-degree stroke is by the lower actuator rotating the valve stem one-quarter turn, and the second stroke is by the upper actuator rotating both the lower actuator body and the valve stem by one-quarter turn.

The other common type of rotary actuator is built quite a bit differently, although it works in the same basic manner as the vane actuator. These rotary actuators are normally designed for very heavy duty, and coincidentally are also often designed for low-maintenance requirements, since one of their early markets was for isolation valves on

**Fig. 3-34.** Vane actuator with air accumulator and pilot valve, serving the same function as a spring return for fail-safe operation. (*Courtesy of Xomox Corporation.*)

cross-country pipelines (Fig. 3-36). They are often designed to use the commodity being transported in the pipeline, such as natural gas, as the actuating medium.

The basic components of the rotary actuator are an outer shell, which is roughly the same height as its diameter, and the two ends, which are the pressure-retaining components along with the shell or cylinder. The bottom end is also drilled for the mounting bracket. The whole assembly stands on end like a direct cylinder actuator. Both ends support the central shaft, with O-ring rotary seals to prevent leakage around the shaft. The shaft itself projects through the bottom end for the coupling mount, and projects up into the top end for bearing support and as a convenient means to attach the indicating device to. The shaft is normally cut out to accept the vanes, to provide maximum strength in the attachment. The vanes have a sealing device retained in the three free edges. Normally all these components are machined steel. Increasing the output of the actuator can be done by increasing the actuating pressure, or by increasing either the diameter or the height of the actuator.

There is a second set of fixed vanes in the shell, which act as the other ends of the air chambers. Rotary actuators, by the nature of their

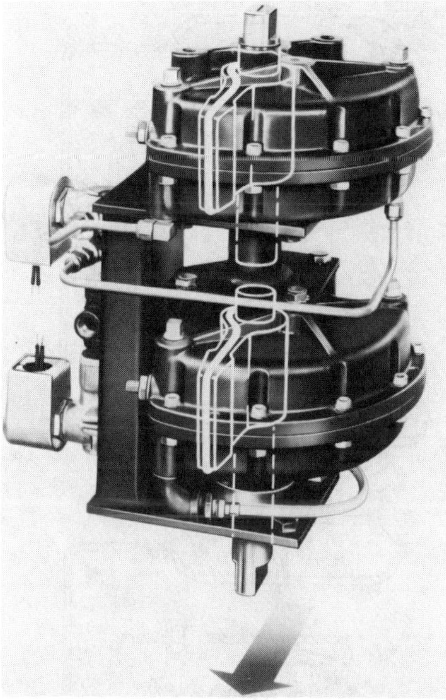

(a)

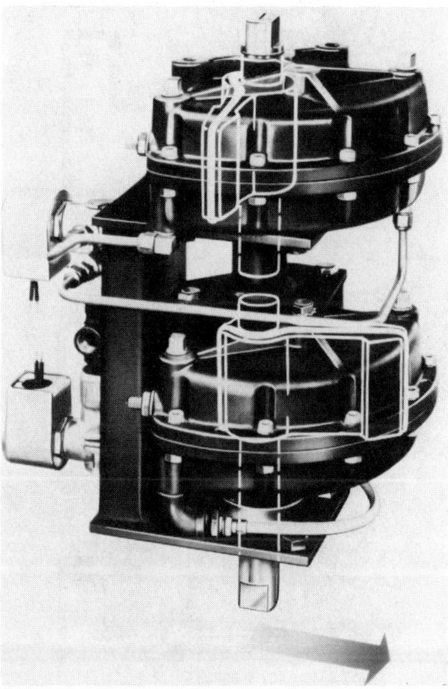

(b)

(c)

**Fig. 3-35.** Rotary vane actuator for three-way valve. (*Courtesy of Xomox Corporation.*)

**Fig. 3-36.** Rotary actuators. (*Courtesy of Shafer Valve Company.*)

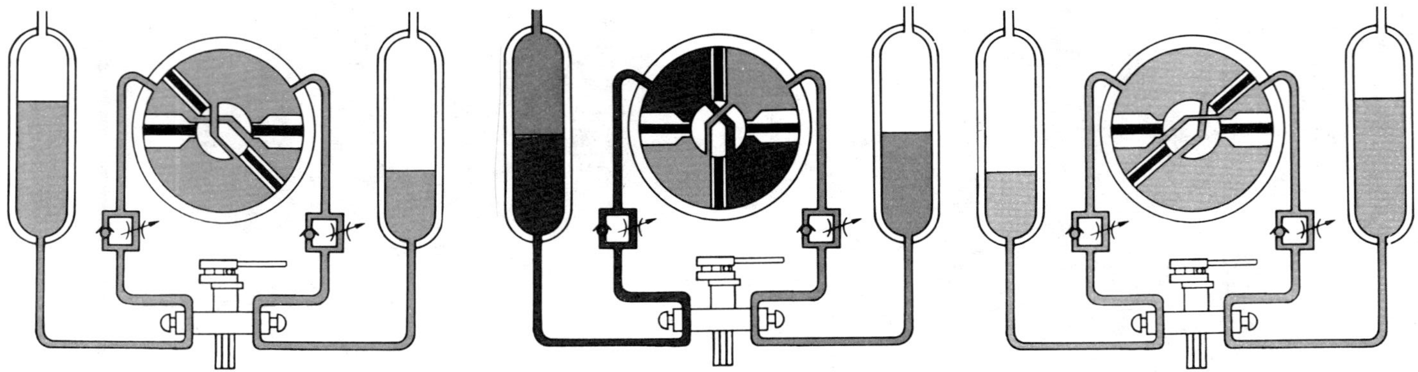


**Fig. 3-37.** Operating sequences of rotary actuator. (*Courtesy of Shafer Valve Company.*)

geometry, are always built with two active "cylinders," each one occupying 180 degrees of the shell arc. The fixed vanes are attached by cap screws to the outer edge of the shell and to the top and bottom, with a sealing device along the inner edge to permit the shaft to rotate under them. Stationary soft seals are provided around the other edges of the fixed vanes, as well as at the end-to-shell joints. The backs of the fixed vanes could act as the stops, but normally a set of adjustable stops are provided. The adjustment is by means of cap screws that pass through the shell from the outside and into the stops, which can move radially in and out. Moving the stops farther in means that the moving vanes contact them sooner, thus reducing the radial travel distance.

The air inlets and exhausts are in the shell, directly adjacent to the edges of the fixed vanes. Usually only one inlet and one exhaust are provided, with the shaft drilled with flow ports from one cylinder to the other. The ports crossing the shaft are located right next to the leading and trailing edges of the moving vanes. This ensures that the pressure in each of the two sets of air chambers at any given time is the same, for even force distribution (Fig. 3-37).

These actuators are most often designed for high actuating pressures, ranging upward from the 80 to 100 psi ( 5 to 7 bar) of standard air supply to the 2200 psi (210 bar) typical of nitrogen bottles to higher pressures for specialized applications. All external components including the tubing and the control block are steel or stainless steel for the pressure. Actuating gas can be supplied from a standard supply header, from bottles, or from the line itself if the gas is clean enough. The use of line pressure is possible because of the mechanical advantage inherent in the actuator's configuration.

## Solenoid Actuators

The group of actuator designs known as solenoids are among the physically smallest of all actuators, and the physically smallest of all actuated valves (Fig. 3-38a,b). Yet they are quite significant, since many of these valves are incorporated on larger control valves as accessories as well as being used independently.

A solenoid is an electromagnetic device in which a coil of wire surrounds a steel rod and induces a magnetic field that moves the rod into and out of the coil. When the current is varied, the rod can be made to move in and out of the coil with varying force. When a spring is added, the variable current can move the rod in and out by a specific distance until the spring force and the magnetic force equalize. The rod can be used to produce work, such as moving a valve disc up and down. The conventions used with solenoid valves are that normally closed

(a)

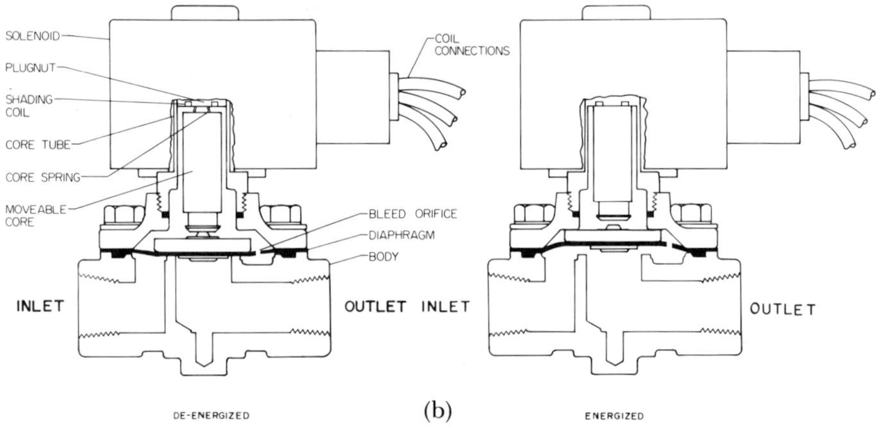


(b)

**Fig. 3-38.** Solenoid valves for in-line mounting. (*Courtesy of Automatic Switch Co. (ASCO).*)

valves are closed when the solenoid is not powered (deenergized) and open when the solenoid is energized. Normally open valves are energized to close the valve and deenergize to open. The spring moves the valve disc to its normal position, and the solenoid works against it.

The solenoid valve as used independently and as used as part of a larger control valve are similar in operation, although not always similar in appearance. For stand-alone solenoid valves, the uses are generally similar to those of small hand-operated needle valves. They find their application in small board-mounted piping arrangements, or controlling very small flows in small lines, generally in 1-in. pipe size (DN 25) and smaller down to ¼-in. size (DN 6). Basically, a solenoid valve is like either a globe or a diaphragm type, but is considered here separately because it can also be classified as an electrical device and the solenoid is really an integral part of the valve.

The body of the solenoid valve is generally cast brass or die-cast aluminum, but stainless steel and some thermoplastic bodies are also used. The disc is sometimes the same material as the body. The portion of the stem that is in contact with the line medium is also sometimes a more corrosion-resistant material. The solenoid rod is steel, generally a special magnetic steel, and is attached to the stem proper. The magnetic coil of the solenoid is in most cases not exposed to process fluid, and its mechanical components are not normally designed to be resistant to the fluid handled. The coil is generally encased in a solid molded-plastic block for protection and electrical insulation. The valve body and parts that are within the magnetic field of the solenoid cannot be steel or any magnetic material. The solenoid case is sometimes plastic, but usually metal, often pressed steel or die-cast aluminum with appropriate gasketing for electrical isolation as required, including weatherproof or explosion-proof designs.

Power into the solenoid is any standard electric power, usually 120 VAC or 12 VDC. The current flows through a coil of hundreds of turns of wire, sufficient to generate a magnetic field capable of moving the solenoid and the attached valve disc up and down against line pressure. In general, the solenoid valves are not required to produce a large amount of closing force directly (Fig. 3-39a,b). Their value lies not only in their being used to control some small flows directly but in their ability to control some very much larger flows in other valves by acting as pilots (Fig. 3-40).

A number of body designs are available. Besides the common globe-type or two-way valve, there are a number of three-way designs. The three-way valves are available in several body configurations, but basically they provide for flow from one port to a second port when de-energized, and from the first port to a third port when energized (Fig. 3-41). The different configurations are determined by which way the flows occur with respect to the location of the solenoid. A different three-way valve, which is often installed on control valves as the air control valve (Fig. 3-42), has a larger port for exhaust and a specially shaped disc that opens wider in the exhaust mode. This allows a higher $C_v$ and a faster flow rate, so that the air volume that passes through the valve passes through it in one direction, from the pressure port to the cylinder port (inlet cycle), in perhaps 5 seconds, and the same air passes through it in the exhaust cycle, entering the valve from the cylinder port but leaving from the exhaust port, in perhaps 2 seconds.

There are several four-way configurations available, which find their application as pilot valves as well as for flow switching. These are built

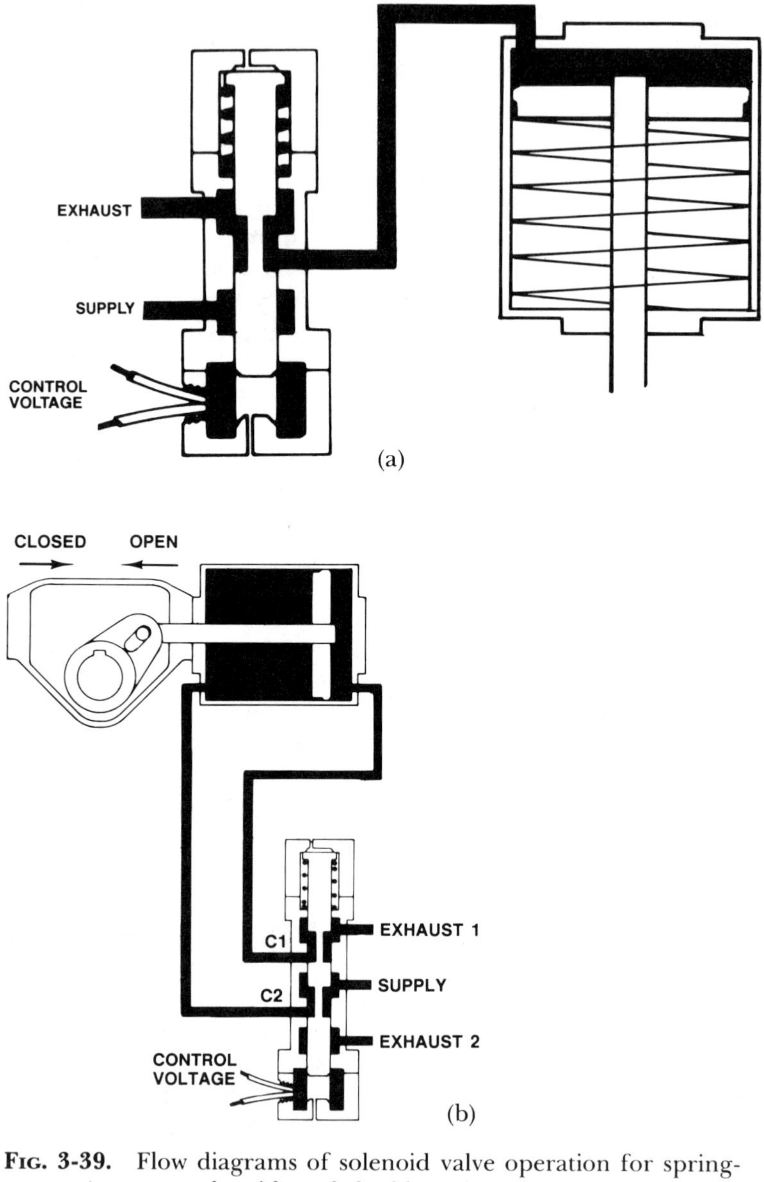


(a) (b)

**Fig. 3-39.** Flow diagrams of solenoid valve operation for spring-return (two-way solenoid) and double-acting (three-way solenoid) actuators. (*Courtesy of Fisher Controls Intl.*)

like manifold valves, with several inlets in a flat plane. The solenoid shuttles the disc, which is known as a spool, back and forth to cover and uncover the various ports and connect the different flow passages to each other.

Other options available concern the actuation. For some applications indoors, a special low-noise valve is available that produces a much lower

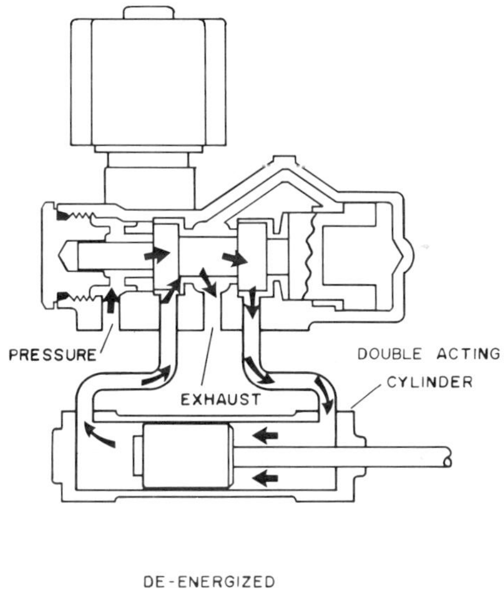


**Fig. 3-40.** Flow through solenoid-operated pilot valve. (*Courtesy of Automatic Switch Co. (ASCO).*)

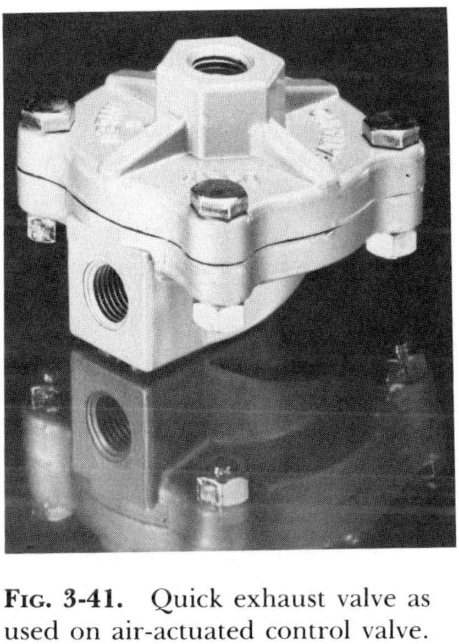

**Fig. 3-41.** Quick exhaust valve as used on air-actuated control valve. (*Courtesy of Automatic Switch Co. (ASCO).*)

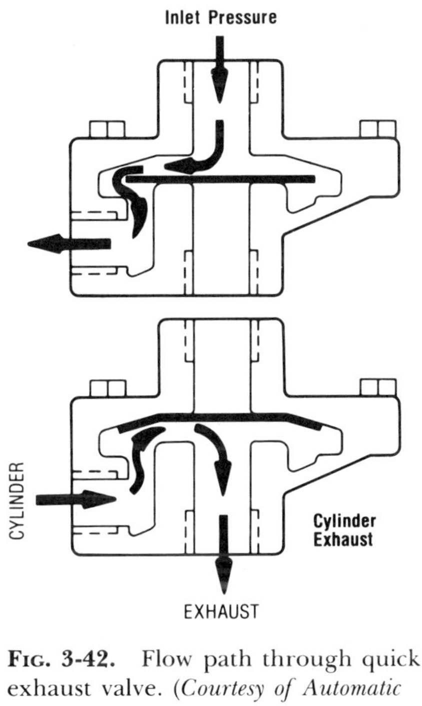


**Fig. 3-42.** Flow path through quick exhaust valve. (*Courtesy of Automatic Switch Co. (ASCO).*)

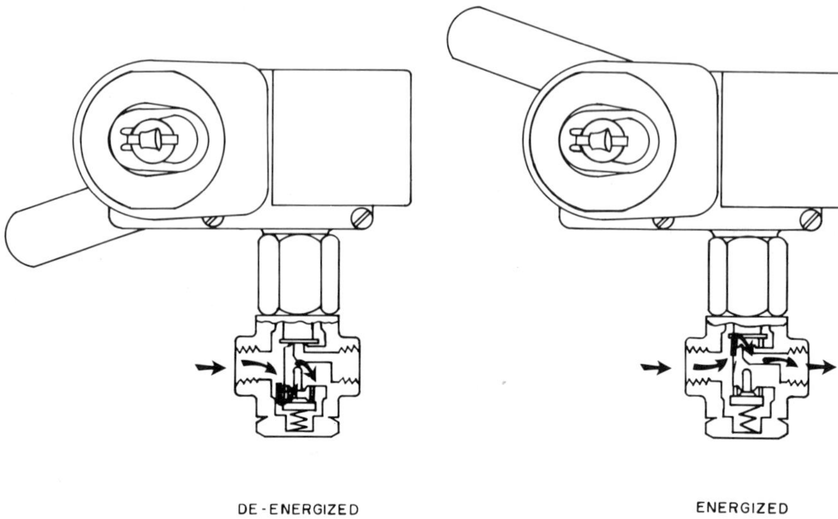


**Fig. 3-43.** Manual latch on solenoid valve. (*Courtesy of Automatic Switch Co. (ASCO).*)

level of sound (known as hum) than the 60-Hz electrical operation of a normal solenoid. This valve also is designed to seat more quietly. Another option that is often used for safety in process plants is a manual reset or latch (Fig. 3-43). It prevents the valve from opening again, once it closes, until a person goes out to the valve and manually moves the handle back to the actuating position. Most of these manual latches also allow for a manual closing of the valve without electrical power, which is useful during maintenance to isolate the piece of equipment that the solenoid valve feeds. The whole point of manual reset is to give a human operator the ability to assess the situation after the valve has closed for some reason, to determine if the situation is safe for operations to resume. They can be designed either to latch open or to latch closed.

## Motor Actuators

### *Electric Motor Actuators*

Many different types of valves can be actuated by electric power. There are the types that use an electromagnetic force to directly move the valve stem, known as solenoid valves, and there are the types that use a conventional motor to supply rotational energy (torque) that is transmitted to the valve stem by mechanical means (Fig. 3-44). This section will

**FIG. 3-44.** Large multi-turn motor actuator on pressure-seal gate valve. (*Courtesy of Pacific Valves, a unit of Mark Controls.*)

discuss the motor operators, primarily electric motors but also air- and hydraulic-powered motors.

Motor operators have been in existence in basically their present form since the 1920s. The motor efficiency, the quality of the components such as insulation, and the compactness of the design have all improved. In fact, there are now two basic approaches to motor actuator design just as there are to air actuator design. One consists of assembled components in a steel, cast-iron, or aluminum housing and is often designed for larger and special valves, and the other is a compact unit that looks like a black box that drops onto the valve.

In the component type, the motor actuator consists of several basic parts (Fig. 3-45). There is the motor, which ordinarily is constructed specially for valve service. The electrical controls are an essential part of the operator, just as the pilot valves are on an air actuator. Supporting the motor and the on-board portion of the wiring is a gearbox that transmits the motor torque through the yoke nut or other gearing to the valve stem or to a separate actuator shaft that is coupled to the valve shaft. Both rising-stem valves such as gates and globes and rotary-stem valves such as butterfly, plug, and ball valves can be actuated by a motor (Fig. 3-46). It is more common for nonmodulating valves to have motors, but with electronic feedback it is certainly possible to use a motor for modulating services.

The motor itself, like any good-sized industrial motor, consists of an

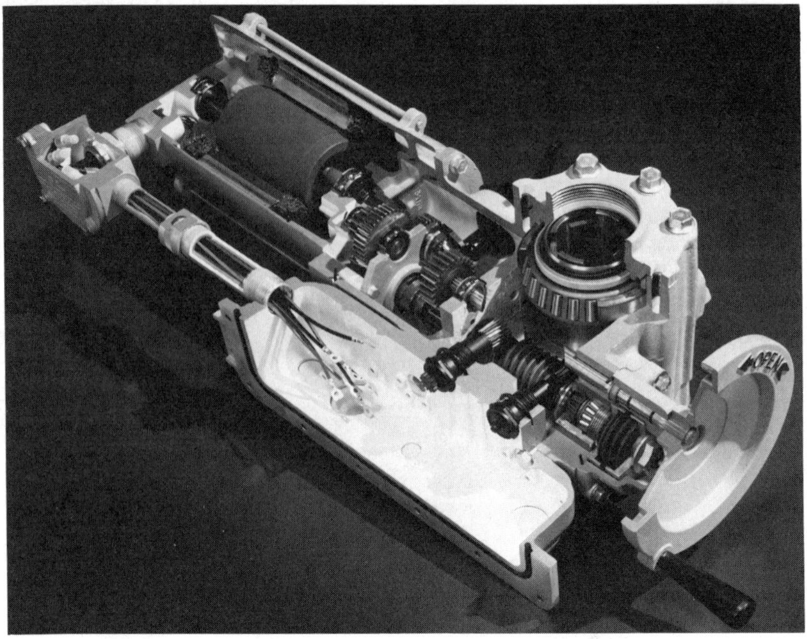


**Fig. 3-45.** Cutaway of motor actuator for multi-turn rising stem valve. (*Courtesy of EIM Controls, Inc.*)

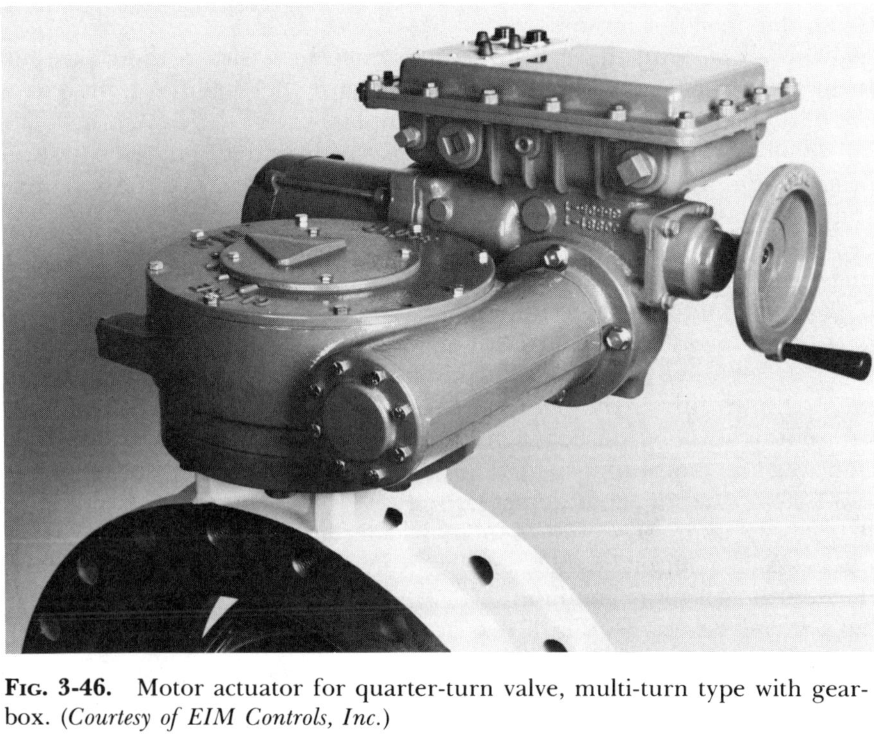

**Fig. 3-46.** Motor actuator for quarter-turn valve, multi-turn type with gear-box. (*Courtesy of EIM Controls, Inc.*)

armature with windings, which is driven by another set of stationary windings enclosed by a motor case. The shaft has bearings at each end, usually ball bearings, a commutator to power the armature windings (in DC and single phase AC motors), and a gear keyed to the one free end of the shaft. Many motor designs have a can-shaped motor case, where the end away from the gearbox is integral with the sides. The end that adjoins the gearbox has a gasketed joint. The amount of sealing provided by this and other mechanical seals in the actuator is determined by the location of installation of the actuator to a degree that the corresponding design of an air or manual actuator is not, because of the danger of moisture to the electrical components and also the danger of the electrical currents, and sparking to the atmosphere and to nearby personnel. Electrical and electronic devices installed in industrial locations are required to comply with National Electrical Manufacturers Association (NEMA) standards that determine their degree of separation from the environment according to the severity of the hazard involved. Thus an electrical device that will operate while being occasionally splashed with liquid or one that will operate in a potentially explosive atmosphere is designed differently than one that will operate in an environment comfortable for people.

The current feed to the motor is through heavy-duty cable, which is run to an electrical box or integral control compartment on the actuator (Fig. 3-47a,b). The electrical connections to the power source are made in this box, which is also sealed according to the degree of hazard represented by its location. Also in this box are the other electrical components, including limit and torque switches and other controls. The operating controls are either on the actuator itself, located on the junction box, or at a remote site such as the control room or a spot a safe distance away from the valve. Sometimes both types of controls are provided. The controls generally consist of three switches or push buttons, one for opening the valve, one for closing the valve, and one for stopping the motor. A red and a green light are normally provided. One light is illuminated when the valve is in one position, and the other is illuminated when the valve is in the other position. The red is often used to indicate closed and the green to indicate open, but it can be wired the other way. Almost always both lights are illuminated while the motor is operating, as a means of showing that the motor is energized.

Both the open and close buttons actually start the motor, one for each direction. The stop button is mainly for emergencies, to allow the motor to be reversed after it has opened partway, or to allow the valve to be left in the partially open position, because the switches that are normally provided are designed to stop the motor when it has reached the end of the stem travel. As an option, on the junction box or on the motor case a set of meters can be installed, for current and voltage, displaying the electrical conditions of the actuator. Generally, the elec-

(a)

(b)

**Fig. 3-47.** (a) Motor actuator for quarter-turn valve, showing controls and indicator lights. (*Courtesy of EIM Controls, Inc.*) (b) Other side of motor actuator, showing mounting arrangement, position indicator, manual override latch, and conduit connections on wiring box. (*Courtesy of EIM Controls, Inc.*)

trical box is rectangular in shape and the motor is cylindrical, often with cooling fins in larger sizes, but almost invariably the electrical compartment is bigger than the motor.

The motor itself is designed to operate in both directions, which is known in electrical terms as a reversible motor. Since motors operate much more efficiently at higher speeds, reduction gearing is almost always required between the motor and the valve. The power output of a motor begins at zero as the motor begins to turn and peaks at the motor's maximum speed, while almost all valves require the greatest power input at the beginning and end of the opening or closing cycle and quite a bit less power in the middle of the stroke. This requires some device that allows the motor to come up to full speed before the load is fully imposed. For motor-operated gate valves, a hammer blow yoke nut assembly serves this function very well. When very large, very high speed, or very high temperature service valves are actuated, the yoke nut assembly may have Belleville springs to allow it to float up and down slightly.

Because the motor provides torque to operate the valve, regardless of the valve type, it is easy to provide a manual override. Manually operating a motor valve requires the physical presence of a human operator near the valve, because the actuator will have a clutch to disengage the motor and engage the handwheel. Standard handwheels appropriate to the torque required to operate the valve are provided, mounted somewhere adjacent to the motor in an easily accessible positon. The clutch is located right next to the end of the motor shaft, so that in manual operation mode the gearing is still in use. Small motors may not have any gearing on the manual handwheel, and it is usually located directly atop the actuator and driving the yoke nut assembly directly (Fig. 3-48). The clutch also keeps the handwheel from revolving when the motor is driving the valve, and keeps the handwheel from being able to operate the valve when the motor is running. Most actuators are designed to disengage the handwheel when electrical power is applied to the motor.

The current required to operate the motor varies with the size of the valve and the motor installed. The supply voltage is also a factor. Many smaller actuators are designed to operate on 110- to 120-V, 60 Hz, single-phase power just like standard house current. The power consumption might be from about 3 to 30 A. Larger actuator motors are normally designed for 240 to 480 V, three-phase 60-Hz and 50-Hz power or 125 V or 250 VDC. Heavier current draws might range up to 300 A. The largest actuators produce up to 240,000 inch-pounds (30,000 Nm) of torque, or translated into vertical force for gate valve operation, up to 500,000 lb (225,000 kgf) of thrust. Current ratings are what is known as locked-rotor current, meaning the maximum current the motor will draw if it is loaded down so heavily that it stalls.

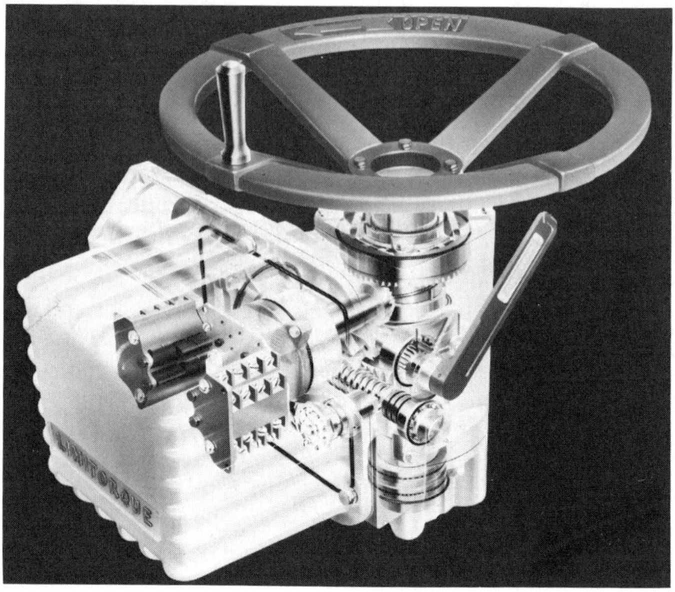


**Fig. 3-48.** Motor actuator and controls for rising-stem valve with handwheel override. (*Courtesy of Limitorque Corp.*)

Some discussion about electrical characteristics is in order. Motors produce heat, which limits the amount of time the motor can be operating during a given time span by the rate at which the heat can be dissipated. This is known as the duty cycle. A motor with a 100 percent or continuous duty cycle could operate continuously at full load torque without overheating, while a 50 percent duty cycle means that the motor can be in continuous operation at full load torque without overheating only 30 minutes of every hour of elapsed time. Overheating is determined by the electrical insulation in the motor, but common motors are often designed for a 144°F (80°C) temperature rise. Another electrical characteristic is the frequency, which is almost universally 60 Hz in the United States. An induction motor, which is the typical power motor for this kind of use, will operate only 83 percent as fast if powered by a 50-Hz current of the same voltage, but will develop just as much torque. The duty cycle will decrease by about 17 percent (Fig. 3-49).

The means of mounting the motor to the valve depends on the type of valve being actuated. For a rising-stem gate valve, the actuator normally has the same bolt pattern on the bottom that a gear or any other type of actuator has, for mounting onto a standard pad on top of the yoke (Fig. 3-50a,b). Many valves now have a standard yoke nut also, with a pair of lugs designed to be the driven half of the hammer blow operation feature. The actuator gear case is generally cast iron or alumi-

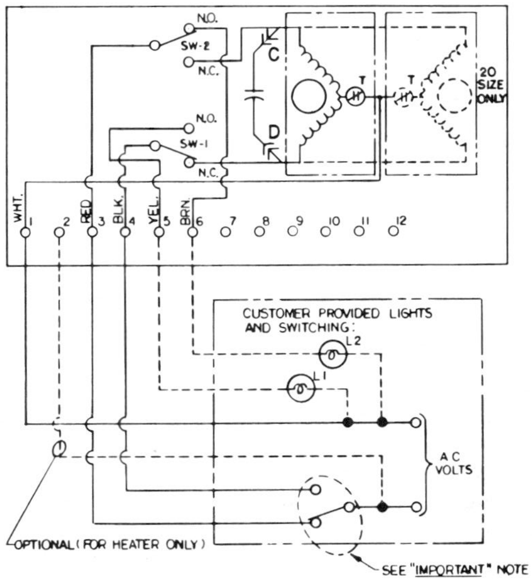


**Sizes 10-30**
**AC Power**

**IMPORTANT!**
EACH ACTUATOR SHOULD BE ELECTRICALLY POWERED THRU ITS OWN INDIVIDUAL SINGLE POLE SWITCH CONTACTS TO ISOLATE THE UNUSED WINDING.

**NOTE:**

1. ACTUATOR SHOWN IN COUNTERCLOCKWISE EXTREME OF TRAVEL, OR "OPEN" POSITION.

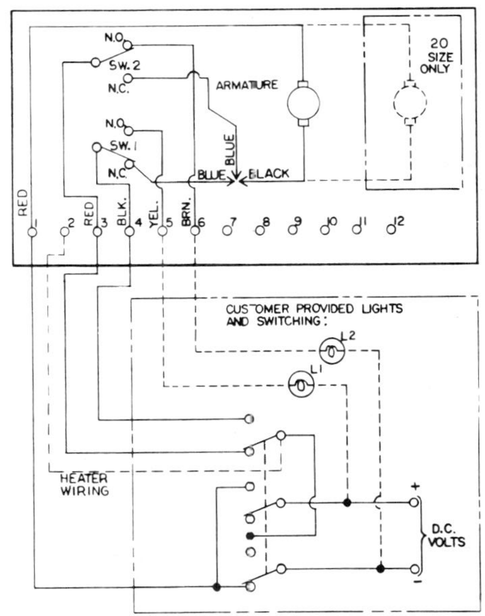


**Sizes 10-20**
**DC Power**

**Fig. 3-49.** Typical motor actuator wiring diagrams. (*Courtesy of Worcester Controls Corp.*)

(a)

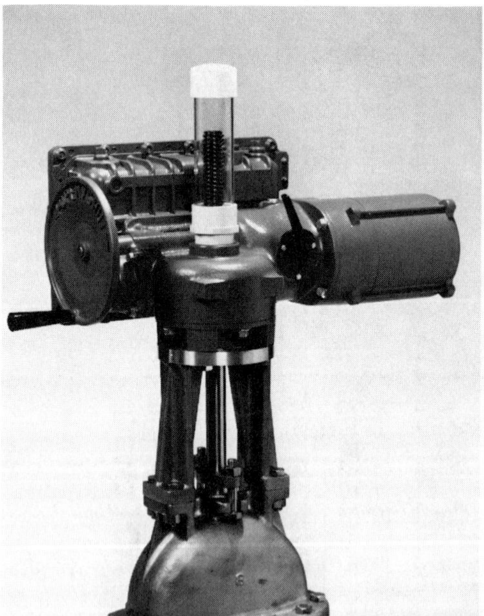

(b)

**FIG. 3-50.** Motor actuator for rising-stem gate valve, with local position indicator. (*Courtesy of EIM Controls, Inc.*)

num like a gear case, only somewhat sturdier to take the increased loads associated with the motor actuation. Several different combinations of intermediate gearing may exist, but in almost all cases there is a considerable reduction from the 900 to 3600 rpm of the motor to the 25 to 70 rpm of the yoke nut, corresponding to a stem travel speed of 5 to 12 in. per min (2 to 5 mm/sec). These are common speeds, and greater or lesser speeds can be obtained for special applications. Speeds at the lower end of this range apply to, for example, globe valves closing against high differential pressures.

The limit switches and other accessories installed on a motor-actuated valve are much like those on any other valve. One of the features of an electric motor actuator, unlike many air and other more mechanical actuators, is that the motor is often capable of exerting so much torque that the stem or the valve body can be severely damaged if the full stall torque of the motor is applied. Cases have been reported of the disc punching through the bottom of the valve or of the stem buckling from column overstress in the closing mode, and of stems pulling in two in the opening mode. Therefore the torque and limit switches are very valuable in the successful operation of a motor-operated valve.

The switch used to control the motor shutoff in the downward, or seating, direction on wedge gate and globe valves is generally a torque switch located inside the actuator. It senses the increase in torque due to the increased resistance to motion as the valve disc seats, or as it encounters some foreign object in its path. Sometimes this is done by detecting the lateral movement of the worm gear as it moves laterally against the torque spring pack. This torque switch can also be supplied for the upward, or opening, stroke, to detect the point at which the stem comes to rest against the backseat. It is much more common to have a limit switch on the upward stroke, to stop the motor just before the stem backseats. Limit switches can also be provided on the downward stroke, to stop the motor just as the valve comes to the fully seated position. Most limit switches are internal also, and they are geared so as to count the number of revolutions of the stem nut or the worm gear shaft, since that has a fixed relationship to the distance the stem has traveled up and down.

In addition, a position indicator can be provided on the motor or the electrical box (Fig. 3-51). The position indicator is an electromechanical device consisting of a potentiometer driven by the output shaft, which gives a signal that varies according to how far the shaft has turned from the zero position on the potentiometer. The output signal is a current, generally the standard 4 to 20 mA used for electrical analog signals.

The other type of motor actuator is the drop-in type (Fig. 3-52), very similar in concept to the aluminum air actuators. This prepackaged actuator uses a die-cast metal housing, with the complete set of motor,

**Fig. 3-51.** Position indicator for motor actuator. (*Courtesy of Limitorque Corp.*)

**Fig. 3-52.** Integral motor actuator and controls for quarter-turn valve with position indicator. (*Courtesy of Worcester Controls Corp.*)

gears, and controls mounted inside it. These actuators are, of necessity, for lighter duties than are the assembled motor actuators described above, but they fulfill their intended service quite adequately.

Like the larger actuators, these drop-in motor actuators still have position indicators, manual overrides with handwheel, and conduit connections for power and signals. Like the small prepackaged air actuators, they have a simple standard bolt pattern and a shaft end that couples to many types of valves. Also, like the air actuators, they are often manufactured by the companies that make the valves they are designed to fit on. But there are certain important differences. It is not possible to simply turn one upside down to reverse its action as it is with the small air actuators. It is also not necessary, since the motors are fully reversible, but it is also impractical to provide spring return.

There is an exception to the fully reversible design. Motors designed to operate small ball valves are occasionally built for unidirectional operation. In ball valves, this is feasible because the conventional ball and seat configuration lends itself to movement in either direction. When the valve opens, it is turning in the standard counterclockwise direction. When it closes, it continues turning counterclockwise instead of reversing direction. Thus one seat contacts first one side of the ball, then the opposite side. This design cannot be used for any ball valve design that is not completely bidirectional. Gearing and controls are much simplified with this design.

The motors are small open-frame designs, often off-the-shelf motors, and generally operate from either 110 or 120 V, single phase, or 220 to 240 V, single phase. They will not normally be designed to need three-phase power. However, these small actuators can also be designed to run (with the appropriate motor) from 12 or 24 VDC. This makes them suitable for powering with the same circuits as are used for instrumentation, like a pneumatic control system with air actuators. Alternating-current motors draw from 0.3 to about 3 A, while the corresponding DC motors will draw perhaps 1.5 to 6 A. Their torque output ranges from about 100 to 3000 inch-pounds (12 to 375 Nm).

The housings, which are often aluminum for light weight, are generally provided with corrosion-resistant baked-epoxy finishes. The motor shaft is often parallel to the actuator output shaft, with several reduction gears and a clutch for manual override. The starting capacitor and position-sensing devices are also located in the housing. A mechanical brake is also available, for applications like butterfly valves where the fluid flow through the valve will tend to turn the shaft. It is also possible, with electronic controls, to infinitely vary the speed of actuation from a few seconds up to several minutes. Another option commonly supplied on actuators for severe environments is a small heater, designed to keep the interior of the housing at no less than 80°F (25°C) or so, to maintain low humidity and avoid low temperatures that hamper the efficiency of some of the electronic components. Many other electronic options are available, including a timer circuit that can automatically open the valve, leave it open for a specified time, then close it and open it again at a specified time, without having input from a remote control system.

Most of these prepackaged actuators are designed for any rotary-shaft valve. The same actuator can as easily drive a three-way valve that rotates 180 degrees—it just runs twice as long. The only modifications needed are three-position switches with a center-off position, if the plug or ball needs to be capable of being oriented in a center position.

### *Air and Hydraulic Motor Actuators*

Although all the preceding discussion was about electric motors, it is also possible to equip valves with air motor or hydraulic motor actuators (Fig. 3-53). These designs are probably more a specialty now than a major force in the market, but the air motor has a fair bit of use in remote locations where electric power is not available or the hazards from sparking are considered great enough to prohibit electric operation. The development of electrical devices safe for explosive atmospheres has put a dent in those uses of the air motor.

An air motor is mounted in much the same way as an electric motor, with a gearbox and a means of allowing the motor to come to speed

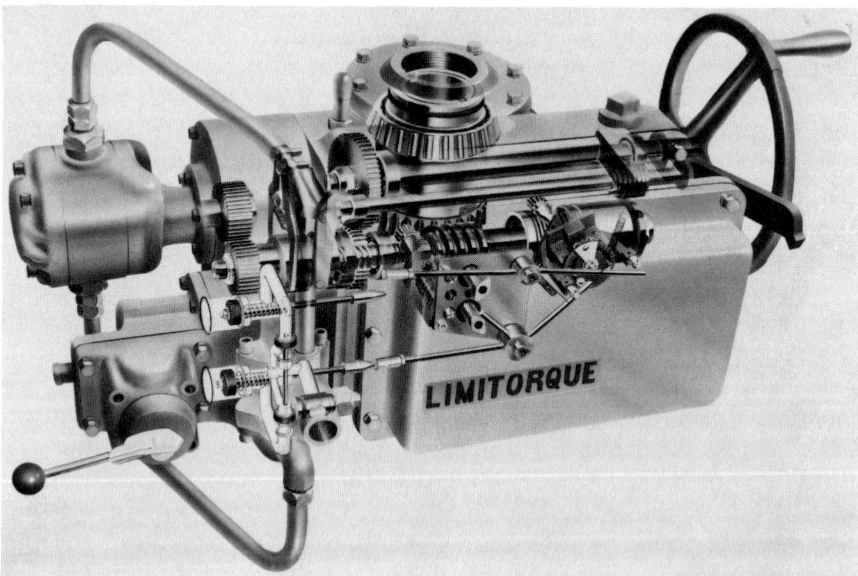


**FIG. 3-53.** See-through view of air motor actuator, showing air motor on left. (*Courtesy of Limitorque Corp.*)

before loading it with the valve unseating force. Because it will consume large amounts of air, ranging from 20 to 600 standard cubic feet per minute (SCFM) (33 to 1000 $m^3$/hr), a dedicated air supply is usually required, and the exhaust must be directed somewhere safely. An air accumulator can also be used to store the volume of air required to cycle the valve as a fail-safe feature. The air motors are generally of the vane type, meaning they spin like a small turbine as air passes through them. For proper lubrication, an air oiler is normally installed to convey small amounts of lubricating oil along with the air supply. Air actuators are provided with mechanical limit switches that operate like their equivalents on an electric motor, including the torque seating feature. Pneumatic position signal capability is available. These units can also be built to operate on the high-pressure gas being transported in the lines they control, as is commonly done for cross-country pipeline work (Fig. 3-54).

Hydraulic motors can be used also, with hydraulic oil supply and return lines. They are more commonly used on small and specialty valves, in areas where hydraulic power is being used extensively for other applications.

## Manual Operators

### *Handwheel Operators*

In the simplest types of actuators, the actuating force is provided by a human being. This generally means that the mechanical parts are simpler than those in other powered actuators, since a human being can adjust the magnitude and direction of applied forces and can supply additional power (a bigger person or a longer wrench) as it is needed. It is, however, often possible for the force required to exceed that which the human being is capable of producing. Therefore, although human-powered actuators are very versatile, they do have their limits.

There is no question that human-powered actuators are occupying a smaller share of the market than they occupied at one time. The entire actuator system, including the human operator, is definitely more expensive in the long run than a purely mechanical or mechanical-electronic system, and is not capable of responding nearly as fast as, for example, an electronic control system. Yet the human operator is capable of receiving input from sources that purely automated systems cannot adequately deal with.

The above discussion summarizes, in a tongue-in-cheek fashion, the trend toward more automation and fewer people in process plants. The entire breadth of this trend is bigger than this book, but there is no

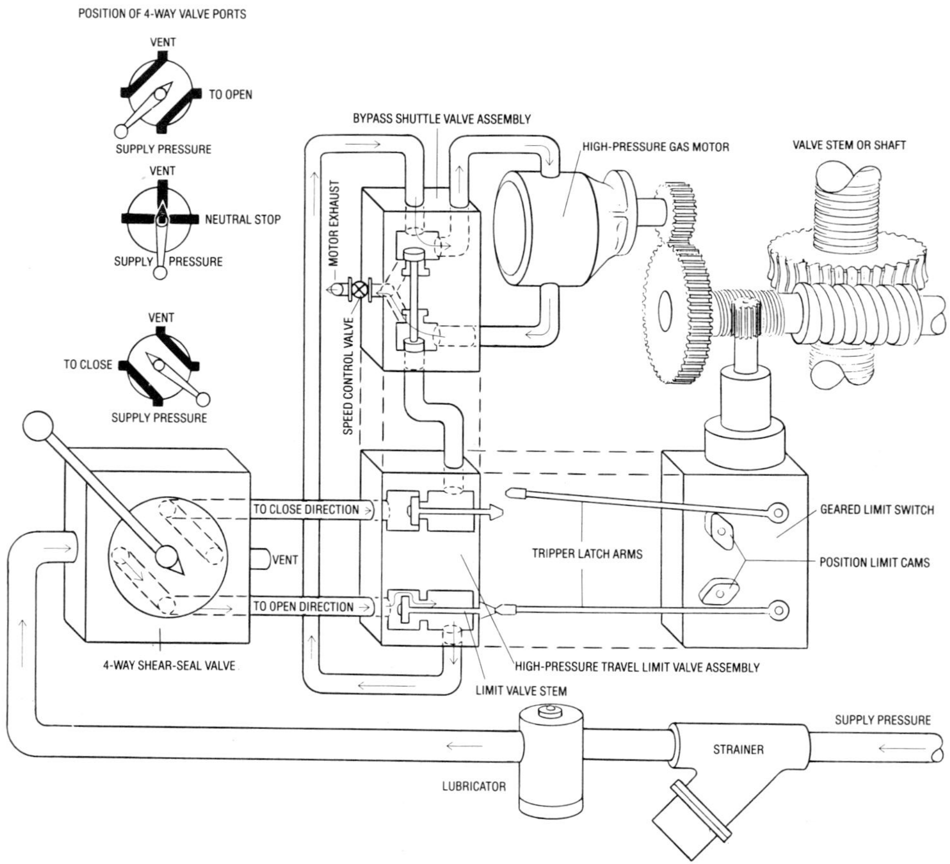


**Fig. 3-54.** Diagram of gas flow and operation of air motor actuator. (*Courtesy of Limitorque Corp.*)

question that it has affected and will continue to affect the details of valving and valve actuators.

What has been meant by an actuator in this chapter has been any device that causes the moving parts inside a valve to move. This is the definition commonly employed among engineers concerned primarily with "control valves," or valves that operate in response to a signal. Among engineers concerned primarily with "manual valves," or valves that are operated by a person's hands, the terminology is a little different. There, an *actuator* is any device that operates the valve is response to a signal, while an *operator* is a device that moves the valve internals by translating the forces applied by a person to the operating device.

The simplest means of operating a valve is by grasping the valve stem and pushing or pulling. This method is still found today in a few wier- or ditch-type valves, but is otherwise unfeasible because of the forces generally required. Many hundreds of years ago, someone applied the principle of the screw to a device with a fixed, but rotatable, operating nut surrounding a shaft threaded with short, wide, sturdy threads. This device was probably first used on wine and olive presses, but it also provided the power behind the first printing press and eventually found its way into valves.

This is the source of the term used to describe valves where the stem moves the disc or seat up and down. These valves are generically referred to as "multiturn," because the operating device makes multiple turns around the valve stem in order to open and close the valve. Notice that it does not make any difference whether the stem moves up and down or not, or whether the stem rotates or not as it moves the valve open or closed. This term is in contrast to "quarter-turn," where the valve stem rotates the seating device 90 degrees from the open to the closed position and the valve stem itself also moves 90 degrees.

Most valves with threaded stems use the Acme thread, and have for many years. A few valves designed for close regulation of flow or light duties have V-notch threads. The Acme threaded stem normally is double threaded, where there are two threads side by side, and each revolution of the operating nut moves the stem up or down by a distance of two thread pitches. This is really relevant only to the valve designer, because all it does is influence the speed of valve opening and the force required to turn the operating nut. The threads are normally cut into the stem; that is, the portion of the stem that travels through the packing gland and into the valve is the same diameter as the *outside* diameter of the threads in the threaded portion of the stem.

On valves that have nonrotating stems, such as gate valves, there is an operating nut, or yoke nut, that bears against the threads cut into the stem and is normally made of a softer material. In cases where the stem is chrome steel or a stainless steel, the yoke nut is normally a brass or bronze. Valves built to API standards are required to have the yoke nut made of a material with specified minimum melting point, to keep the nut intact when the valve is moderately exposed to fire. The yoke nut must be retained on top and bottom to keep it from moving up and down instead of the stem. This is generally accomplished by a portion of the yoke nut being larger than the main body, with one surface bearing against the yoke itself and the other side bearing against a retainer that is attached to the yoke. Current API standards require that this retainer be fixed by a pin or a tack weld to prevent it from accidentally coming loose, which would allow the stem to come out to its full travel if line pressure were to push against it unexpectedly. The operating device, such as a handwheel, is then attached to the operating nut.

On valves where the stem is designed to rotate, such as globe valves, there is typically a nonrotating component that looks like an operating nut, but acts simply as a bearing or as a softer inlay into the yoke. This part is normally threaded into the yoke or otherwise solidly affixed. Only very cheap valves do away with this part altogether and just thread the top of the yoke to accept the stem. Regardless, there is a short section of the stem that projects above the top of the yoke and it is here that the operating device is attached. Normally, the stem is machined to a square or square-tapered shape, and a corresponding shape is cast into the handwheel. The end of the stem is then tapped for a short screw, which holds a light washer over the top of the handwheel, effectively securing the handwheel to the stem. When the valve is opened, the stem, handwheel, and all rotate and rise as one.

This difference between the handwheel attachments of gate and globe valves is one of the easiest ways of telling gate valves and globe valves apart in situations where the bodies are similar in shape and size.

The operating device used on most multiturn valves is the handwheel. Most handwheels have knobs, ridges, or just the spokes to aid in grasping the wheel with gloved hands or in oily or slippery conditions. Occasionally there is a vertical member to grasp that projects up from the handwheel rim. Sometimes the spokes project out from the rim like the spokes on a ship's wheel. The handwheel itself, on standard valves, is often cast malleable iron, sometimes with elaborate spokes. Larger handwheels are often welded steel, which is considerably stronger and cheaper. Some handwheels are pressed steel sheet. On smaller and lighter duty valves, the wheel is sometimes die-cast soft metal such as zinc or aluminum. Needle and instrument valves often have a metal or plastic bar-type handle instead of a handwheel. Malleable iron and steel handwheels are usually painted, sometimes with a high-build epoxy paint for extra durability, and are often a distinctive color. Occasionally, on special alloy valves, it is possible to get stainless steel or other corrosion-resistant handwheels to resist atmospheric corrosion or leaks.

Some industrial valves are equipped with quick-opening operators, although these are not common anymore. One place where this device

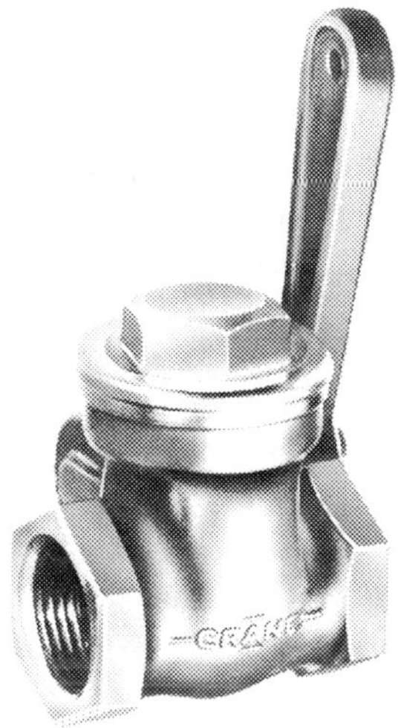

**Fig. 3-55.** Quick-opening lever on gate valve. (*Courtesy of Crane Co.*)

was once popular was on the inlet valve for a steam whistle. This was a globe-type body with a bracket that held a lever. The disc was spring-loaded into the closed position. When the free end of the lever was pulled, the other end would push on the end of the stem and hold the valve open as long as tension was maintained on the end of the lever. Releasing the lever allowed the spring to push the disc shut.

There are, or were, two common types of quick-opening gate valves also (Fig. 3-55). One was basically a cam, with a specially designed gate that rotated up instead of lifting straight up. This required a special valve body. The other type had a standard gate and stem, with a bracket cast into the outside of the valve body. A lever attached to this bracket held another lever that, when the end was pushed up, would pull the valve stem up and open the valve. They were intended for steam and air lines, since the fast shutoff could produce water hammer in liquid services.

### *Lever and Wrench Operators*

Quarter-turn valves are operated by an entirely different and much simpler method than are rising-stem valves. Since it is not physically possible for the human hand to grasp a bare quarter-turn valve stem and rotate it, various methods have been devised to get more leverage.

Simple quarter-turn valves, such as gas cocks, may have the handle cast as one piece with the stem. These valves often have a handle that resembles the key used to wind a clock, where half of the handle is on each side. Others have a lever-type handle, where the entire length of the handle is to one side of the stem centerline. Sometimes, the handle is detachable.

On larger and more sophisticated quarter-turn valves, the predominant operator is the wrench or handle. The difference between a wrench and a handle is that the wrench is not solidly attached to the valve and is often moved from valve to valve, while the handle is attached to the valve stem with a nut or other securing device. Wrenches are more common on plug valves and special ball valves, while handles are more common on most ball valves and nearly all butterfly valves. Wrenches are also more likely to be preferred on valves that are infrequently operated and/or located in remote areas but are exposed to the general public, such as on pipelines.

Most handles and wrenches are cast malleable or ductile iron, or occasionally fabricated steel. On small ball valves the handles are pressed from steel or stainless steel plate. There will be a soft plastic coating on the end of the lever to make it more comfortable to grasp with the bare hand. Sometimes the levers are cast or fabricated in distinctive shapes as part of the manufacturer's effort to make its brand unique and more visible than the competition. Some in fact, are difficult to categorize as either handwheels or handles by their shape. The size of the handle should be in proportion to the torque required to operate the valve. The stem end has some means of attachment to the valve stem, either a deeper section in the case of wrenches or some means of attachment, including a heavier reinforced section, for a handle.

Small ball valves normally have their handles attached with a nut that threads onto the top of the stem. Even relatively large ball valves occasionally have the top of the stem threaded to accept a castellated nut or other retaining device. The stem normally has two flats machined to line up with the open position (which is perpendicular to the stem-ball attachment slot). The handle thus can be installed only so that it lines up with the open position of the ball, making it possible to tell the valve position by looking at the handle.

Most handles are fixed with respect to the valve stem, but a few have the handle portion made from round steel bar that slides through an operating nut much like the handle on a vise (see Fig. 1-21). This design has the advantage that the handle can be positioned midway through so that you can simultaneously push and pull on it, or it can be adjusted to avoid an obstruction on one side without taking the handle completely off and reattaching it.

Valve wrenches have an opening that matches the top of the stem. The stem sometimes has just two flats machined into the side, sometimes four flats for a square section, but almost always there is some mark on the stem to indicate where the open and closed positions are. Generally, the wrench can be attached in only two ways, so that it lines up with the open position on either one side or the other. Some wrenches have a four-sided or eight-sided pattern so they can be put on to line up with the closed position, or at a 45-degree angle with the open or closed position. This is an advantage where space restricts the swing of the valve wrench, but here a stem mark to show the open position is mandatory.

Sometimes the "wrench" is actually just round bar, and there is a cast wrenching device that is actually attached to the stem like a handle (see Figs. 2-42 and 2-43). The wrenching device is not large enough to operate the valve without inserting a bar through it, although it may be equipped with an extra hammer arm to allow wrenching the valve around with the bar and simultaneously providing hammer force. Usually several bar slots are provided, since access around these valves may be limited, as on offshore platforms.

Butterfly valves also normally use a lever, but since the butterfly is often used for throttling, a different lever design is often used. The lever has a latching or locking device pivoted below the lever itself (see Fig. 2-144), which is spring-loaded so that the outboard end must be grasped and lifted up into the lever, which pulls the opposite end of the latch down, disengaging a locking device that looks like a segment of a gear. Each tooth of the locking device provides a spot for the lever to be latched, making it possible for the disc to be positioned and locked into one of about nine or ten positions. The first one of these notches is at the fully open position, and the last one is at the fully closed position. This device also aids in ensuring that the valve is fully closed, since a fair amount of force is required to seat any butterfly valve, and especially a rubber-lined valve. The operator knows that the valve is fully seated only when the locking mechanism drops into the last notch on the locking plate. This whole assembly is generically known as a lever-lock handle, and it is very rarely needed for anything other than a butterfly valve.

Quarter-turn valves do not get very large before they require more torque, which means more force applied to the lever by the human operator than the average human can supply. These valves normally must be fitted with gear operators. It is possible to avoid gear operators by using longer wrenches, but eventually the wrench size gets impossibly large. The largest valve wrenches may be 3 ft (1 m) or a little longer, but at this size the wrench is even hard to pick up, let alone operate a valve with. In terms of valve size, the largest feasible lever-operated valves are about 4-in size (DN 100) in a sleeved plug, about 6-in. size (DN 150) in a lubricated plug, 6-in. size (DN 150) in a soft-seated ball valve, and 8-in. size (DN 200) in a butterfly valve. In higher pressure ratings, even smaller valves require gears. Some high-pressure metal-seated ball valves may require a gear operator in size as small as 2 in. (DN 50).

In smaller sizes where the torques are not high, many quarter-turn valves can be fitted with round or oval handles that look like handwheels (Fig. 3-56). These handles have the same stem attachment, with a sym-

**Fig. 3-56.** Oval handwheels reduce the risk of snagging levers on ball valves. (*Courtesy of Worcester Controls Corp.*)

metrical piece of steel bar bent upward to attach to a rim section. The portion of the bar that attaches to the valve stem still provides the same visual indication of the valve's position. Although it is somewhat harder to apply a large force to an oval handle because it does not have as great a moment arm as an equivalent lever handle, in the last 5 to 10 years there has been a great demand for these handles because they are much safer than a lever handle. It is much more difficult to open or close a valve accidentally with an oval handle than with a lever. Not only are the oval handles safe from being snagged by clothing or elbows of passing workers, but they are symmetrical and not prone to drooping or vibrating around. Any valve near machinery is potentially vulnerable to vibrations, and any valve mounted with the stem in a horizontal position is potentially vulnerable to the force of gravity pulling the lever down if the valve is not tightly seated into position.

There are limits to the sizes of valves that can be fitted with oval handwheels. Generally the largest size of ball valve that can be equipped is 2 in. (DN 50) regular port or 1½ in. (DN 40) full port, and plug valves can be equipped only in ½- and ¾-in. sizes (DN 12 and DN 18). Most oval handles are about the same size, and most are made of pressed steel or welded wire with plastic coating. Most are available in a standard height, roughly the same height as a lever handle and just high enough to get an operator's knuckles comfortably under the rim of the handle.

Some are also available in a taller model that allows room for insulation to be applied completely over the valve body and still leave room to grasp the handle. Valves located out of reach overhead can be operated by chain levers arranged so that a pull on one side opens the valve and a pull on the other side closes it (Fig. 3-57).

There are many accessories that can be provided with handle operators, other than the lever-lock described above. Many of these accessory devices can be used on almost any type of quarter-turn valve. Some means for preventing accidental or deliberate opening can be provided by either a latching mechanism or a locking device (Fig. 3-58). The latching handle has a clip that fits around the inclined portion of the handle, just back from the stem attachment area. The bottom of this clip

**FIG. 3-57.** Chain handle for ball valve, with position indicator, for operation of overhead valves from grade. *(Courtesy of Swagelok Co.)*

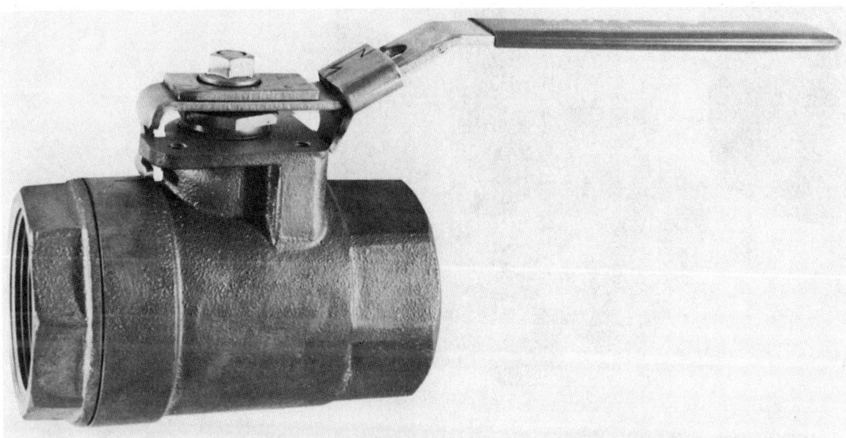

**FIG. 3-58.** Ball valve lever showing latching mechanism that requires positive action (lifting the latch) prior to opening. Also capable of accepting a locking device. *(Courtesy of Watts Regulator Company.)*

engages a boss or finger on the top of the valve body. Moving the handle requires the operator to grasp this clip and slide it back a ways toward the end of the handle to clear the boss. These latches are often used to prevent vibrations from jarring the valve out of its last position, or to make it more difficult for a person moving past the valve to brush up against the handle and open or close it accidentally. It can also be equipped with a locking configuration in which an additional piece of steel sheet with a hole in it lines up with a hole on the handle. A padlock can be slipped through these holes to prevent operation by anyone not equipped with a key.

There are other ways of providing locking capability for handle operators. One is simply to provide an arm attached to the valve body, with a hole, and another arm as part of or attached to the handle (Fig. 3-59). Again, a padlock is simply slipped through the two holes when they line up. When locking capability for both open and closed positions is required, then two locking holes must be provided on either the handle or the valve body.

When a valve needs to be operable but it is very desirable to keep it in one position, a spring-return handle can be furnished (Fig. 3-60). In a sense, these are like spring-return devices on actuators. They mount on the valve in the same way that an actuator would, and consist of a housing containing a shaft and a spring that holds the shaft in one position. The bottom of the shaft couples to the valve stem, and the top of the shaft on top of the housing has a standard valve lever handle

**Fig. 3-59.** Locking device with padlock. (*Courtesy of Worcester Controls Corp.*)

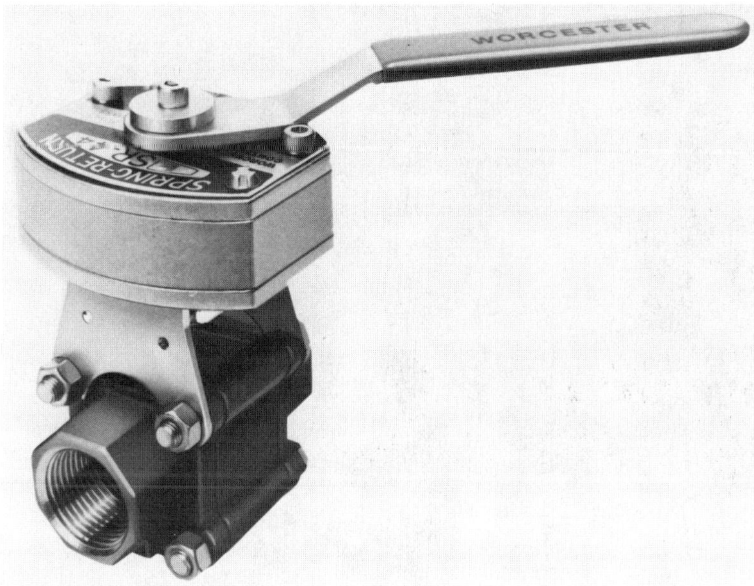


**Fig. 3-60.** Spring return handle for manual ball valve. (*Courtesy of Worcester Controls Corp.*)

mounted on it. An operator can open the valve with not much more force than is required to open a standard valve, but when hand force on the handle is released, the spring closes the valve again. This assembly is used often on sampling or dispensing valves, and in places where leaving a valve open could be a safety hazard. They can be considered as fail-safe devices in that sense. Like all mechanical actuating devices, they must be sized to deliver at least as much torque as is required to close the valve reliably and fully.

A further improvement to the spring-return actuator is a version that has fusible links retaining the spring (Fig. 3-61). The operator is completely usable in the manual mode, but in the event of a fire, the fusible links melt, allowing the spring to close the valve if it was not already closed. This device could be added to air actuators also.

Valves that are completely insulated can be fitted with different types of extensions to put the handles comfortably above the outside of the insulation. One is a stem extension, which is just a shaft that couples to the valve stem and has a handle connection on its top end. Some more sophisticated versions of this device are bolted to the body like an actuator bracket, and include an extra bearing to make the handle stiffer and avoid imposing side loads on the top of the stem and the bearing and packing areas (Fig. 3-62). These extensions also make it easier to insulate around the valve without getting in the way of the handle travel.

**Fig. 3-61.** Fusible link spring return actuator that overrides manual gear operation. (*Courtesy of Fisher Controls Intl.*)

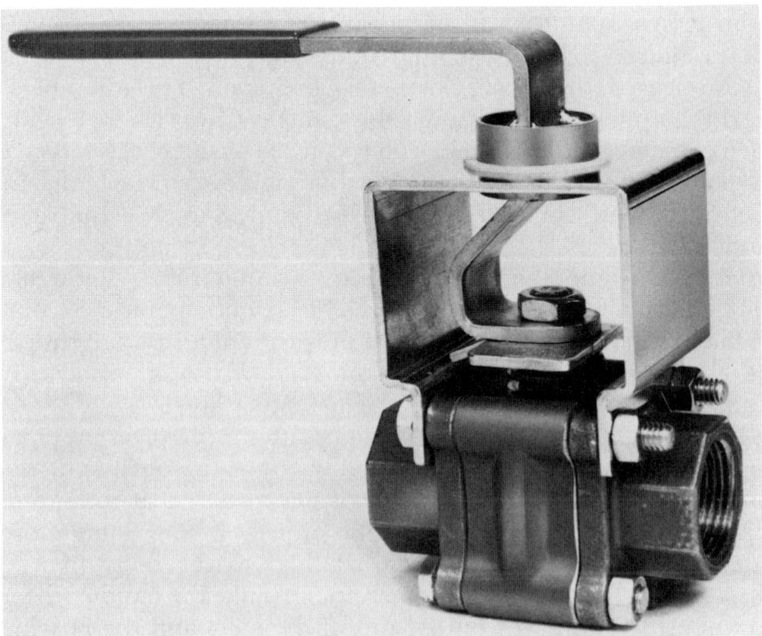

**Fig. 3-62.** Extension stem with outboard bearing and insulation protector for hot insulated lines. (*Courtesy of Worcester Controls Corp.*)

The same limit switches that are available with actuated valves can be mounted on valves equipped with handles. Generally, this is needed only for interlocking or when a remote indication of the valve's position is required. The limit switches usually mount above the handle, in the same place they would be on an actuated valve, and they really detect the position of the handle, not the valve internals. Therefore, it is even more important for these valves than for standard valves that it be impossible to put the valve together with the handle out of phase. Since the limit switches require power and send signals, a conduit connection must be provided.

For valves that are used in services where leakage must be kept to an absolute minimum, it is possible to build a bellows seal lever operator (Fig. 3-63). Examples of these services include not only toxic or hazardous services, but also high-vacuum services, high-purity gases, and expensive fluids. The lever works like any other quarter-turn lever, in the sense that you grasp it and turn it 90 degrees, but it is not built the same way. The heart of the lever consists of a stem section in one continuous piece, bent in a 90-degree arc. The bottom of the stem has a ball tongue or other configuration required to interface with the internals of the particular valve being used. Outside the stem is a bellows

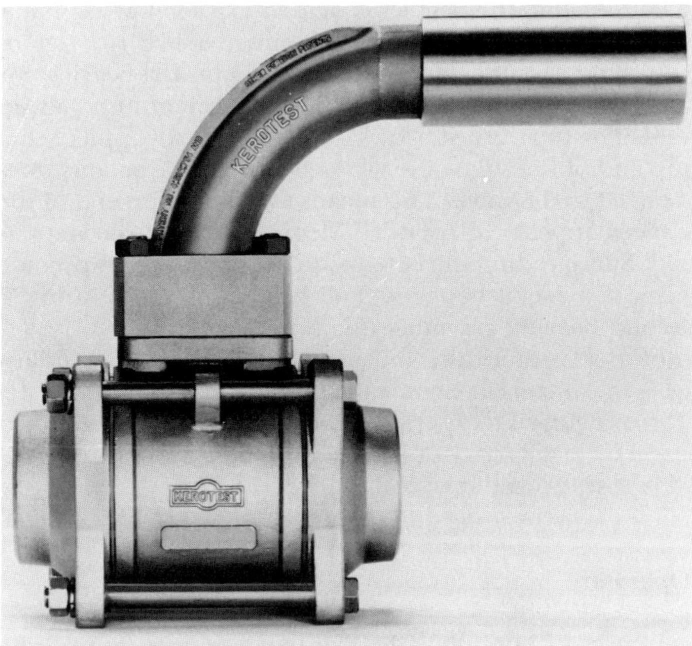


**FIG. 3-63.** Bellows seal lever operator for quarter-turn valves, shown on butt-weld-end ball valve. (*Courtesy of Kerotest Manufacturing Corp.*)

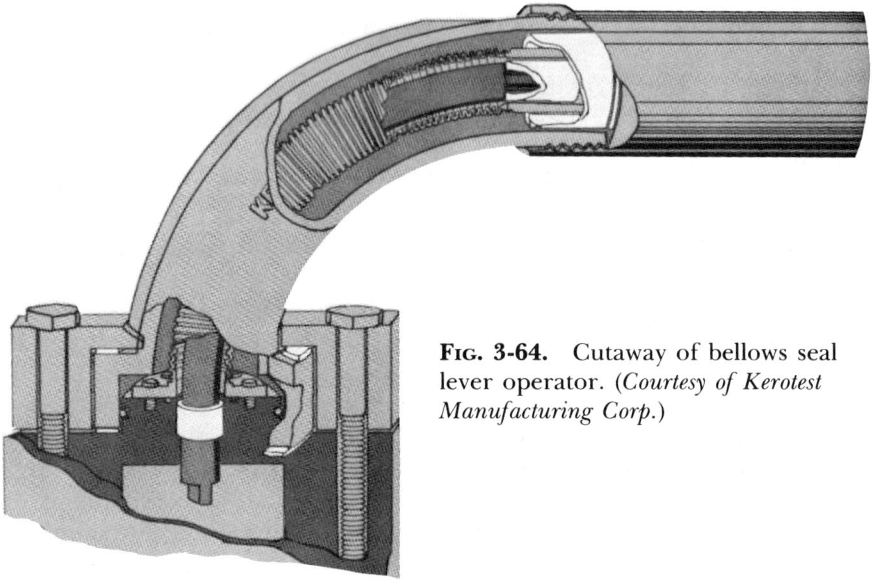

**Fig. 3-64.** Cutaway of bellows seal lever operator. (*Courtesy of Kerotest Manufacturing Corp.*)

consisting of a convoluted thin sheet of stainless steel or high-alloy material. At the bottom end, the bellows is welded to a solid piece that is bolted to the valve bonnet. Under this attachment piece is a set of O-rings or other sealing arrangement, and in the center is a bearing for the stem. There are other guides for the bellows spaced at intervals up the arc of the stem. The top of the bellows is welded to a cover piece that surrounds the top end of the stem, also with a bearing. The bellows does not rotate with respect to the valve. This means that at the top end of the stem, the cover piece appears to rotate 90 degrees around the stem as the stem is turned. Surrounding the bellows, to protect it and to provide a surface to grasp, is a metal housing. The bottom of the housing is retained by a second bearing assembly (Fig. 3-64).

This lever is, of course, quite a bit more expensive than any standard lever, and its mechanical complexity means that it contributes a fair bit by itself to the force required to operate a valve that it is installed on. It currently is not widely available. But it does provide a bit of additional leak resistance for cases where it is really needed.

### *Manual Gear Operators*

The standard practice on any manual valve that requires more force than one person can conveniently apply is to increase the available mechanical advantage by using a gear. Here, as elsewhere, the approaches taken for multiturn valves and quarter-turn valves are different. Gearboxes, or gear operators as they are commonly known, are built with bevel or spur gearing for multiturn valves since they are required to drive the operating nut through multiple revolutions, often dozens of revolutions to open a large valve. Gear operators for quarter-turn valves are the rack-and-pinion or worm gear type since the driven member rotates only 90 degrees along with the valve stem.

Gears have been used on valves for many years. Originally, when almost all industrial processing was carried out inside buildings, the gears and extension stems and other devices required to make valves operable from the floor were considered as general mechanical details and were not too heavily engineered. Because of these indoor locations and because of the lower degree of worker protection that was once considered adequate, most early gears were just held in brackets or bearings in the open air. Gearing of valves was often done as much to make the valves accessible as to reduce the force required to operate them.

By the 1940s, and probably earlier, standard gear combinations became available. These gear operators were designed to fit onto valve yokes in specified directions and were sized for specific models and ratings of valves. Some were even capable of being mounted in any one of several standard orientations and were designed for easy field mounting onto existing valves. These gear operators were offered by valve manufacturers to fit their valves. They were still generally open-gear assemblies on cast mounting brackets. Many combinations existed, in both spur and bevel gear designs.

Later, weatherproof or enclosed gearboxes became available and have in recent years become the standard design in domestic production. The enclosed gearbox serves to protect the gearing not only from the weather but from airborne corrodents and mechanical interference, and it also keeps operating personnel from coming into contact with the gears. Sometimes the gearboxes are built by valve manufacturers themselves, but often they are outsourced like other specialized work.

The typical construction of an enclosed gear operator (see Figs. 1-35 and 2-1) is a cast-steel or cast-iron housing in two pieces, with bearings on the bottom and top for the drive sleeve and a pair of ball bearing races at the side for the handwheel shaft. There is a bolted cover that retains the handwheel shaft. The bottom of the gearbox normally has a standard bolting pattern to fit a standard drilling on the valve yoke. Since gearboxes mounted by other than the original manufacturer generally have to be adapted to a non-gear-compatible yoke casting, an adapter bracket may have to be welded or bolted to the yoke before the gearbox can be bolted on. The bolt pattern has four, sometimes eight bolts, so that the gear can be positioned in any one of four orientations.

At the top of the gearbox is an opening with female pipe threads. The threads are for attaching a stem protector, which consists of a section of pipe threaded onto the top of the gear and a cap threaded onto the end of the pipe. The stem rises up through the inside of the protector, keeping it safe from damage and keeping bystanders safe from the stem's movement. However, the protector also keeps the stem from being observed to determine the valve's position. This problem can be cured by drilling and tapping a small hole in the end of the stem (there is normally a center hole left over from the stem machining anyway) and installing a rod that can rise through another hole drilled in the end of the stem protector. It is a good idea to install a plastic end cap on the end of this rod, for visibility and to prevent impaling anything with the telltale rod.

The gears are generally hardened steel, with a gear ratio of between 3:1 and 7:1. For multiturn rising-stem valves, the almost universal standard now is a bevel gear, with the driven gear facing teeth upward and bolted directly to the drive sleeve that turns the operating nut. Generally this drive sleeve surrounds the operating nut, with a hammer blow design of two lugs that interlock with the operating nut. These lugs are sections of a circle, generally about 45 degrees of arc in width. The yoke nut has two lugs that project up, generally about 90 degrees of arc in width, so that the drive sleeve moves about 45 degrees in either direction (one-quarter to one-half a turn on the handwheel) before it contacts the operating nut. The stem rises through the center of the gear. The driving pinion is mounted on a shaft, with the handwheel mounted on the other end of the shaft outside the gearbox. Handwheels on gear operators vary widely in size, with some being quite small since the operating torque no longer is as high as the corresponding nongeared valve would require, while in others the wheel is as big as the original handwheel. Sometimes, as in retrofits, it is the same handwheel.

In place of manual handwheels, it is possible to provide a special handwheel that has the chain teeth and retaining rim for remote chain operation. For ball and butterfly valves, this configuration is the best solution for the operation of valves located beyond the operator's reach. It is even reasonable to install gears on valves that might require only a lever if they were accessible because the gear makes it possible to chain-operate the valve. Regardless of the gearing, it is inherently more difficult to operate a valve with a chain because it is much more awkward to grasp a hanging chain than a solid handwheel, and so some allowance must be made in sizing gears with this consideration. It is also possible to provide a bolted-on chainwheel (Fig. 3-65), for gears or directly onto handwheels, that uses small U-bolts to clamp on to the handwheel spokes. These chainwheels are available with integral hammer blow devices, to compensate for the slightly more difficult operation.

For large valves where the gear is accessible but because of its size takes many minutes of work for an operator to open and close, the handwheel can be furnished with a standard 2-in. square operating nut as part of the hub (Fig. 3-66). An air or electric wrench can then be used to spin the handwheel through its travel. This is a time-saving measure, but the seating and backseating of the valve should be done manually

**Fig. 3-65.** Chain wheel with hammer device, designed to clamp onto existing handwheel. Chain wheel allows operation from grade. (*Courtesy of Roto Hammer Corp.*)

because the air wrench can overtorque the valve and cause damage from overtightening, either to the valve or to the gear or both.

If the output torque required is too great for a standard gear ratio, a second gear operator can be piggybacked on the first. This is known as a compound or double-reduction gear (see Fig. 2-149). The second gear can be another bevel or a spur gear type. The use of a second bevel gear also allows the handwheel to be located in the horizontal just as if there were no gear at all, or to one side. A spur gear operator simply moves the handwheel out and upward a little from the position it would have occupied on the original gear operator. These double-reduction gears are sometimes specified not because of the torque requirements, but simply to allow a change in the handwheel orientation to suit the conditions around a particular valve. In these cases, the second gear may have a 1:1 gear ratio to only change direction and not torque or operating time.

The considerations required for determining if a valve needs a gear are a little different for a rising-stem valve than they are for a quarter-turn valve. On the quarter-turn valve, the range of sizes for which there is a question about whether a gear is required is very narrow. In other words, you usually either need a gear to operate the valve or you do not. For a rising-stem valve, since the distance that the input force (at the handwheel rim) has to move is dependent on the number of threads, or the number of turns required to open, and since that force is exerted over a long distance, the instantaneous force is not that big (Fig. 3-67, page 363). An operator will get very tired opening a large gate valve without a gear, though. A common rule of thumb for the use of gears is

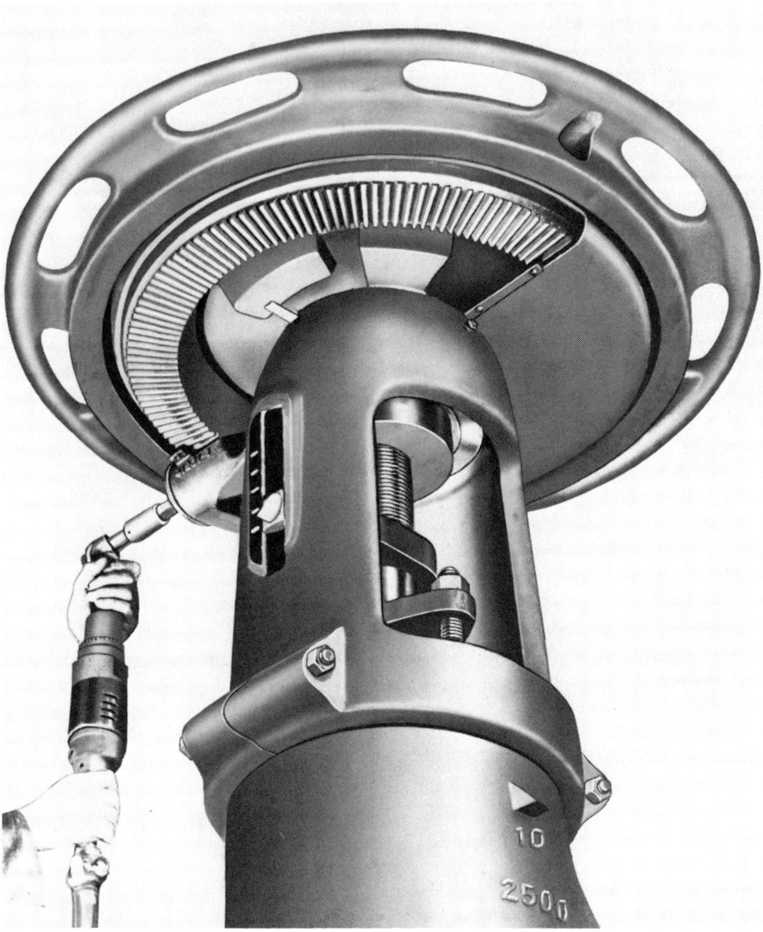


**Fig. 3-66.** Operating large handwheel-operated valve with air wrench. (*Courtesy of Edward Valves, Inc., formerly Rockwell International Inc.*)

that a gear operator is required on gate valves above 12 in. to 16 in. (DN 300 to DN 400) for class 150 (PN 16), above 10 in. to 12 in. (DN 250 to DN 300) for class 300 (PN 25), above 8 in. (DN 200) for class 600 (PN 40), and so on. Globe valves require much greater torques, so the corresponding sizes are 8 in. (DN 200) for class 150, 6 in. (DN 150) for class 300, 4 in. (DN 100) for class 600, and so on. These sizes assume full differential pressure across the valve. Where a lower differential, and hence closing force, exists, the same size valve may not need a gear. Conversely, a frequently operated valve that is impractical to actuate might benefit from a gear even in smaller sizes.

Clearances on the valve itself permit the use of almost any size handwheel, but other clearances may represent a problem. Piping layout

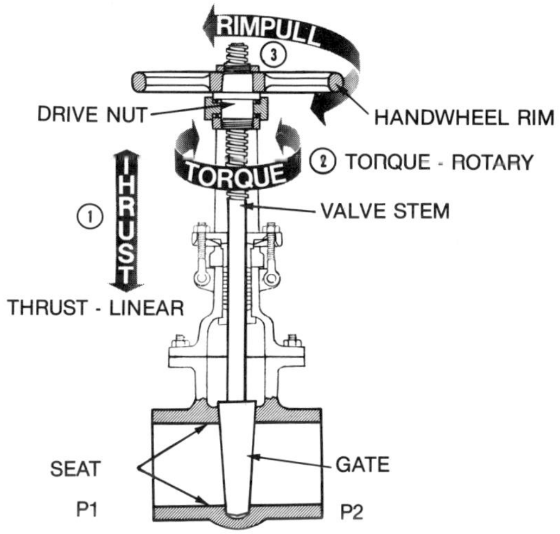


**Fig. 3-67.** Diagram of forces and moments involved in rising stem valve operation. (*Courtesy of Pacific Valves, a unit of Mark Controls.*)

is often a problem with gear-operated valves. Since the gears have to be accessible, they often have to be reoriented to suit each location, and handwheel sizes very widely enough that sometimes the handwheels themselves do not fit into the available space. Without a stem protector in place, the handwheel generally is the limiting topmost dimension of the valve. When an auxiliary gearbox is used to turn the handwheel to the horizontal position, where the handwheel shaft is parallel to the valve stem, the handwheel must be far enough out that the stem will clear the edge of the handwheel as it rises.

Gearboxes for quarter-turn valves (Fig. 3-68) are built similarly to the rising-stem designs, except that they normally consist of a partial gear mounted to a shaft with a coupling to turn the valve stem, with a single- or double-threaded worm gear mounted on a shaft with bearings at each end and a handwheel on one end. The worm shaft is often reversible by opening the case and turning it end for end. Reversing the worm shaft does not change the direction of rotation of the driven gear. The partial gear generally consists of about 120 degrees of arc with gear teeth, since it only moves back and forth and never rotates completely around. Stops are normally provided for the driven gear, which are adjustable to 5 degrees of arc on each side. As with any other actuator, it is much preferable to adjust the travel of the valve internals by adjusting the *actuator,* not the valve. The worm gear drive also means that the gear is self-locking, that is, that torque on the valve shaft cannot turn the handwheel and rotate the valve.

Some manual operators are built with a Scotch yoke arrangement rather than a gear segment. The Scotch yoke permits higher torque at the ends of the travel than in the middle, for easier seating and unseating. The driving mechanism is a lead screw, which looks very much like

**FIG. 3-68.** Enclosed worm gear operator for quarter-turn valve, showing travel stops and position indicator. (*Courtesy of Fisher Controls Intl.*)

the stem from a rising-stem gate valve with an Acme thread along much of its length. The rotation of the lead screw pushes a drive nut back and forth along its length. The drive nut has a pin that engages the yoke. Generally the drive nut slides across the bottom of the operator housing, which is often the only guide required to keep the drive nut in its proper plane. Sometimes an additional top guide is provided.

Handwheels on quarter-turn gear operators can be quite small, sometimes as small as 4 in. (100 mm), but occasionally can be 2 ft (600 mm) or more in diameter. In place of a handwheel, many smaller operators have a crank made from bent bar or cast iron, or a three- or four-armed handle like a drill press. Because of the small overall or envelope dimensions of most quarter-turn valves, the gear can be oriented only at right angles to the pipe, on one side or the other. The bottom of the handwheel sometimes falls below the centerline of the valve. The gear normally is mounted to a bracket made of steel plate or box section that then bolts to the top of the valve or valve bonnet. Output torques of gear operators for rotary-shaft valves may range from 2000 inch-pounds to 750,000 inch-pounds (125 to almost 100,000 Nm). The number of turns of the handwheel will range from 10 to 600 or more, the latter being with the addition of a double reduction gear for the high torque requirement. These gears usually require an input of no more than 1000 inch-pounds (125 Nm), which is about the maximum a person can exert.

Some valves, especially ball valves, are large enough for the mounting bracket to be bolted to the bonnet or a mounting pad around the packing gland. Plug valves and smaller ball valves generally have mounting pads provided as part of the end flanges. Butterfly valves usually have a small mounting pad at the level of the top of the packing, which can take either a lever-lock manual operator or a gear operator. Occasionally on top-entry valves four of the cover bolts are extra long with double nuts for securing the mounting bracket. This arrangement is not permitted by some users.

There is also a worm gear equipped with a bellows seal device for extra high integrity leak proofing (Figs. 3-69 and 3-70). The purpose of the bellows seal, like that of a bellows seal on a valve, is to provide a flexible metal seal that is welded at the ends to keep toxic or valuable fluids from escaping. On worm gears, the application is for cases where a leak could come through the valve packing up into the gear area. The bottom of the gear is gasketed to the top of the valve. Leaks into the gear are up into the area around the pinion shaft and surrounding the worm

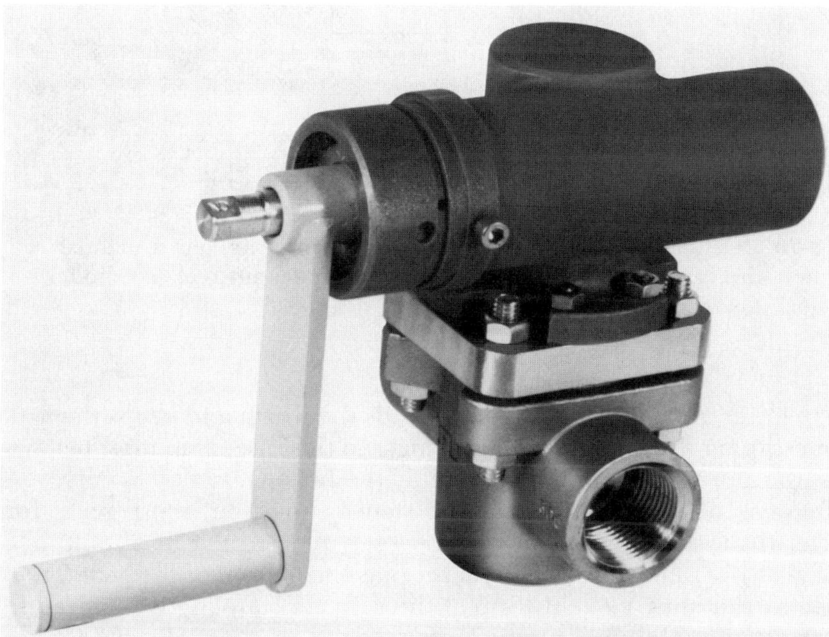

**Fig. 3-69.** Bellows seal worm gear operator mounted on plug valve. (*Courtesy of Xomox Corporation.*)

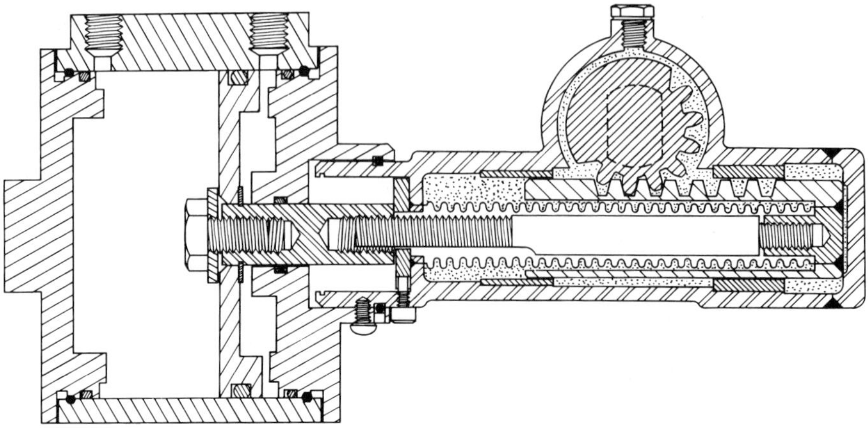

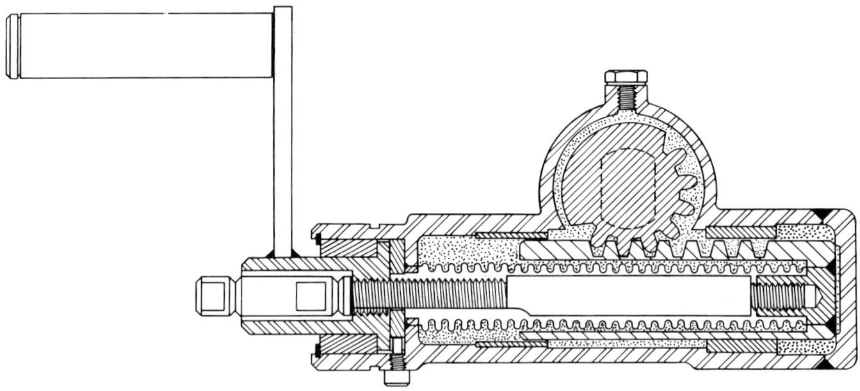

**Fig. 3-70.** Cutaway drawings of bellows seal worm gears, one a manual operator with crank handle, the other an air-to-open, air-to-close cylinder actuator. (*Courtesy of Xomox Corporation.*)

and follower area. The bellows surrounds the worm and keeps the leak from escaping through the end bearings. In this case, as in most bellows seals, the pressure is on the outside of the bellows.

Because of the geometry of the stem connection, most gears for quarter-turn valves do not mount as closely to the valve as a multiturn valve gear operator does. The bracket puts the bottom of the gear 2 to 6 in. (50 to 150 mm) above the top of the valve. A coupling is required to join the stem to the shaft on the bottom of the gearbox. However, since this space exists anyway, it provides room for limit switches and other devices to be mounted. Sometimes the coupling can be marked or fitted

**Fig. 3-71.** Nonenclosed spur gear operator on globe valve, showing damaged sheet metal housing. (*Photograph by R. C. Merrick.*)

with a small pin to help show the position of the valve. This is in addition to the position indicator normally furnished on the gearbox, which is attached to the gear and is visible on the top of the gearbox. Most gearboxes are designed so that the position indicator can be installed only in positions that line up correctly with the position of the valve itself, but this is not always true.

The enclosed gearboxes described above are almost always equipped with grease fittings. A standard gearbox will be furnished fully lubricated and will never need additional lubrication. Not for lubrication but for corrosion protection and to help keep water out, some users will prefer to fill the gearbox completely full of grease.

Some foreign-made valves with factory-installed gear operators may have a considerably cheaper design. These are similar to the old open gear brackets with a sheet metal enclosure built around them (Fig. 3-71). This is in no way equivalent to the enclosed gearbox described above, except that it may keep operators from becoming entangled in the gears. Many of these sheet metal enclosures do not even survive the trip to their final destination.

## Signal Interface

Any valve actuated by a mechanical device requires a signal to tell it what to do. Many of these valves also have to provide a signal to indicate what position they are in. It is the job of the control system to give and to receive all these signals, interpret them, and pass them along to other devices that it controls. The control system also must provide some interface to human operators, giving them knowledge of what is happening in the process. The theory and practice of control systems are much too complex to treat here. Instead, the focus will be on how signals are given, received, and interpreted by valves.

Originally, when valves had to be actuated from a distance or from a central location it was done by rodding, that is, by manipulating one end of a long rod or a rope that mechanically applied an actuating force to a valve, or by opening and closing a valve that directly controlled the flow of air or steam to a valve actuator. For example, all of the flows on a steam locomotive were controlled in this manner. To this day, the hydraulics on mobile equipment such as road graders or on machine tools are often directly controlled, with valves in the operator's console and the cylinders located at the points where force is applied.

Two generations of advances in control systems have occurred since then. The first was the use of individually controlled, remotely actuated devices. The second was the distributed control system that united all the available inputs and signals and a set of control logic into a highly automated system for receiving, processing, and distributing the information required for process control. Here the concern will be not with the methods of processing information (signals), but with the valve and its actuator as it is connected to the signal line (the tube or wire coming from some point external to the valve, through which a signal will come).

Most actuators do basically the same things, regardless of the type of signal they receive or the type of power they use. All of the mechanical devices described in the preceding sections are the means of applying power to the valve-operating mechanism. The receiving of signals from the outside, the response, and the feedback are performed by the small auxiliary equipment attached to the actuator.

The various signaling systems all do almost the same things, usually with analog signals. Digital signal transmissions exist, but usually are used for long-distance controlling where the signals have to be transmitted by telephone, radio, or microwave, such as for pipelines or for objects in space. (There are a large number of valves in a liquid fuel rocket motor.) Even where digital signal transmission is used for more prosaic applications, most of the inputs and outputs are analog so an analog system is easier. Common inputs used for valve controlling are fluid temperature, fluid pressure, liquid level, vapor density, and fluid flow rate, all of which are generally analog. Many signals that are of the true-false or on-off type, such as a liquid level that has or has not reached a predetermined point, are sent as analog signals with 0 percent of signal scale for one condition and 100 percent of signal scale for the opposite condition.

Most signals in a process plant are sent by either pneumatic or electronic carriers. There are others, including hydraulic or liquid signal systems, but these are less common. Pneumatic signaling was once the dominant mode, with mechanical devices to do the detection and signal generation. The development of electronic circuitry, especially semiconductor and circuit board technology, allowed the introduction of devices that were more accurate, more reliable, smaller, lighter, and eventually cheaper than their pneumatic counterparts, as well as being capable of some things that were unfeasible or impossible with mechanical pneumatic equipment. In present-day installations, electronic controls have captured virtually the entire market for conventional control systems. Where pneumatic control systems have continued in use, some of the sensing devices are electronic with converters to pneumatic signaling.

There are several standard signal strengths in common use. These signal standards must be specified for any item that connects to the signal transmission system in order for the components to "talk" to each other properly. For pneumatic systems, the common ones are 3 to 15 psig (0.2 to 1 barg) and 3 to 30 psig (0.2 to 2 barg). For electronic systems, the common ones are 4 to 20 mA at either 12 or 24 VDC. For special purposes and some industry standards, there are other conventional signal levels.

## Positioners

Most actuated valves that are intended for on-off service respond to signals in a quite simple manner. When the signal to close or open is sent, and received by the valve, it begins to close or open and does not stop until some mechanical interference occurs, such as seating or hitting the stop. For valves in modulating service, much more is required. There must be a feedback method and a control device that senses the feedback signal and controls the power to the valve in such a way as to keep the valve position in proportion to the signal being sent at any given time.

In some cases, the feedback method can be purely mechanical, through the resistance of a spring, without depending on a special feedback device. This generally works only in specific cases, such as on many diaphragm actuators. This method is not applicable at all to many types of actuators and in any application where the actuator is double acting, that is, in any situation where there is no spring return. Generally, a dedicated feedback device, known as a positioner, is either required or desirable for optimum functioning of a valve intended to modulate flows.

Each type of signal and power supply calls for a different type of positioner, but each performs in much the same fashion (Figs. 3-72 and 3-73). The diagrams that explain how a positioner works can be drawn using a mechanical, electrical, or hydraulic analogy, and each can ex-

**Fig. 3-72.** Internal view of pneumatic positioner (for pneumatic signaling with air actuator). (*Courtesy of Worcester Controls Corp.*)

**Fig. 3-73.** Internal view of electro-pneumatic positioner (for electronic signaling with air actuator). (*Courtesy of Worcester Controls Corp.*)

plain adequately how it functions (Figs. 3-74 and 3-75). The valves that use the same type of transmission for signal and power have the simplest positioners, and generally the choice of positioner is related to the type of power used. When the type of signal transmission is different from the type of power supply, then a transducer or converter is used, either external to or integral with the positioner.

Regardless of the type of positioner, essentially what it does is to compare the position of the valve with the signal being received from outside. This can be either from the control system or from a specific device. The positioner then modifies the power (current, air pressure, or other) being fed to the actuator, increasing or decreasing it until the signal showing the valve's position is proportional to the signal coming from outside.

Several concepts in control system technology are useful here. One is span, which is the total range from minimum to maximum input. Others are linearity, which is how closely the ratio of output to input approaches a straight line; repeatability, which is how closely the output comes to being the same each time the same input is supplied; hysteresis, which is how much the output varies when a given input is arrived at on the increase versus on the decrease; resolution, which is the minimum amount the output moves as the input increases as slowly as possible; and sensitivity, which is the ratio of the change in output to a given change in the input. Many of these terms have official definitions established by ISA. See the glossary for more complete details.

One common positioner type as used on pneumatic signal diaphragm actuators is illustrative of how many of the types function (Fig. 3-76). The actuator shaft coupling has a bar attached to it, which moves up and down as the actuator shaft moves. The end of this bar is linked to a lever arm that rotates back and forth as the actuator shaft moves. The other end of this shaft is attached to a cam that also rotates, pushing another lever back and forth slightly. Attached to this lever (which is D-shaped, for easy adjustment of the lever arm ratio) at one point is a bellows that moves as the signal pressure changes. At another point on the lever is the mechanical relay that controls how much air is sent to the diaphragm. As the signal pressure increases, the bellows expands and moves the feedback lever outward. This causes the mechanical relay to open farther, admitting more air into the actuator. The pressure rises, causing the diaphragm to move downward, pushing the indicator lever downward, and rotating the cam. This moves the feedback lever, halting the flow of air into the actuator. If too much air has been admitted, the lever is tilted too far and it allows some air to escape from the actuator, depressuring it slightly and allowing the spring to push the diaphragm upward slightly. This continues until the forces acting on the lever arm are equalized, or at least drop below the point at which friction forces are greater than the magnitude of the total forces. When the control system

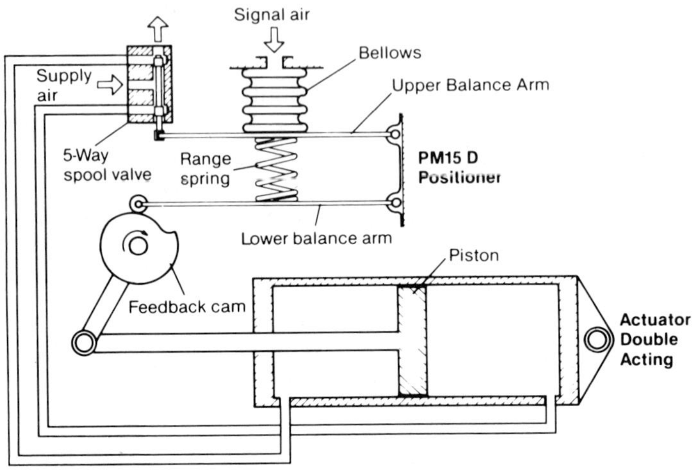


**Fig. 3-74.** Operational diagram of pneumatic positioner. (*Courtesy of Worcester Controls Corp.*)

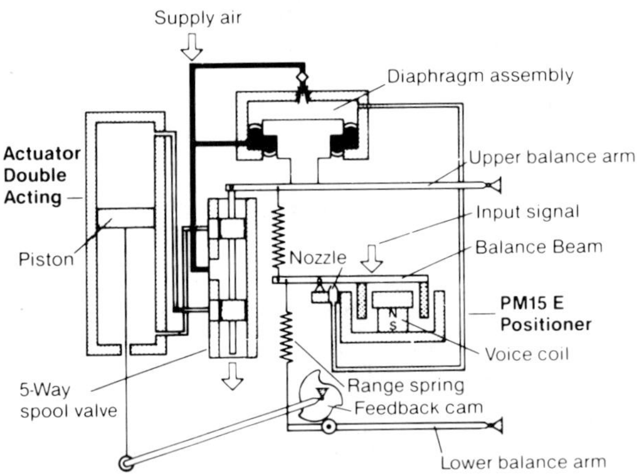


**Fig. 3-75.** Operational diagram of electro-pneumatic positioner. (*Courtesy of Worcester Controls Corp.*)

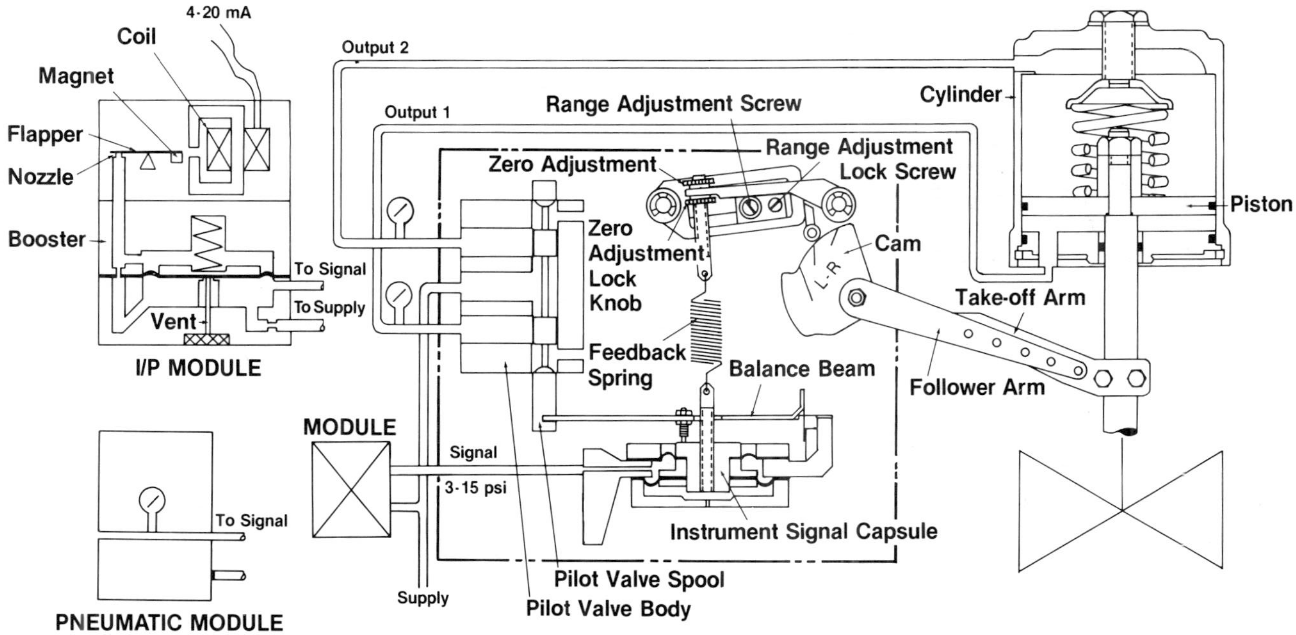


FIG. 3-76. Operational diagram of positioner. (*Courtesy of Voltek, Inc.*)

is very stable, very little back-and-forth motion occurs. As stated earlier, in a stable enough system or one in which the signal moves decisively and fairly fast, a positioner is less needed than in any of the cases where the foregoing is not true. When the positioner cycles back and forth quite a bit before coming to rest, that is a symptom of instability or overcontrol. This cycling back and forth is known as hunting, and is undesirable because it increases the amount of time the mechanical parts are in motion and increases wear, and keeps the positioner "busy" longer so it has more difficulty responding to new signal changes.

Purely electrical devices, such as motor actuators, can have mechanical positioners, but most now have purely electronic feedback systems in which the physical position of the shaft is detected by a potentiometer, which varies the resistance in the feedback circuit and controls the amount of power being fed to the motor. The use of motors for modulating rising-stem or sliding-stem valves is rare, but is quite common for rotary-shaft valves. Hydraulic actuators have systems that operate in very much the same way as the pneumatic one described above.

# 4

# Valve Procurement

## Bid Development

In the preceding chapters there has been discussion of valve design and the choices of valves available, the philosophy and the mechanics of choosing and specifying valves, the tricks of the trade, the mechanics of inspection and meeting requirements. Up to this point, all the discussion has been related to valve engineering and design. None of this specification work is really worth anything, though, except in terms of billable hours to the specification engineer, until the procurement phase is entered. Procurement of material and services is where the rubber meets the road, so to speak. Even negotiating with manufacturers about capabilities, production time, and test results is essentially an academic exercise until the procurement phase. In the course of procurement of materials, which covers the period from when the specifications are first made available for use by others until the finished product arrives at the site, all of the specifications written by the engineers are actually put to the test.

Methods and practices used in valve procurement vary widely from company to company, and industry to industry, and depending on whether the owner is buying the material or a contractor is buying material for an owner. The procurement practices used in private industry of any sort are radically different from those in government agencies on any level. Of course, these practices are rooted in the reason for being of each of these respective organizations, and apply to procurement of goods and services of almost any type, not just valves or piping material.

There are several essential factors in procurement of material, all of which vary widely depending on the organizational circumstances mentioned above. The production of specifications and the input into them is the first factor, more fully discussed in Chapter 1. Another is the bidders list, which is the means by which the manufacturers and/or sellers who are invited to bid on a given item are selected. Still another is the actual bidding process and the practices involved in the evaluation of bids and the selection of successful bidders. Then there is the purchase order itself, with its terms and conditions of sale, and the administration of the order. Finally, there is the inspection of the product at the supplier's location, shipping/receiving, inspection at the receiving site, job site storage prior to installation, and the manufacturer's warranty. All of these activities involve technical considerations, even though they seem to be of purely commercial or administrative nature.

The two entities, procurement and engineering, that are involved in the procurement cycle are inextricably intertwined. Some organizations attempt to rigorously separate the two and have very rigid boundaries, while others tend to lump the two together or consider the whole process as one. The ideal situation is somewhere between these two poles, probably closer to the latter. An example of the first extreme might be a government agency, and for the opposite end of the spectrum, the process you go through when you go to the hardware store and buy a valve, perhaps a kitchen faucet. Here you are acting as the specification authority, the engineer, the procurement force, the finance organization, and the installer. Your spouse probably acts as the inspector.

A "typical" procedure can be illustrated in the steps of buying a valve. Assume that the specification activity performed in Chapter 1 has already occurred. The process of reviewing and issuing approved specifications has occurred. (This process may be occurring simultaneously with the beginning of procurement activities. In fact, the bidding process often can be used to assist the development of final specifications. But for now, assume that the specification development has been completed.) All the mechanics of printing specifications and making them available to the people who require them are somewhat outside the scope of this book, but the point to remember is that the specification-writing effort is not complete until the documents are published in some form and available to all the personnel in the organization who need them.

The procurement cycle really begins with the preparation of a bidders list. A bidders list may also be known as an acceptable manufacturers list, a vendors list, a manufacturer reference list, or any one of several other names. This may be a bidders list for one single item, a list of bidders who have performed satisfactorily in the past on similar material, or a corporate bidders list that includes all companies known to be in the business of supplying any of the items that the company uses. It may just be the phone book. But to target the appropriate manufacturers and distributors efficiently, it is wise to maintain a bidders list. Some government organizations have procurement rules that more or less disallow the concept of a bidders list, and instead are required to advertise the letting of bids on a particular item in the newspaper. At least some of these, though, use a "prequalification" stage for technical material to weed out the unqualified before serious requisitioning begins.

Who decides on the makeup of the bidders list? That depends on the organization. Sometimes it is seen as a procurement function. Sometimes it is considered purely an engineering or technical area. The best organizations make use of input from both groups, since both technical and commercial considerations are involved. It also depends on whether only manufacturers are listed, or only distributors, or both.

Where do you look to find bidders? That is not a worry because most of them will find you. This is perhaps an overstatement, and perhaps some of the prospective bidders that find you and want to sell you something will have things you could never use. But since manufacturers and distributors make their living by selling things, they usually try diligently to find people and organizations to sell them to. In the author's experience, about one-third of all eventual sales of specified or engineered material result from a first contact made by the manufacturer or the manufacturer's representative.

When you have to find manufacturers for an item you wish to buy, there are several sources. For general industrial products, nationwide, there is the *Thomas Register*. This publication, issued annually, is a large-format 14-volume list, by category of material, of virtually every domestic company that makes or sells a given product or service. It covers almost the entire spectrum of manufactured finished or semifinished products, from the machinery to steel. Within a category, it lists manufacturers alphabetically, by state. If anything, its scope is too broad. It can take time to page through the books just to find the correct category out of 50,000 or so product categories. But if your responsibilities include any sort of material beyond just valves, or just piping material, it is an invaluable office resource.

If your primary need is for manufacturers or distributors located in your area, or in some other very specific geographical area, another resource is the Yellow Pages. Distributors tend to congregate where the business is, whether in a major "chemical alley" or even in a fairly remote town where there is even one plant or fabricator that uses significant amounts of material. In smaller, more remote locales the selection will not be as good, and since one company will be covering quite a large area, it could be that the closest one is several counties away. But local distributors, of which there are many more than local manufacturers in most cases, are familiar with the area, and sometimes familiar with your plant or the types of industries common in the area. Working your way through the Yellow Pages, or course, may take some investigation. Some distributors may even show in the ads what manufacturers they carry or can buy from. Most do not; so if you are trying to set up or expand your bidders list you may have to visit a few of them, or have them visit you, or request a line sheet, or at least interview them over the phone.

More specific lists of manufacturers and distributors are maintained by various private publishers that cater to different industries. On the Gulf Coast, because of the concentration of refiners and chemical manufacturers, and suppliers and equipment manufacturers, several companies specialize in printing buyer's guides or listings, organized by category of material, and often cross-referenced between manufacturers and distributors. Many similar guides exist for other regions of the country and other industries. These are usually published ad hoc, depending on the business cycle, every year in good times and skipping a year or two in bad times.

In foreign countries where the piping and general industrial business is dominated by American and European companies, such as the Middle East and Far East wherever there is extensive oil and chemical work, there will usually be a business directory, in English or bilingual, listing importers and the domestic manufacturers that have sprung up to cater to this work. Most of these countries encourage domestic investment in service and supply businesses as one more way to move wealth from the natural resource companies into the local economies. They therefore tend to encourage local purchases as much as possible.

Regardless of how you have obtained names of possible bidders, the next step is to decide how many of them you really want. This is especially true in several cases. One is where there is a corporate or plant bidders list that you are working from. There may be companies on the list that you have never heard of, or rarely do business with. If you have several on the list that you know well, and a few that you do not, it is probably safe to go with the ones you know. This depends a lot on company policy. You may be required to inquire of everybody on the list, whether you want them or not. More usually, corporate procurement policies require three or four bidders, or as many as six, depending on corporate philosophy. Government agencies, especially local ones, may not have any choice about the matter. You may simply have to advertise for bids and evaluate every one that responds.

There is a practice known as "single sourcing." This is also known as "no equal" or "no substitution." Briefly, it means that the item in question is to be purchased exactly as specified, by the manufacturer specified, with no other product from any other manufacturer considered to be equal or equivalent. This practice simplifies the job a great deal, but it is a powerful tool that should be selectively applied.

A physician has the choice of "no substitution" as opposed to authorizing a generic equivalent on a prescription, just by checking a box on the prescription form. Engineers can do that too, provided it is justified. Again, this practice is subject to the procurement regulations of the organization. First, invoking a "no substitution" clause almost always leads to increased cost, especially if the supplier knows that its products are a "no substitution." Second, it takes considerable engineering judgment to arrive at a conclusion that some item has no equal. Sometimes, depending on circumstances, the evidence is there for all to see. In a manufacturing plant, sometimes it is obvious from observing failures that one product works or stands up to its required duty and that others, seemingly equivalent, do not. All it takes here is for the engineering group to convince the procurement group of that fact. Sometimes, for spare parts inventory or just general warehouse inventory reduction reasons, it can be shown that it is cheaper to continue buying the item that already is in use, rather than switch to a competing product. This last example, of course, is an economic decision as well as one based on engineering logic. Another example that occurs quite often is with fire and safety equipment. This is equipment that must be operated quickly, and reliably, at times when the personnel using it do not have time to play with it to make it work. "No substitution" provisions are fairly common in many industries on fire and safety equipment.

Another question you may have is about how much trouble is bidding worth. Most corporations acknowledge this, in that the labor involved in competitively bidding items is worth something, regardless of the size of the bid package. Many companies have minimum limits, of $500 or $2000 or $5000 or so, whereby if the items to be purchased are below that value, competitive bidding is not required. Sometimes it may be that a price inquiry is made on a less formal basis, to still allow for the best deal.

Some organizations' procurement regulations require competitive bidding on everything. In other cases, the value of the equipment you wish to purchase is above the minimum value established for competitive bidding. In such cases, it may be possible to get competitive bids *on the same item* by choosing two or three distributors who each represent the same manufacturer (most common in areas where there are many sources of supply) or by soliciting quotes from the manufacturer directly as well as a couple of distributors (assuming that the manufacturer will quote directly—some do and some do not). This really does represent a legitimate business practice, because you are still obtaining competitive bids and choosing whichever one comes in lowest, and thereby getting the best value for your company's dollar.

The last few paragraphs have discussed the selection of bidders from the viewpoint of trying to control or limit the selection. Most of the time, limiting the selection is not desired at all. Usually, the goal is to get the widest possible selection of quotes, within reason, and within the confines of the engineering requirements. A good example would be the procurement of rather ordinary flanged steel gate valves. These valves might be for a fairly ordinary service, such as low-pressure steam or light noncorrosive hydrocarbons. Assume that you want valves made to a very common standard, like API 600. The number of manufacturers in the world that make such a product is in the hundreds. Assume that you are used to dealing with many of the larger supply houses in the country. These companies generally stock, or represent, one to three of the major world manufacturers. In other words, you can be confident that when you solicit bids on such items, you will generally get "name brand" material quoted to you. Almost any distributor you go to should be expected to provide a valid quote, one that should not take too much analysis to evaluate and approve. Material like this, which might constitute three-quarters of the volume of valve purchasing that most companies do, and 95 percent of the volume of total purchasing of construction or maintenance material, would almost always be expected to be put "out on the street" for competitive bidding and would almost always be expected to result in at least one bidder who has adequate delivery, a good enough price, and is technically acceptable. In other words, such material would never be considered to be of such critical nature that a "no substitution" would be considered.

The engineer or specifier should be the one who, with some careful procurement advice, makes the judgment whether or not to sole source such material. Such decisions in the real world will always be tempered with the financial and other considerations that come from the procurement side. Certainly, it is in your organization's best interest to put all material out for competitive bidding except the items that absolutely will not work as intended if furnished by another vendor than the one you envision when you write your specifications. Any engineer who has a budget responsibility can appreciate that.

It should be noted that competitive bidding does not have to be performed on each item every time you need it. Again, this is governed by procurement practices and guidelines, but many plants and entire corporate purchasing groups, and sometimes large construction projects, operate on blanket orders, pricing agreements, or maintenance or repair order (MRO) contracts. Here, an order is placed for an indefinite amount of material, but with the scope of supply and pricing structure clearly specified and the price for any variation in the material spelled out. Such orders typically are placed with a contract duration of 1 or 2 years, sometimes with a fixed price per item to be renegotiated at the end of the contract, or sometimes with constant multiplier and price increases based solely on manufacturer's price structure. Often these have maximum escalation clauses included. Such orders can be in place for years at a time, sometimes after being competitively rebid every year, sometimes simply being renegotiated. Sometimes there are hard-fought battles, with the contract changing hands every time it comes up for renewal.

The scope of these orders vary. Sometimes there may be one for forged-steel valves only, another one for forged stainless steel valves, another one for cast-steel valves, and so on. In other cases it may be for all valves, or perhaps one for all carbon steel piping material, valves, flanges, and everything, with a separate one for stainless steels. (Carbon steel and stainless steel pricing, market conditions, and distribution systems are quite different from each other.) For smaller plants, it may be all piping material plus even other material. Generally, though, distribution of electrical equipment, plate/structurals, and machinery hardly ever overlaps with piping material except in very remote locations such as oil patch work.

All of the work in developing the bidders list and determining the type of inquiry to be used is leading up to the actual act of soliciting bids. Preparing a bid package to go out the door is a fairly critical activity. Generally, this is a shared responsibility in most of private enterprise, with the engineering department or engineering representatives providing the information or the finished paperwork on the technical requirements, and the procurement department or the purchasing representatives providing the terms and conditions and the other business-related or commercial requirements. Some organizations are structured such that almost all of this activity is centered in engineering, and commercial terms are considered as an addendum to the engineering package. Some organizations are the other way around, where a package of material requirements is prepared and the procurement department takes it over, solicits bids, evaluates, places orders, and arranges for delivery all on its own. As is usual in these cases, the ideal case is probably somewhere between these two extremes.

There are a number of components in a bid package. The preparation and assembly of the set of specifications were covered more fully in Chapter 1. Broadly speaking, the responsibilities of engineering personnel extend as far as to prepare, if not the complete technical package for a bid request, at least a checklist of items required to be included. Besides the specifications, there has to be a listing of the items to be bid, with the quantity required of each (inquiries for blanket orders may or may not have quantities, and the quantities may be either "hard" or approximate), and a set of the commercial terms and conditions of order. Sometimes a copy of a sample purchase order is included, to show the bidders an example of the information that would appear on the actual order. If delivery is required by a specific date, that information must be included in the bid package also. Any other special conditions that could affect the price, such as the presence of inspection requirements, special cleaning or preparation, and material certification requirements, also need to be stated. Some of these items can be left open, if necessary, and agreed upon with the bidders as the evaluation process proceeds, but as more of these items are left open or omitted, the evaluation process becomes longer, more intense, and more expensive.

Sometimes this information is gathered into a package and sent by the purchasing group directly to the bidders chosen to receive it. Since the bidders are going to do the work of pricing out these inquiries, and they understandably would like to get this information as soon as possible, many bidders prefer to come out in person to collect their bid package. For government-style open procurement, the packages are made up ahead of time and made available for anyone who presents the appropriate intention to bid. These packages are sometimes known as formal inquiries, or requests for quotation, or requests for proposal. They are often intended to be received back as sealed bids. Often, but not necessarily, the minimum dollar limit that requires competitive bidding also requires sealed bids to be returned. Sometimes, for more routine work or for situations where time is more critical, the step of preparing packages is dispensed with or reduced to its bare essentials, namely, the engineering requirements (bills of material or data sheets) and commercial terms, which are then sent by facsimile or teletype to the bidders. For that matter, it is perfectly common for the buyer to call a prospective bidder and read the items required over the phone. Sometimes this action is followed by a request for quote, sometimes not, but in almost all cases the bidder is expected to follow up with a written quote.

All of these activities are carried out with no obligation, as the advertisements always say. When paper requests for quotation are sent out, it is common practice to put a statement on the document somewhere that says "this is not an order," in case there could be any misconstrual.

For centuries, this bid request information has been prepared on paper, but there have been a number of experiments in recent years on ways to transmit the data electronically. There is a standard interface that has been developed, known as electronic data interchange (EDI), to send description data back and forth. The EDI interchange has to be arranged for in advance by both parties; that is, it is still not like putting an inquiry in the mail and getting prices back. At present EDI is most suited for interchange between companies where commercial and financial relations are already set (for instance, in MRO situations). But the full cycle of data is provided for in the EDI data sets—from inquiry, to bid, to purchase order, to change order—with scheduling data, shipping notices, and other information already configured for electronic interchange.

EDI is an ANSI standard, or more precisely a set of ANSI standards, ANSI X12.1 through X12.22. As part of the same general development of standard information interchanges, there now exists an ANSI standard relational data base with a structured query language (ANSI X3.135). Together, these standard ways of handling, retrieving, and interchanging information have the potential for revolutionizing, again, the ways of handling the business dealings concerning engineered products. As yet, these systems are still in their infancy, and although there are many companies looking at ways of doing business that can take advantage of these standards, not many programs or practices are fully in place and proven performers yet. Probably within a few years, though, there will be many good systems and a few will be dominating the market.

## Bid Evaluation

The bid development stage of material procurement ends with the buyer's representative, usually meaning the purchasing department, sending out all the inquiry packages to the prospective bidders intended to receive them. At this point the focus of the action shifts to the organizations planning to provide a quote on the materials or services described in the bid package. The buyer's organization typically shifts into low gear, or, more accurately, turns its attention to other work such as other bid packages, while the pricing and bid preparation activity is going on.

In the bidder's organization, there will normally be a department or a group responsible for pricing of bids. Since the success or failure of the company often depends on how well these people do their jobs, their activities are usually in the spotlight within their organization. This pricing activity takes various factors into account, such as the actual cost of the manufactured material (raw material, purchased components, labor and machining time), the prevailing price structure (what the market value of these goods are), and the importance of getting the job. Pricing is generally lower on similar material in markets where goods are plentiful and sellers are plentiful, and higher where either or both of these factors are less present.

Pricing will also have to be higher when such factors as import duties, freight, and taxes are considered. Generally import duties are considered to be the importer's responsibility and rolled into the actual cost of the material, while freight and taxes are normally required to be priced separately for a number of reasons. For one thing, the actual work of shipping the goods is usually done by a third party such as a trucking company, and also there are many different ways of billing freight such as direct bill, third-party invoice, or prepay and charge. Taxes are also almost always required to be stated separately since the owner's organization knows exactly whether and how taxes are due on the goods being bought. Although both freight and taxes do ultimately end up as part of the installed cost of the plant being built, accounting convention requires both to be treated separately and the government requires a separate accounting of taxes in addition.

The pricing structure of most manufacturers generally revolves around a list price sheet. List prices are published on individual items of the company's product, usually with incremental pricing for options. This method is exactly the same for valves as it is with automobiles, although the sales effort is not usually the same. However, on top of the list price structure are two other variables, one being a volume discount and the other a "competition" discount that, taken together, yield the final price. The volume discount accounts for the efficiency of scale in that more business to the same customer results in very little added overhead cost to the seller. The "competition" discount, which may be made up on the spot by the salesperson in negotiations or may be the results of a very carefully crafted study of market conditions, is how the seller responds to the effects of the market. There have been cases where a company that did not want a particular job for some reason but bid it anyway out of some other motive, or that discovered after getting the bid package that it was not one that they wanted to do, in effect came up with a negative discount by quoting a deliberately high price in hopes of losing out in the bid evaluation. Those cases have been known to backfire.

The very existence of companies is ultimately determined by the price of the goods they sell, relative to the cost of producing them. In a controlled economy this is not true, or not as true as it is in the U.S. economy, depending on what the purpose of the company is seen to be. In the United States the purpose of most companies, and of most individuals for that matter, is to make as much money as necessary. In other economic environments the purpose of a company may be seen as to provide gainful employment for the populace, or to make money from outside the national economy by export sales, or any of a number of other things. In free-market economies the world over the point is still, of course, to make money. The difference is that some nations have, by nature of their economic realities, a very strong export drive. There are lots of products made in these countries that are intended for nothing but export, products for which there is very little demand at home. In the United States, most products are made primarily for internal consumption, and if there are others around the world who also want them, that is a side benefit. The United States has been able to get away with this approach for many years, first, because American products for many years were the best in the world, and second, because the United States also exported the technology that required these products. To a large extent, these factors are both still operative, but the technology side of the picture is more active today. The technology export is not merely plant designs and engineering, but codes and standards as well.

Not only do many ISO standards agree with ANSI and other domestic American standards, but the use of U.S. codes and standards in plants worldwide has exported the demand for U.S. standard material. This has also led to the familiarity of valve and other material manufacturers worldwide with U.S. codes and standards and has helped the process of internationalizing the manufacture of material.

The point is that the price of goods, and the profits earned therefrom, are relative to what the competition is. In many products the competition is worldwide. Where this is not true, it is because the cost of the products and their potential earning power are less than the cost of the freight, such as in sand and gravel, to the point that it is not worth the trouble to bring them in from very far, or because the market is restricted to a given segment by some other reason.

Some basic industrial products, which in the last century were on the fringes of technology, are now "mature" in the sense that the markets are quite static and the improvements in technology are mostly refinements that create slight decreases in the cost of manufacture or slight improvements in product quality. Steel is a good example. There are now blast furnaces and melt shops and rolling mills virtually all over the world. Valves are somewhat more specialized, but the principle is much the same since the technology of making valves consists of basic foundry and forge processes plus machining technology. This refers to mass-produced valves, such as the ones you can buy at the hardware store for $5, not the ones that are on the fringes of technology. There will always be fringes, no matter how far out they are, and there will always be engineering and manufacturing that are on the cutting edge. But there is a continuum between these two extremes, not a sharp demarcation, and many valve plants around the world and many industrial economies occupy a considerable portion of the spectrum.

Here is another analogy between valves and automobiles. Regardless of personal opinions, or the value of national self-reliance, or any other outside factor, it is a fact that the valve market is international and that the lower cost producers in general are not in the United States or in North America. It is also a fact that the spread of technology will continue. Trade barriers and artificial restraints will not stop or slow this trend for very long or very effectively, any more than the spread of technology from England (textiles) or China (silk and porcelain) could be stopped centuries ago. This means that more products, and better products, will be available around the world from more places in the future. It means that a nation that had more of the world's manufacturing capacity than any other nation in the world at one time will not enjoy that position again.

What does this discussion of economic evolution have to do with the price of valves? Anyone who has been involved with the procurement of valves over the last 15 or 20 years already knows the answer. Sourcing of products, and not just in valves, has changed dramatically. Procurement practices and engineering acceptance practices have had to change also to meet the realities of the market. "The Market," of course, varies all over the map depending on the commodity being purchased, the local or microeconomic climate, and the taste and knowledge of the consuming organization. For instance, federal government procurement practices that require 100 percent domestic content of material are drawing from a vastly different pool of manufacturers than much of private industry does, where foreign and quasi-foreign producers dominate some segments of the valve market. The pricing structure for material going to the federal government is much different, also, probably higher than for similar material bought by private industry. There may be several reasons for this, such as volume of purchases and the more rigid structure of government procurement, as well as the fact that federal and military engineering requirements may be more severe than commercial or industrial requirements.

As a general rule, the greater the engineering content of a piece of material, the more likely it is to be made entirely or partly in the United States. This is true partly because most engineers feel most comfortable working with companies they can see and can discuss requirements with in great detail, which is more true when the two parties to a discussion are closer together and can speak the same engineering language. Also, more high-tech products have more design engineering time in them proportional to their total cost, while simpler goods have a greater proportion of labor and manufacturing cost. For the latter products, there is more of an incentive to move manufacturing to a place where labor is cheaper.

There has been a great deal of such outsourcing of castings, assembly, and complete manufacture in the last 10 years, and there will probably be more in the future. But U.S. manufacturers doing outsourcing is not what caused the internationalization of the valve market. The original impetus was foreign producers observing the U.S. market and deciding that they could sell products there. Again, as with the first foreign cars, the product was not that good and the manufacturers were not very understanding of domestic practices and requirements. Over time the product improved while domestic producers had labor cost increases, production difficulties, and perhaps some lack of recognition of the true magnitude of the potential worldwide competition. Eventually some domestic producers failed, some successfully fought back, some developed a mixed line of domestic and imported products, some simply became shell companies doing a totally import business under the name of a former domestic producer, and many new players entered the game. Some of the first new players were purely foreign entities doing export business in the United States. Now there are a number of American and Canadian companies building valves at various locations around the world to domestic designs and engineering standards, such that the engineering content is essentially domestic but the material and labor content are imported. In effect, the pipe, valve, and fitting market is now totally internationalized, with a broad mix of domestic and foreign content that varies literally from one end of the scale to the other. It also varies from month to month and item to item. There are no product categories made for the industrial market that can be said to be purely domestic, but there really are not any product categories in which there are no domestic manufacturers either.

What, then, can you expect to see on your bid evaluation? That depends on what you asked for, and why you wanted it. Consider first the case where 100 percent domestic content is required. If this is a hard-and-fast rule, for example, in an organization that is required by law or as a condition of funding to obtain 100 percent domestic content, or in a unionized plant, you probably would want an affidavit from each bidder certifying the domestic content. But what do you mean by domestic content? Anything that contains chromium, for instance, represents at least a small amount of imported raw material. If you mean 100 percent domestic labor content, that is quite easy to certify. The situation is not as depressing as it may seem at first glance. There are literally thousands of valve manufacturers actually making product in the United States. Most of them are small, since the largest volume producers are generally international. Most American companies that produce the bulk of their product overseas are completely capable of making valves with 100 percent domestic content. But, in some cases, domestic content means these are special order products.

Now, suppose you have the more common case, where you are the specification engineer or the valve procurement specialist for a private company, say a manufacturing entity or a utility. Many such entities had in the past a procurement policy that dictated the purchase of only domestic materials and supplies. Most companies now have a policy that provides for priority to goods made in the United States where possible or feasible. Many other companies have no policy at all regarding domestic content, simply expecting the purchase of lowest priced material that meets specification.

Your bidders list may be restricted to companies headquartered in the United States. That, in and of itself, is no guarantee that you will get domestic material. Some manufacturers have a coding or figure number system that indicates the origin of a particular product, as by having a different number or a suffix for imported product. More usually, there is nothing to distinguish foreign and domestic product anywhere on the bid. You can ask the manufacturer to state exactly where the material is produced and by whom. Remember, the country of origin is required by law to be marked on the material. Although it is unlikely that any bidder will fraudulently state country of origin, there is enough room for interpretation in current statutes on country of origin and percent of domestic content that you may not necessarily be reading the same thing that the bidder is saying on the quote. Just do not assume that because your current quotes show the same manufacturer and figure number that you have been seeing for 10 or 15 years, that you are going to get the same valve, made in the same place by the same people, because odds are that that will not be true.

For that matter, since it really is up to you to decide what is acceptable, you may feel that products engineered in the United States but produced elsewhere are every bit the equal of material made here. You may prefer to check out every item to be sure that all the material you receive is domestic. There are enough people who prefer foreign cars that for some products you may have arrived at a conclusion that the foreign valve is better made and has more desirable features, even at a higher price, than the domestic competition. All three of these approaches are present in the marketplace today.

To return to the mechanics of bidding and bid evaluation, let's look at this set of bids that has been prepared in response to our request for quotes. The quotation request should have included a date, known as the bid closing date, that the bids are expected to be returned. Sometimes this is stated as an exact time, such as at the end of the business day. The length of time allowed for the bidders to respond varies with the complexity of the bid and the time constraints of the buyer organization. For routine purchases of commodity material, it may be a couple of days to 2 weeks, while for highly engineered products it may be 4 or more weeks. It depends also on whether completed design drawings are required with the quote, or as is more common, preliminary drawings are required with the quote and finished drawings later, for the successful bidder to do after purchase order award.

Many organizations that do open procurement, where anyone who qualifies can submit a bid on a particular quotation request, will have a bid box that is set out for receiving those bids. On occasion, the organization's own cost or estimating group will be among the entities depositing a proposal into the bid box. Sometimes, if no bidder's price beats the cost estimate provided by the estimators, the project does not proceed and is put out for rebid until someone does come in under the estimated price. Generally, the bid closing date is hard and fast, but in private industry, not having enough bidders is a worse problem than not meeting the date, so sometimes the quotation request in effect stays open until the person in charge decides that there are enough bidders to proceed. Usually this is a day or two, when it occurs, and then usually it is because a bidder has requested an extension for a good reason.

Bid opening procedures vary just as bid requests do. Some government organizations that require sealed bids also require that all the bids be opened in an open session, often the regularly scheduled board meeting. These procedures are often written into an open meetings law. The apparent low bidder is then announced, after which the engineering department gets its chance to review the quotes to determine conformance to specifications. Private business sometimes has bid opening meetings, often with a representative from each of two or three departments present. The purpose is the same, to ensure that no person from the buyer organization gets a look at the bids before they become public knowledge, or at least before they are known to be available. This, plus other rules that most organizations have about discussing bids, is also intended to prevent competitors from gaining knowledge about each other's bids. The difference between the government and private industry approaches is that in government the bid prices become a matter of public record after the bid opening session, while industry generally prefers to keep all information about costs and pricing on a need-to-know basis for its own competitive reasons.

Where sealed bids are not in effect, quotes are normally just mailed or delivered to the buyer in time to meet the bid closing date, and accumulated in a file. Where faster response is the norm, a lot of quotes and proposals are now sent by facsimile. It is assumed that only clerks and disinterested people will be receiving the documents on the other end.

Someone is responsible for collecting the data contained in the quotes into an organized statement of comparisons. This is usually referred to as a bid tabulation (Fig. 4-1) and is generally done by the person who is named as the addressee of the quotes. Most companies have forms for this purpose, or a computer bid tabulation system, but because the bid tabulation is really just a spreadsheet a piece of paper or a spreadsheet program will do almost as well.

A bid tabulation contains several sections. First is the general portion that describes what was put out for inquiry and when the closing date was. Then each bidder's quote is summarized, to include the price of each item and the total price, and usually the delivery date for each item. For some items such as valves where the products of many manufacturers could be offered, most bid tabulations will show at least the names of the manufacturers, and often the exact manufacturer and figure number quoted for each item. Sometimes all this work is done by the same individual, and sometimes the commercial portion is done first and the technical summary is done later.

At some point in the cycle there should be a technical review of the bids. Generally the technical review is done by the engineer who was responsible for preparing the technical content of the request for quotation (RFQ) in the first place. This is the most reliable method of ensuring that requirements are actually met. Both the purchasing representative and the engineering representative may have questions and clarifications to request from the bidders. Most of the time it is normal practice

COMMERCIAL QUOTATION SUMMARY

* CHECK RECOMMENDED ITEMS

PROJECT: ______________________________

MATERIAL/EQUIPMENT ______________________________

| | | | |
|---|---|---|---|
| BIDDER: | | | |
| DATE OF QUOTATION: | | | |
| QUOTATION REFERENCE NO.: | | | |
| BID VALIDITY DATE: | | | |
| LOCATION OF MANUFACTURER: | | | |
| NAME OF MANUFACTURER: | | | |
| | | | |

| TAG NO. OR ITEM CODE | QTY | DESCRIPTION | DEL. WKS. | UNIT PRICE | EXTENSION | * | DEL. WKS. | UNIT PRICE | EXTENSION | * | DEL. WKS. | UNIT PRICE |
|---|---|---|---|---|---|---|---|---|---|---|---|---|
| | | | | | | | | | | | | |
| | | | | | | | | | | | | |
| | | | | | | | | | | | | |
| | | | | | | | | | | | | |
| | | | | | | | | | | | | |
| | | | | | | | | | | | | |
| | | | | | | | | | | | | |
| | | | | | | | | | | | | |
| | | | | | | | | | | | | |
| | | | | | | | | | | | | |
| | | | | | | | | | | | | |
| | | | | | | | | | | | | |
| | | | | | | | | | | | | |
| | | | | | | | | | | | | |
| | | | | | | | | | | | | |

| | | | |
|---|---|---|---|
| COST OF MATERIAL QUOTED | | | |
| FOB POINT | | | |
| EST. SHIPPING WEIGHT | | | |
| EST. SHIPPING CHARGES | | | |
| TERMS OF PAYMENT | | | |
| CASH DISCOUNT (DEDUCT) | | | |
| FIRM PRICE/MAX ESCALATION | | | |
| | | | |
| | | | |
| EST. TOTAL COST | | | |
| | | | |
| ACCEPT. OF FLUOR DANIEL TERMS & CONDITIONS | | | |
| MEETS SPECIFICATIONS – YES/NO | | | |

**Fig. 4-1.** Typical bid tabulation form. (*Courtesy of Fluor Daniel, Inc.*)

INQUIRY NO: ______________________

SHEET NO.: __________ OF __________

DATE: ______________________

BY: ______________________

| | | | BID CLOSING DATE: |
|---|---|---|---|
| | | | NUMBERS OF BIDS SOLICITED: |
| | | | VENDOR(S) NOT EVALUATED: |
| | | | |
| | | | |
| | | | |

| EXTENSION | * | DEL. WKS. | UNIT PRICE | EXTENSION | * | DEL. WKS. | UNIT PRICE | EXTENSION | * | RECOMMENDED VENDOR(S): TOTAL: |
|---|---|---|---|---|---|---|---|---|---|---|
| | | | | | | | | | | |
| | | | | | | | | | | REASON: |
| | | | | | | | | | | |
| | | | | | | | | | | COMMENTS/ATTACHMENTS: |
| | | | | | | | | | | |

| | | | APPROVALS: | (FULL SIGNATURE) | DATE |
|---|---|---|---|---|---|
| | | | PROCUREMENT | | |
| | | | PROCESS | | |
| | | | DISC. ENG.: | | |
| | | | PROJECT MGR. | | |
| | | | CLIENT | | |

to allow clarifications of bids, wherein the buyer's personnel ask the bidders to respond to specific questions. Some care must be exercised here, especially if there is a concern that one bidder is allowed to provide more information or to provide additional opportunities for pricing improvements that other bidders are not given the opportunity to provide.

Occasionally the technical review is conducted entirely by procurement personnel. If the requirements are clearly enough stated, and the material is unambiguous (that is, anyone can tell just by reading the quotes whether the items meet spec or not), this method is probably adequate. Any questions that arise in this review process are sometimes referred to engineering for resolution. In some organizations, arrangements have been made in advance that items A, B, or C can be considered acceptable and that any reviewer who finds those items that match the criteria can deem the bids acceptable without further review.

At this point, the bid item is ready for final commercial evaluation. Acceptance of buyer's terms and conditions is an important point, and once that hurdle is cleared, the next step is to determine lowest price and/or best delivery. If delivery is relatively unimportant, as it should be for MRO orders whose purpose is to maintain warehouse inventories, or if this is a case where the delivery date would have no impact on the schedule, the problem devolves to the lowest price. There is a practice known as "cherry picking," in which the lowest price per line item is determined, and the order split up among the bidders depending on who had the lowest price on each line item. In other cases, the bidder with the overall lowest total cost is selected regardless of how the pricing is broken out among individual items. Many companies follow a middle ground. One way that this can be calculated is to figure how much more would be paid if each of several combinations of item splits is selected. This premium is then compared against some reference number that represents the theoretical cost to issue a second or third purchase order to a second or third vendor in order to choose the lower priced items from each one. If the premium is less than or roughly equal to this number, then it is cost-effective to split the order among the bidders involved.

In cases where delivery is significant, it next becomes necessary to determine whether selection is to be made solely on the basis of delivery or whether some combination of improved delivery for a slightly higher price is acceptable. Splitting orders on the basis of delivery promises is practiced sometimes, although delivery promises are just that and are thus somewhat "softer" numbers than prices usually are. Overall price, by the way, is not necessarily the number that is quoted. First, pricing can be either firm or subject to escalation. Escalation can be by a fixed percentage or a "not-to-exceed" percentage, or be open (price in effect at time of shipment). The best bid tabulations are the ones where these numbers can be tied down as firmly as possible, although poorly defined engineering requirements can make this goal difficult to achieve.

Once all these activities are performed, the bid tab is signed off. Signatories are normally the purchasing person(s) authorized to spend money on behalf of the organization, the engineer(s) responsible for the technical content of the goods being bought, and usually some level of management. Once these signatures are obtained, it is possible to award the order to the bidder(s) selected, who then become vendors.

Awards are sometimes simply phoned to the successful bidder. This is most often done when time is valuable. Sometimes, there may not be a formally entered purchase order until almost time for material shipment, or not even then, depending on the supplier-client relationship and the company's policy on paperless transactions. Normally, though, the buyer's organization will issue an order, on paper or by electronic means, giving an order number, some statement of technical requirements and quantities ordered, with a delivery date and location expected. Unsuccessful bidders are usually notified as a courtesy, although it usually does not take very long for most unsuccessful bidders to figure out that they did not get an order.

It is possible that the material quoted by the successful bidder, or perhaps all the bidders, was not exactly as described in the RFQ. This can legitimately occur for several reasons. Of course, one common reason is that the engineering personnel who developed the requirements were not fully aware of exactly what materials and designs were commercially available. Perhaps the bidder or bidders chose to offer an item that was different in some respect from the original engineering requirements, but was still considered suitable by the people who did the technical review. Perhaps the bidders improved on the engineer's ideas. Regardless, when the successful quote does not match the engineering specifications, the best course is to change the specifications. After all, these specifications are going to be written into the purchase order, which is a legally binding document. Any discrepancies between what is in the order and what is actually provided, no matter how benign, form the basis for potential legal troubles somewhere.

From the time of order placement on, the problems become ones of order or contract administration; seeing that promises are kept; referring technical problems to the appropriate person for solution; expediting the order to see that it stays on schedule and does not get lost; arranging for inspection; ensuring that the data requirements such as outline drawings, parts lists, material certifications, and other technical data are produced and received; and arranging for shipment of the finished material to the free-on-board (FOB) point where the owner takes delivery of it. Then there are the financial considerations, being sure that the invoicing is received and is correct, that invoicing is reconciled with the quantity and identity of goods received, and that the bills are paid. Finally, there is the manufacturer's guarantee period, which is typically 18 months after shipment or 1 year after installation. Only after all these activities have faded into the past is a purchase order for material really complete and closed.

## What to Look for in Bids

The details of performing a technical evaluation of bids were glossed over in the previous section on the process of bidding and awarding an order. This subject merits a much more detailed look, because the review of bids from a technical aspect is second in importance only to the preparation of the engineering requirements package itself. In fact, the evaluation could be more critical, since here is where the engineering requirements come face-to-face with the manufacturing capability.

Sometimes a good manufacturer or manufacturer's representative can produce a good quote from a poorly stated or uncertain request for quote. Sometimes the manufacturers can produce more reliable descriptions of what is required than an engineer can, if the engineer is responsible for many facets of the plant requirements and not a specialist in any one area. More usually, a poorly written technical package means that important aspects of the engineering requirements may be left out of the bid preparation or improperly understood. The occasions when the salesperson can be expected to know more about the requirements than the engineer does are rare.

There can be a number of reasons why engineering requirements are not adequately or completely stated. Sometimes it is a matter of the experience level of the people involved. Stable personnel development situations, where experienced people can train newer people over a period of years, are getting less common, and not just in engineering. Work requirements in a design office often result in people working over their heads, which naturally results either in inadequate attention being devoted to RFQ preparation or in people doing things that they have never been properly trained to do. Those are not desirable situations, but their occurrence is a fact of life. In fact, one of the reasons for the publication of this book was the belief that engineers and users of valves could use more help and background information in the performance of their jobs.

In Chapter 1, there was a discussion of the preparation of a package for requisitions and what items and details need to be included in order to properly convey the full scope of the technical requirements. Here, the evaluation of the quotes received in response to that request for quote will be discussed, as well as how to determine if the bidders understand your requirements and what to do when they do not (which, for most engineers, happens more often than the case where the bidder gives you what you want even though you did not ask for it).

The person who does the technical review of bids, in most cases, is the person who prepared the technical package. There are organizations in which the person who reviews the bid packages for technical acceptance is deliberately chosen separately from the one who does the engineering. Sometimes this is done for convenience, as in cases where the procurement organization takes total responsibility for providing goods for the plant's use, at the obvious disadvantage of not having proper technical review and the risk of buying equipment that is not at all suitable for its intended service. The person who developed the requisition package is usually the one most knowledgeable about what the requirements are and is the best equipped to judge whether those requirements are met by the various bids. In any case, it should be a person familiar with the general requirements of the job. For plant work, if you have spent any time working in the plant, it should not be hard to visualize the places where these valves will be installed. Think about what services they will see and what abuses they will be subjected to. That will help focus your attention on whether the items you are evaluating will perform as you expect. Of course, a great deal of engineering work is performed in office locations far removed from the job site. There, almost all you have to go by are the specifications and your experience on similar projects, or what you have been told by the clients.

There is no denying that experience helps in performing a good bid review. There are many factors that influence the selection of acceptable bidders, and selecting the successful bidder. Some of these factors are explicit and are demonstrably true, such as cases where bidders state that they cannot meet certain of your requirements and either offer an alternate or decline to quote an item. Certain factors are simply the result of previous experience and are not easily documented, such as knowing that a particular manufacturer's product is less reliable than another's—perhaps a certain brand of valves seems to leak more than another. Another example might be a design for a proposed valve configuration that has never been built. One proposed design might look flimsier and less likely to operate successfully than another, but of course both bidders say they meet all your requirements and are willing to guarantee their work as being suitable for its intended service. What do you do then?

The answer is that there are a number of things that you can do, depending on your organization's procurement policies and your skill at documenting the technical review. But let's start with an overview of what you should be doing when performing a technical analysis of quotes.

Most of the time, quotes will come in addressed to the buyer who solicited them, usually over a span of a couple of days but sometimes all in the afternoon of the bid due date. That person will compile the quotes and separate out a copy for engineering review. (Some procurement policies require an unpriced copy for engineering review and a priced copy for procurement, so as to—theoretically—keep the pricing from influencing the engineering evaluation.) Sometimes the quotes come to the engineers just as they come in, and the person doing the evaluation has to track them and determine if all of them are accounted for. Sometimes there is a transmittal from procurement to engineering of a complete quote package, including everything received from all bidders, all at once. Sometimes a draft of the bid tabulation summary is included. Sometimes the technical evaluator prepares a summary of the analysis separately from the commercial bid tabulation. Sometimes, the engineer will disqualify all or portions of certain quotes, and the ones remaining are all considered technically acceptable, with procurement personnel making the final selection from among these, with price and delivery being the criteria. Sometimes, less often, the engineer may select the successful bidder, based on some combination of technical and commercial criteria, possibly including price.

However it is that the quotes have come to you, the person responsible for engineering evaluation, your first task is to read them carefully, one by one. Look for a statement of whether or not the bidder has taken any technical exceptions. Read each page of the quote, looking for notes that the bidder has made about the material. Sometimes the body of the quote, where the prices and the details of individual items are located, is the same bill of materials or requisition form that you sent out, but with pricing and delivery information filled in. This is the easiest way for both sides to work. The quote may be redone in a different format, though, and this is more difficult to follow since if the bidder chooses not to bid on some items and simply leaves them out of the proposal, it can be hard to keep your place while reading the quote, let alone figure out what items are omitted.

If your data sheet or valve description has a model number or figure number referenced, check each one to see that the figure number being proposed is equivalent to the one you listed. If you listed several acceptable figure numbers, this is a little easier. To be certain of a true correspondence, you should look up each figure number in the manufacturer's catalog and determine if the features of that valve truly correspond with the valve you referenced. If you choose not to list any representative figure number, you will still need to do this to ascertain that the bidder understood your description correctly.

There are published tables of equivalent figure numbers that show several manufacturers' products and the figure numbers that match up for each one. In general, these equivalent tables are not reliable enough to use, because they are not capable of defining each valve as specifically as you can in your description, and because these tables are not capable of taking into account some of the differences that exist in certain manufacturers' product where there is no *exact* equivalent and the approximate equivalent may or may not be suitable for what you plan to build. It is much better to check each and every one of the items against catalog data, or the data provided with the quote if it is sufficient enough. At this point you may find it worthwhile to look at the same item in each quote in turn, which may cause you to notice differences that you might not notice if you dealt with each quote in order.

After you have read each quote, and looked at some items across from quote to quote, you are ready to begin formulating your evaluation. Commodity materials generally can be evaluated on the basis of either meeting or not meeting the specification. Valves that are being designed specifically for your application, especially if they are a truly custom design, may take a much longer evaluation. Here you will have to determine if each design is truly suitable, especially if the design approaches are different from one quote to the next, and give more of a dissertation on the advantages and disadvantages of each design. For commodity valves, the presumption is that any item you deem to be acceptable could then be purchased for your use, and the engineering evaluation stops at this point. The commercial evaluation of acceptable delivery and lowest price can then occur. For special valves, evaluation may have to continue beyond considering whether or not the design is suitable. For instance, competing quotes may have valves that are different sizes, yet capable of the same flow capacity because of some design advantage. The cost consideration is then not just the cost of one valve versus the other, but also includes the cost of the extra reducers or perhaps the cost of modifying piping that already exists. A lower cost valve option may not be the overall least cost, in other words. If you have the ability to quantify these extra costs, by all means make note of them on the evaluation. These costs may not all appear in the same budget item or cost center, but since they are certainly costs that the company will have to pay, they are costs that need to be addressed, no matter how irrelevant they may appear to some individuals.

Your completed technical evaluation may take the form of, "These quotes are technically acceptable except for the following items . . ." or it may, in the case of highly engineered valves, be better to state that "quote X is the best technical proposal, followed by quote Y as second choice, followed by. . . ." In either case you have presented options for the people involved in the remainder of the procurement process leading up to order placement.

It may be that none of the quoted items that differed from your original description is acceptable. It can be a matter of great debate as to how close is good enough to qualify as "acceptable." For most commodity valves, a different manufacturer should be perfectly acceptable, as long as all the features you need are present in the other manufacturer's product. Generally, if you needed a valve that had to fit in a specific dimension, a valve that performs the same functions but is 2 in. longer is not acceptable. If there are other variables out of your control, such as the orientation of check valves, then a valve that works in horizontal lines only is not an acceptable substitute for one that can function in any orientation. Any item that you can think of that will cause a valve not to be capable of its intended service, regardless of whether you had posed it as a criterion in the original bid request or not, is fair game for rejecting an item. However, these items must be explained sufficiently on the technical evaluation for the readers of your evaluation, usually the procurement organization, to accept your reasoning. It is generally not considered legitimate to reject an offering just because it is different from what you asked for. For an item to be considered not acceptable, it generally has to be demonstrated to not be totally equivalent, or totally interchangeable, with the specified item as in the examples above. Experience on previous orders, either good or bad, may be cited as long as the experiences can be expressed in a concrete manner, such as, "These valves had a lot of packing leaks" rather than, "These valves were no good."

As a practical matter, it is better not to overevaluate a set of quotes. Where you have listed certain items as not meeting specification, it is better to list just one good reason, or two or three if there are that many, why they are not acceptable. But it is necessary to provide at least that much reasoning on each one because someone else in the organization (an auditor, a management level person, your counterpart in procurement) may quiz you on exactly why something is not good enough, if it is a lot cheaper than some of the other quotes. The documentation you provide should be good enough for others to agree that the items that are not acceptable are truly not equivalent and can not be substituted. It is possible to provide a conditional acceptance, to say, for example, that X item is acceptable provided that this and that are changed. It is better to obtain confirmation that "this and that" will indeed be changed, and whatever pricing adjustments required have occurred, and then release the item with an unconditional acceptance. These conditions for acceptance have a way of not happening if they are not dealt with during the evaluation.

# 5

# Modification, Repair, and Troubleshooting

## Valve Modification

This chapter will discuss valve modification, valve repair, and the ways you can determine whether your valves need repairing. The sections on valve modification and repair come first because the modifications and repairs that are done on valves usually do not have anything to do with the problems of particular valves. Most of the time, the valve *modifications* that you see are those performed by others on valves you are buying. The valve *repairs,* on the other hand, are in the nature of a general overhaul, as if you were sending them in for their 30,000-mile checkup.

The first time you will encounter valve modification is at the time you buy new valves. This may seem odd, but it is a fact of the valve business unless you need only very basic, simple valves. Most valve distributors, if they are at all reputable, will tell you in their quotes that they are planning to furnish valves that are modified to your requirements, and that someone is doing the modifications locally. Sometimes, though, distributors treat the modifications so routinely that they do not always mention that fact.

Local modification (i.e., not done at the factory) is a fact of life for several reasons. The valve business, in most of its market segments, is a fairly competitive one, and the more efficient producers, regardless of where they are located, are the ones that can crank out production runs of hundreds or thousands of the same thing. There are, by contrast, thousands of variations of body and trim types that could be produced, most of which are in demand at one time or another by the thousands of valve users in the world.

The usual compromise is to mass-produce the types of valves that sell the best, and to allow for local modification of trims and ends and other details. The other compromise that is made is to upgrade the choices of materials and types somewhat. The most obvious examples of these are the material choices, such as 304 versus 316 stainless steel. Although 304 stainless is cheaper in dollars per pound than 316, the material cost is not that big a component of valve cost, and the salability of the 304 is limited in the U.S. market. Therefore, although 304 valves are occasionally available, the economics of production costs are such that 304 valves will normally cost quite a bit more than 316. Or what is more likely is that few manufacturers will quote the 304 grade unless the valves involved are fabricated weldments. Thus, even though the 304 material is available for pipe, fittings, and flanges, there is no point in specifying it for cast or forged valves since the 316 will normally be offered anyway. There are few, if any, cases where the 304 would be suitable that the 316 would not be. In fact, it normally is the other way around, which is why the 304 fell out of favor in the first place.

Another example is body styles. At one time, small steel valves were available with your choices of several different combinations of union bonnet, bolted-bonnet, or welded-bonnet, and inside-screw or outside-screw stems. The present value of steel screwed and socket-weld valves is around $20 to $40 apiece, depending on size, which makes it not worthwhile to offer so many combinations. Therefore, while union bonnet, inside-screw valves may cost a dollar or so less to make than a bolted-bonnet, outside-screw valve, assuming there were production runs of thousands of valves, the size of most customers' orders is such that it is no longer economical to offer the lower quality designs, which is what the union bonnet is. Although it might have been economical to do so until the 1950s or 1960s, the economics have changed with higher labor and inventory costs to make it worthwhile to stock only the one or two varieties that sell the best and upgrade the customers that once would have used the lower grade products.

Another example is threaded versus socket-weld ends. The two end connections have dimension envelopes that almost completely overlap, such that the threaded end has a slightly larger outside diameter requirement (not even including the hex or octagon end used for wrench seating) and a smaller inside diameter, even measured to the bottom of the threads. The only dimension that crosses over the other way is the length of the female threaded portion, which is somewhat more than the depth of the socket-weld end. The implications of this to a valve manufacturer are obvious. Even though socket-weld valves probably outsell threaded-end valves in most markets, many manufacturers produce mostly the threaded-end variety. The ones that produce a lot of socket-weld end valves are the ones that do a lot of project work on a custom basis. Most overseas manufacturers, where the labor cost is relatively less and the freight cost relatively more, produce nothing but threaded-end valves for the U.S. market. The socket-weld ends are done by a machine shop, usually under contract to the distributor, sometimes without taking the valve apart.

The relationship of the distributor and the valve shop is quite important. In areas where a lot of valve sales occur, which are also areas where many plants are located, there are many small shops that specialize in working on valves. Some of these are large, well-established operations that may do some light manufacturing of their own. Some of them are also refurbishing and repair shops. Some are purely valve shops or welding shops that do only modification work for distributors. Many, especially in more outlying areas, are general machine shops or welding shops that might also, for example, fix plows. But most major distributors have under contract a modification shop, occasionally located on premises but usually in a separate location, that performs modifications like producing socket-weld ends from threaded ends and performing trim changes.

Trim changes are also quite common. New valves ordered for a project, where delivery times of 10 to 40 weeks can be accommodated in the design and construction schedule, are often built by the manufacturers exactly as specified by the purchaser. This includes special trims, special end preparation, special packing, and so on. But for plant orders where the delivery time may need to be a couple of weeks, or a couple of days, such practices are impractical. Instead, the valves that most distributors stock for plant orders have standard trims, the ones that are sold the most, such as F6 trim with hardfaced seats and braided packing for gate valves and 316 ball and stem with Teflon seats for ball valves. If you want something else, the most practical way to get it is for the distributor to pull a standard valve off the shelf and send it to their modification shop. There, almost anything feasible can be done. The most common modifications are packing changes, trim changes or seat material changes, flange facing modifications, and automation.

Some of these, like the packing change, are quite simple. All that is involved is unbolting the packing gland, pulling the old packing out, and adding the packing you have specified. Even a packing change, though, changes the pressure integrity of the valve, and you should expect the valves to be hydrotested again after the packing change (Fig. 5-1). The trim change is more involved. Here the valve is disassembled and the existing seating surfaces, usually F6 (13% chrome) are machined back. The seat rings in the body are usually removed and replaced if the new trim is 316 or another high-alloy material, but if the seating surfaces just need hardfacing, the existing surface is machined down and the hardfacing is deposited by a welding process. The seating surfaces on

**Fig. 5-1.** Test stand for flanged valves. (*Courtesy of Dunn's Valve Testers, Inc.*)

the wedge or disc are usually also machined back and the new alloy seating material is weld deposited. This is the most common practice now, since the gates and discs are usually carbon steel with a weld-overlaid seating surface. Only in smaller size valves are the discs usually made entirely of the trim material. Since the stem cannot be modified, most modification shops replace it with a new forged factory-made stem. Not all shops take the trouble to do this. Some may simply make one out of the appropriate barstock. This is a situation where the user must decide whether or not the shop-built modification is acceptable.

Other modifications to the body are common. One is the addition of taps and drains to the body. Most cast-steel valve bodies have bosses cast in at the points where taps are commonly made. The modification shop can easily drill and tap these bosses. Some castings do not contain these bosses, and in some it may not be possible to make the taps because the thickness is not there. Other possible problems include the tap size, especially in smaller size valves. MSS-SP-45 shows the standard tap locations, configurations, and sizes, and generally anything that exceeds the standards is difficult or impossible to do.

Modifications to flanges are fraught with problems. It is fairly common to reface flanges, either because of flange damage or because of the necessity of having finer finishes to accommodate spiral-wound gaskets.

However, especially with 1/16-in. (1.6 mm) raised-face flanges as used in class 150 and 300 (PN 16 and PN 25) valves, you can generally do one refacing. Not only do you have to watch the flange seating surface height (the 1/16-in. or 1/4-in. [1.6 mm or 6 mm] elevation of the seating surface above the balance of the flange face), you also need to watch the end-to-end dimension, which changes slightly every time you remachine a flange face. More than one refacing on a 1/16-in. (1.6 mm) raised face may cause some portions of the flange face outside the bolt holes to contact and exert pressure on the gasket, thus lessening the gasket seating force and increasing the possibility of a leak, but also making it difficult to see into the flanged joint after makeup to inspect the gasket edges. Remachining the outer portion of the flange to compensate for this is not always an acceptable solution, since the flange thickness is prescribed by ANSI and some of the class 150 (PN 16) flange thicknesses, at least, are light enough that it is actually possible to distort the flange during bolt-up if stiff gasketing and high bolt torques are used. Flange distortion is a direct cause of leaks.

A worse problem is the difference between flanged and butt-weld ends. It is possible to make a butt-weld end valve from a flanged-end valve, but generally this practice is more like butchery. A flanged-end gate or globe valve probably has a larger inside diameter at the end than most matching pipe sizes, so some weld buildup may be necessary. Since this weld buildup is precisely where the root pass of the butt weld will occur, it must be inspected thoroughly by dye penetrant or other means to ensure that no cracks are present. The outside diameter contours may not be adequate either. In general terms, it is a bad idea to try to convert a flanged-end valve to a butt-weld end. You can always convert a butt-weld end valve to a flanged-end, just by welding a flange to it, but then of course the end-to-end dimensions are off by several inches. There is no good way to convert a butt-weld end valve to flanged-end and keep the same dimensions, except in very small sizes.

Another common situation is converting a valve with raised-face flanged ends to ring-joint ends. Sometimes this can be done without weld buildup on the flange faces, but usually not. Any kind of major welding work like this could cause body warpage, depending on the relative size of the valve and the welds, resulting in at best a complete remachining of seating surfaces and at worst a valve that never closes completely. And while an average shop could perform such work on a carbon steel valve body, the same work on a chrome or stainless valve might not be as successful. Of course, in emergency maintenance work, there may be situations where you have to do any of the most undesirable things in order to keep the plant running, and a patched-together valve that works somewhat is preferable to no valve at all.

Automation is one of the most common major modifications to valves, although in many cases this entails no modification per se to the valve at all. There are a whole set of companies that do almost nothing but valve automation, and the selling of automated packages that consist of valves, actuators mounted on the valves and tested, with all the accessories required. The valves most often handled here are the basically on-off types that need some type of remote or power actuation, since true control valves typically are intended to be sold as a package and are designed for modular assembly, although they are often shipped as a complete package from the factory.

By contrast to the true control valves, there are a large number of gate, ball, butterfly, and plug valves that, although they are used in applications where an actuator is required, are exactly like any manually operated valve. These valves are generally bought from distributor stock or from the factory by the automation house, or sometimes provided by the customer from a larger valve order or from the customer's warehouse, along with manufactured actuators. The work of automating the valves consists of building brackets to mount the actuator onto the valve, or choosing the correct one, since in the last 10 years more and more manufacturers have designed this ability into their valves from the start, doing the mounting, and testing the complete package to see that it works and works correctly.

Many manufacturers also do this work. Critical or special valves, where the actuator is relatively standard, probably should be automated by the manufacturer, especially if special testing procedures such as kerosene testing or helium leak testing are required on the finished package. But the actuator houses can probably do it faster, especially if the valves and the actuators are available from stock. The only caution is that some manual valves are not good enough to be actuated, in the sense that the loads imposed on the valve in power actuation and the number of times it may cycle are usually in excess of the requirements for manual valves. Also, actuators, with the exception of some motor actuators, do not have the "feel" a person would have for how much force to apply or when to stop. The stops are normally in the actuator itself, and part of the work of automating a valve is making sure that the stops are set properly and that any signals that the valve has to send, like position indication, are being sent properly. Improper setting of stops, or significant oversizing or undersizing of actuators, is hard on the valve and/or the actuator and is likely to be the cause of a premature failure.

All of these modifications, where reputable companies are involved, carry the full warranty of the distributor and, depending on the arrangement, occasionally also carry the manufacturer's warranty. Usually, though, trim changes and other modifications will invalidate the manufacturer's warranty. Warranties are mostly good for replacement or repair of defective work, so in the practical sense for most private industry the distributor's warranty is good enough. Only if there are contractual or policy restrictions on warranties do you have to worry about losing the manufacturer's warranty. In these cases you may prefer to custom order your material from the manufacturer. It also might be better not to attempt unusually complex modifications at a local shop.

The owner of a valve is free to modify it in any way desired, subject only to the limitations of the codes on wall thickness and other basic safety-related requirements. The same modification shops that perform work for distributors engage in this business, as do many more whose business is limited to modification and repair of valves brought in by owners. The occasion for modifying a valve that has been in service might be that the original trim or original design did not work quite right, and that the components in question are being upgraded or replaced with something else. Often when a whole group of valves has problems like this, it is because the material furnished did not perform as advertised. Occasionally, owners modify valves to obtain special trims or other detail changes, like drain taps, because the delivery time on a new valve is unacceptably long. Generally, though, valves brought in by their owners for modification are the exception in the commercial valve modification business. It is more common for owners who are equipped with any sort of decent maintenance shop to do the work themselves, even on fairly specialized ones like ball valves. This depends on the availability and relative cost of outside labor, so plants located in outlying areas are more likely to do their own rebuild work. Specialized parts like balls and seats are usually sourced from the manufacturer. Companies who do their own maintenance work tend to do it on an as-needed basis, rather than wholesale rebuilds of groups of valves. The quality of work may not be as high as it might be with outside rebuilders, at least in part because "nobody else is ever going to see it."

Some companies have regular rebuild programs as part of their preventive maintenance program, wherein a group of valves at a time is sent out to be refurbished. Sometimes these are valves that have been pulled out of service one at a time as a result of failures or leakage, and sometimes this is an entire group of valves "harvested" from a unit that has been shut down or mothballed. These valves are normally not being modified in the sense of being changed, but more likely they are being rebuilt back to original condition. In cases where valve standards have changed over the years, some modifications may be made to bring the valve up to compliance with the current edition of the valve standard. Sometimes these refurbishing shops have contracts with the owner to maintain an inventory of rebuilt valves, wherein the owner sends the valves to the refurbisher, where they remain until the owner requests a given valve back. The refurbisher performs the necessary work and delivers the valve to the owner in "like-new" condition. Some valves that are not in a condition to be rebuilt acceptably may be used as a source of parts, with the remains being eventually scrapped.

For valves that are rebuilt, the process is as follows: To begin, each valve is disassembled (after having been checked over to determine that no harmful fluids remain inside), and pieces are match marked or marked with a code to ensure that matching bodies, bonnets, stems, and discs get back together at the end of the process. Then the major parts are grit blasted to remove all paint and adherent corrosion products. Usually the finely machined surfaces, like flange seating areas, are masked beforehand, or else the entire machined surface of a part is remachined after grit blasting. When only the bare metal is left, the remaining wall thicknesses can be checked to ensure that they at least meet minimum requirements. If any weld repair is necessary, it is performed at this stage, but generally only a small amount of weld repair can be performed before the cost of rebuilding the valve becomes uneconomical. Parts that are machined on all surfaces, such as the stem and seat rings (if removed), are inspected for corrosion damage and resurfaced if necessary, unless the stem diameter ends up below the minimum. Seat rings and disc seating surfaces are lapped flat. The parts can then be reunited, reassembled, and repacked, and the complete valve is then tested to original pressure test standards. Finally, finished valves are repainted and protected for shipment back to the owner.

The economics of refurbishing valves versus buying them new may depend on how much difficulty you have in getting new valves, but most owners consider 50 to 60 percent of the cost of a new valve as the threshold beyond which it is not worthwhile to repair a valve. By these criteria, cast-iron and bronze valves are almost never worthwhile repairing, except for very large or irreplaceable valves. Steel valves in class 150 are uneconomical to repair in the smaller sizes, up to about 4 in. or 6 in. or if they have suffered heavy damage. Stainless steel valves are almost always worthwhile repairing, since they are quite a bit more expensive and often will have suffered no perceptible corrosion damage, even after years in bad services. Valves that have been in a severe fire are generally not rebuildable, because of warpage in the body and possible loss of strength in the material.

A lower order of modification work is in the "junker" market. Here, a modification shop buys valves for its own account and does what refurbishing it feels necessary, then sells the valves to new owners. Often the same companies who do modification work for original owners also do this work. The valves so refurbished are often adequately rebuilt and capable of providing a few more years of service for their new owner. But these valves are exactly like used cars—you get what you pay for sometimes, but not always, and you never get more than you paid for. The refurbishing shop does give warranties on such valves, but some warranties are not worth very much. Very few plants have such a small maintenance budget that it is worth their while to buy such valves. Many companies do not permit such valves under any circumstances, except when they are desperate to get back on line and there is no other way to get the valve they need. Usually, in such cases, an acceptable product can be obtained if the owner's inspector is present in the shop for the rebuilding. While many owners will sell surplus and used valves to valve dealers, some companies will cut up valves deemed not safe to be rebuilt in such a way that they cannot be rebuilt, for instance by cutting through the bolt holes with a torch.

Another aspect of the business is "new surplus." The same valve dealers (they refer to themselves as "junkers") that buy valves for rebuilding will also buy valves from projects or occasionally from distributors that have received surplus valves back from projects. These valves reputedly are in new condition, and have never been installed. Occasionally this fact is obvious, but since valves do not come with odometers, it can be hard to tell without a careful examination whether the valve has, indeed, never been installed. Most valve dealers will be honest about the pedigree of the valves they sell, but they rarely know much. Any valves that are bought as "new surplus" should be assumed to be of suspect quality until examined and determined to be otherwise. Never assume that "new surplus" is equivalent to "new."

A related problem is "bogus" valves. These are valves that have manufacturers' tags on them, or are otherwise claimed to be made by a certain manufacturer, but in fact were modified from something else or even made new by an entity not related to the claimed manufacturer. This problem occurs from time to time, and the product being sold under the other manufacturer's name is always inferior to the genuine article, since it is a product made without benefit of engineering support and intended to be cheaper than the genuine product. The manufacturer being victimized has the right to sue, and usually does.

Generally, manufacturers discover that their valves are being counterfeited when they receive a service call on a leaking or failed valve. Serial numbers on the tag or heat numbers on the body are used to trace the origin of the valve and obtain the original data on valves in question, since the order data are rarely marked on the valve. For a genuine valve, this permits the proper spare parts to be identified and any problems with that model of valve (if the manufacturer knows of any) to be identified. For bogus valves, this is the point where it becomes demonstrable that the valve is not what it has claimed to be. Often, the salesperson making the service call will recognize on the spot that the valve in question is a fake.

Understandably, all manufacturers and factory distributors are opposed to the bogus valve business. What do the valve manufacturers think of the refurbishing business? Generally, they do not think highly of the practitioners of the art. Most manufacturers recognize that their customers have to perform maintenance on their capital assets, and are capable of supplying parts, which often means making parts to the original drawings, for their customers and companies that do valve rebuilding. Yet most manufacturers are not willing to assist valve rebuilders whose business it is to rebuild and sell used valves. In fact, at least one manufacturer has an unwritten policy to buy its own valves that come on the surplus market in large quantities, for the purpose of melting them down and preventing them from being available as used valves.

In most instances, the valve companies promote rebuilding programs by authorized service centers, usually factory service or their own company service centers but also including selected and qualified valve rebuilders. The manufacturers also provide maintenance manuals, and part numbers of commonly replaced parts, to owners for their own maintenance programs.

## Valve Repair

This section will discuss ways that you, the owner, can repair and maintain valves. It may be of only slight interest if your involvement with valves ends with specifying them, but the knowledge of what can be done to valves may guide your thinking in choosing valves in the first place.

The first question to ask is whether you intend to repair valves at all. The alternatives, as discussed previously, are to send them out for repair or to discard them and replace them with new ones. If you can send valves out for repair and refurbishment, you will want to consider the economic limits of refurbishment, which is that a valve is worth refurbishing if the cost of refurbishing it is less than 50 percent to 60 percent of the cost of a new valve. Obviously, other factors can enter into this decision. For one thing, if you have a large investment in valves for which parts are hard to get, and that are hard to rebuild, you probably will want to bite the bullet at some point and find a different valve to replace them with as they come up with problems. If you have valves that are not easily removable, such as butt-weld end, heavy-wall valves, the cost of removing and replacing them is such that it is economical to have a valve refurbisher living in your plant to rebuild the valves in place. At the other end of the spectrum, many low-value valves probably should not even be considered for rebuilding.

Other factors to consider in deciding whether to rebuild valves yourself are the shop space and equipment you have available, and the personnel. You will need at least a vertical mill and a medium-sized lathe, for flange refinishing and making or cleaning up small parts. You will also need at least some means of cleaning bodies, preferably in a blasting cabinet. Taking valve bodies out to the sandblast yard probably will not do, unless most of the valves you will refurbish are only lightly corroded. For refinishing flat seating surfaces, you will need a lapping machine.

While the bodies are being disassembled, be sure to mark each major piece (body, bonnet, disc, and stem) with a code to ensure that each of the pieces goes back into the valve that it came from. This is imperative if you have a mix of manufacturers represented, or if you have different types of trims. Stamp the parts in nonstressed areas, like the flange or bonnet flange, and the top or bottom of the stem. Be sure to mark the wedges of gate valves so that you know which side was which. The same applies to ball valves that have vent holes in the side of the ball. Discard the bolts (they have been stretched while in service), and for optimum gasket seating, the replacement bolting should be new.

After the body, bonnet, and disc have been cleaned up, you will want to do at least a visual examination for corrosion damage and pitting or cracking. Discard any parts that show gross damage or metal loss. The only way to ensure that sufficient wall thickness still exists is by measuring the thickness with calipers and comparing it to the minimum wall thickness charts in the standard to which the valves were originally bought (ANSI B16.34, API 600, MSS-SP-84, or other). Examine the stem for evidence of corrosion. The worst corrosion will probably be in the packing area, and will likely be pitting. The stems can be turned down and repolished in a lathe. Surface finish is very important to the sealing ability of the packing. The final polishing of the stem should be done by hand, with a motion parallel to the length of the stem (for rising-stem valves), to ensure that no ridges or rough spots remain to catch the packing material and dislodge it or score it. Check the stem diameter to see that it is still acceptable. Check the seating surfaces and relap them to ensure that they are flat and parallel. For ball valves, it may be possible to put the ball in a lathe and repolish the surface if it is slightly corroded, but many carbon steel balls are plated and an attempt at polishing may take the plating off, rendering the ball near-useless. Major pitting or seat damage will require rewelding, which is difficult to do yourself. Rewelding of hardfaced surfaces, especially, is very difficult to do, since it requires a considerable degree of preheat and a very careful deposition technique to avoid cracking in the weld deposit. Unless you are very well equipped with skilled welders, it would be better not to attempt this yourself. Many high-pressure globe valves and some others with conical seat designs are designed to be refaced with the seat in place, not by welding but by carefully shaving down the hardfaced surface. Most manufacturers of these valves sell a tool to do this with.

This is the time to check the seating surfaces to see if they are still capable of seating. Use bluing on the seating surface in the body and test fit the disc in place. Drop the disc into the body (you will need to slip it onto an old stem—not the stem you will be reusing), snug fit it into the seats and tap the top of the stem with a mallet to set it into the seats, then pull it out. Anyplace where the bluing still shows after the disc is pulled out is not seating properly. This technique works for any metal-seated gate or globe valve. Soft seats in butterflies and balls are harder to check prior to testing, but they are also more forgiving of assembly. Keep lapping the seating surfaces until they look as if they will seal. Even so, some seats may not pass the pressure test the first time. Check for wear on gate valves, since the wedging action means that as the seating surfaces wear or are refaced, the wedge will seat farther and farther down in the body. If it seats too low down, the top edge of the wedge will no longer contact the top of the body seat and the valve will leak. This is usually considered condemnable, except that you can make new, thicker seat rings to replace the old ones. Gate valves with welded-in seat rings are much harder to rebuild adequately than are gate valves with threaded-in seat rings. Ball valves that have had the ball returned also normally will require shims behind the seats or new, thicker seats to compensate.

Reassembly for gate, globe, and ball valves begins with reinstalling the bonnet bushing, if removed, and inserting the stem in the bonnet. A little grease or lubricant may help, if it will not be injurious to the process. Repack the valve at this time. Repacking is critical. Begin by installing one packing ring at a time, tamping it down, and being sure to install the packing rings in the correct sequence if you have a mixed packing stack. This is almost always the case with graphite packing. With cut packing rings, stagger the cut of each ring about 120 degrees around from the one below it. If there is a lantern ring, be sure to get it at the right height so it will still function. After the packing is almost all in the stuffing box, test fit the gland and tighten it down, then remove it and add the last ring or two and retighten. Do not forget that normally the gland and gland flange are captive—that is, they have to be in place while you slip the stem up through them. Bolt the gland flange down again and position the stem in the halfway open position. The handwheel does not have to be pinned back onto the stem nut yet, but you will need the handwheel to move the stem.

After you have determined that the seating surfaces have been properly lapped together, install the wedge or disc onto the stem end. Many globe valves have a retainer piece or a pin to hold the stem and disc together. Be sure that the wedge in a gate valve is oriented the correct way. (Since the seats and disc have been lapped together, the disc will seat correctly in only one orientation.) Then drop the bonnet and disc into the body, with the correct bonnet gasket in place, and bolt the bonnet down. After the bonnet is bolted down, check the operation to see that the wedge or disc goes all the way up or down without binding. The stem should not be too easy or too hard to turn—that is a symptom of improper packing tightness. Ball valves go together differently. Split-body and end-entry valves are assembled by setting one seat into the larger half of the body, with the seat springs if present, then setting the ball in place over the tongue of the stem (the ball will be in the closed position). Then set the other seat over the ball, and install the tailpiece or other body half and tighten down. Top-entry ball valves require the ball and both seats to be set in place as a unit, then proceeding as with a gate valve. Butterfly valves usually have to be assembled backward, with the disc in place so that the stem can be inserted from the outside and then pinned to the disc.

Add the handwheel or lever, or in the case of actuated or gear-operated valves perhaps a temporary gear operator to allow for operation during the shell test. Ball and butterfly valves can be tested just with a wrench to turn the stem. The testing procedure, regardless of the test standard used, normally begins with a shell test. The valve is mounted in the test fixture or, if a test fixture is not available, is bolted up with a test flange (a blind flange with a small test connection) on each end. Fill the body with water, being sure to vent it thoroughly, then pressurize to the minimum test pressure. Hold for the minimum time required, checking all bolted joints and the packing area for leakage. It helps to blow any remaining water off the outside of the valve with an air hose prior to starting the test. Look at the body for evidence of through-the-body leakage, which should be unlikely. If there are any leaks in the packing or bonnet joint, depressure the body before retightening the bolts, then retest. The disc or ball should be in the half-open position so that all interior surfaces of the shell are pressurized. Especially with high-pressure tests, be careful of the test gaskets so that if one blows out, it will not cause injury. Even water can be dangerous at 2000 psi.

If the shell test is successful, you can go on to test the seats. Depressure and drain the body, but it is not necessary to turn the valve yet. If this valve is in a test stand with one end up, as is the usual case, you can go on to test the seat that is visible (Fig. 5-2). For actuated valves, including gear operators, you will probably need to mount the actuator at this time, if not present for the shell test. Close the valve, then hold the other end of the valve down (or keep the blind flange bolted on that end) and pour enough water into the valve to cover the seating area. Then apply air behind the seat (into the valve body) for the usual air-under-water test. This is the easiest testing method, since bubbles are easy to see forming if the valve has a slight leak. Another way to do it, if the valve is on the floor in the horizontal position, is to be sure the seat being tested has a test flange securely bolted on, and run a tube from the end of the valve into a bucket filled with water. Look for bubbles there. The bubbles can be counted either way, if your test standard allows X number of bubbles per minute. If the leakage is not acceptable, back off the disc or ball and reseat again. It is a good idea to move the valve all the way from full-closed to full-open and back again. You can use as much seating force as you might expect an operator to use in the field, but not more. Gears and actuators will probably need the stops adjusted. If no amount of playing with the valve seating will get it to shut off, it will be

(a)

(b)

**Fig. 5-2.** Test stand for check valves. (*Courtesy of Dunn's Valve Testers, Inc.*)

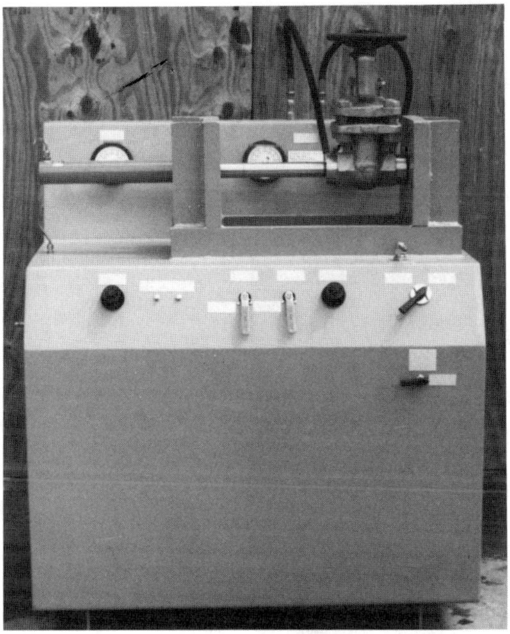

**FIG. 5-3.** Test stand for socket-weld valves. (*Courtesy of Dunn's Valve Testers, Inc.*)

necessary to pull it off the test stand and break it down to see why it does not seat. Once the first seat has passed the seat leakage test, take it off the test stand and turn it over to test the other seat. Single-seated valves will not need this, of course. Testing of socket-weld end valves requires special rams or fixtures to hold an O-ring seal in each end for the shell and seat tests (Fig. 5-3). Be careful of ball valves, especially, if it is necessary to disassemble the valve, because in the open position there will be pressure built up around the ball in the body cavity.

Actuated valves can be shell tested without the actuator being mounted, but the true test of shutoff ability is by seat testing with the actuator mounted (Fig. 5-4). The idea is to see if the *actuator* can successfully close the valve to the seat tightness specified.

If you have a number of valves to test, be sure you can keep track of which ones have been tested and passed and which ones have not. If this is not done carefully, it can lead to problems later (Fig. 5-5). The ones that have passed can then be prepared for shipment or return to the warehouse. Even if these valves will remain in the confines of your own plant, it is still a good idea to apply end protectors. If you have a bad atmosphere, it is a good idea to put a little grease on the gasket seating surfaces of steel valves prior to applying the end protectors. The best are ⅛-in. (3 mm) thick plastic ones retained with separate lock plugs, but the one-piece molded polyethylene end protectors with integral lock plugs are good also. If at all possible, paint the exterior of carbon and chrome

**Fig. 5-4.** Test fixture for butt-weld-end valve, with motor operated valve wired up for seat tightness testing. (*Courtesy of Pacific Valves, a unit of Mark Controls.*)

**Fig. 5-5.** Pressure testing actuated ball valves at jobsite. (*Photograph by R. C. Merrick.*)

valves to keep rust down, even if it is just one coat of paint. Put a cardboard wrapper over the stem of rising-stem valves prior to painting, and leave it there to keep dirt off.

If all of the above description seems daunting, be aware that manufacturers will provide you with repair manuals, some rather comprehensive, that guide you step-by-step through the repair procedures. This is especially true of valves other than regular gate, glove, and check valves, the primary ones discussed in the previous paragraphs. Many of these other valves have very specialized repair techniques, sometimes requiring special tools. All of these requirements, along with a list of recommended spare parts, can be obtained with the initial valve orders. Many accounting practices do not allow you to buy the spares and the tools with the valves, however, since they might be considered expense rather than capital items or because of a similar accounting nicety. Just be aware that, once you have bought valves from X manufacturer, the parts have to come from that manufacturer or someone capable of making parts to the original dimensions, as if you were buying parts for your '57 Chevy.

## Troubleshooting Valves

In the course of running a plant, there are always problems. The preceding section discussed how to fix valves when they are in the shop for repair, but there is another step before that—namely, the problem of how to decide if there is something wrong with a valve, and if so, what is it that is wrong, and what to do about it. The most important parts of this problem are the first and second, because the third part, what to do about it, is generally a matter of having to fix it now or later. Many valves that have something wrong on the inside, though by no means all, have to be either removed from the line and taken away to be worked on, or opened and worked on on the spot. In either case the line the valve is in has to be shut down, and if it contains hydrocarbons or hazardous fluids, purged until it is safe to enter. If it is a high-temperature line, such as a steam line, it will have to cool down also.

Some examples of valves that do not always have to be opened for an internal problem to be fixed are lubricated plug valves and ball valves with sealant injection fittings. These designs are common on pipelines and in a few other places. A few other types of valves, such as through-conduit gates and some specialized gate and globe valves for, say, hydrofluoric acid service, also have these injectors. Their purpose is to add a sealing material in the area where shutoff is supposed to take place, and if the valve has been closed but it does not shut off the flow completely, an injection of sealant may solve the problem.

Leaks to atmosphere are a perpetual problem in any pressurized system. With end flange leaks, if it is a nonhazardous fluid you may be able to use one of the sealant injection devices made for flanges. Packing leaks may or may not be stopped by tightening the gland. Not many packing glands come loose by themselves, or if they have done so, the leak may have already run long enough by the time someone spots it to have disrupted the packing. Either way, it may not be possible to pull up on the gland bolting enough to stop the leak.

With most valves, you will have to decide whether or not to leave a problem alone until there is an easy or an economical time to fix it. That sort of decision is really economic, up to a point. You will have to do something if the problem represents a safety hazard. The germane question is whether or not the piping has been configured so that you can block in and depressure the line that the bad valve is in without bringing the unit down. Assuming you can, then you will want to see if there is a spare valve available to replace the bad one, so that you can diagnose or fix the other at leisure. Economics may not permit such a thing, but be prepared for the worst if you pull a valve out without having a spare. Here is where it helps to have a valve rebuilder standing by.

You might be able to tell what is wrong with the valve before you take it out of line. If the problem is that the valve will not close all the way, there could be something caught under the disc, which is easy to remedy, or there could be scratches or gouges in the seating area, which is not. Soft-seated valves especially are very prone to cuts and gouges in seats. You should easily be able to obtain replacement seats for most soft-seated valves.

If the valve is making noise, you will want to listen to it in operation. If it is making a loud banging noise, the likely problem is that the disc is moving around because of a flow problem (undersizing or oversizing). If it is making a gravelly or grinding noise, it likely is cavitating (pressure drop problems). These problems will not be solved by fixing the valve, but by replacing or modifying it to a more appropriate size. This may take some modifications to the piping.

# 6

# Valve and Actuator Sizing

Valve sizing is a subject that is often neglected, even by valve specifiers who should know better. At times it can be difficult to predict correctly the size that is required. This is not necessarily because the equations are difficult—in fact the equations are generally based on rule of thumb or on flow test data. The difficulty in valve sizing lies in the fact that good data are often hard to come by. Data on variables such as flow rates are unreliable, especially when the valves being sized are located in a new unit under design. Much of the data available to the valve engineer then are based on the theoretical performance of many other components that have not been tested or even built. However, since the appropriate course of action is to work with the best available data, the discussion here will be how to size valves and for which valves the sizing concerns are of any importance.

Most manual valves do not need sizing. The user simply chooses the valve size that matches the pipe size. The only caveat here is that the valve size, meaning the end connection size, may or may not be exactly what you want, considering the fact that reduced ports are standard in many types of valve. In the general case, the reduced-port valve is perfectly acceptable and in fact even desirable, since (a) they cost less than full-port valves and (b) they generally open and close with less torque/force than would be required for a full-port valve. If pressure drop is of concern, the reduced-port does indeed produce a greater pressure drop (it has a smaller $C_v$) than a full-port. In most cases, however, the difference is not that great and pipe routing (length and number of bends) has at least as much influence on overall pressure drop, usually more, than the choice of full- or reduced-port valves. The type of valve also has a bigger influence on pressure drop, since globe valves, butterfly valves, and plug valves all are fairly high consumers of $\Delta P$, more so than reduced-port gate valves or ball valves.

The use of full-port valves is normally required for certain reasons. One is if some object must pass through the valve. A good example would be a fouling service where connections or the line itself must be rodded out periodically to clear the buildup of deposited material. Another example would be a line that must be pigged, whether for product separation or cleaning or inspection. Even check valves in such lines must be full-port. Another case for using full-port valves would concern the inlet lines to safety relief valves, which are normally subject to the requirements of ASME section VIII. Although the code is a little vague in its definitions, it in effect requires the use of full-area (which is almost the same as full-port) valving in the lines from a code vessel to the relief valves protecting it.

Other services that require full-port valves are ones where slurries or entrained solids are carried in the flow stream, and the reduced port provides a place for material to settle out. Entrained material or high-velocity flow may cause erosion damage to any place where a reduction in cross-sectional area occurs, either at a reduced-port valve or at a reducer. Occasionally, a positive drainage may be required, where the line must be sloped to ensure that no liquid remains in it. In such lines, the valves must be full-port to provide a flat flow profile.

Almost all types of valves are available as reduced-port. Some types are essentially not available as full-port. Gate valves in forged steel (API 602 pattern) are normally reduced-port, while API and ANSI cast-steel valves and stainless steel valves are generally full-port or close to it. Most ball valves are reduced-port in all sizes, although in smaller sizes a full-port valve is quite common. Full-port plug valves are difficult and expensive, while a full-port or full-area butterfly valve is basically impossible by definition. Some globe valves are close to full area, but the convoluted flow path makes a globe unsuitable for almost all the reasons cited above for full-port valving. Globe valves built for throttling are almost invariably considerably reduced in port size, both for acutator sizing and for accuracy of control.

Beyond these considerations for the choice of full versus reduced port, the sizing question for manual valves reduces to just a few specific cases. The most common are the sizing of boiler stop-check and blowdown valves and the sizing of special check valves. All of the cases where special sizing procedures are called for are ones where the correct valve size may be smaller than the line size, and if a line size valve is chosen without resorting to a sizing investigation, the valve might just self-destruct. In other words, oversizing can be every bit as harmful as undersizing.

The steps required to size a manual valve are usually simple. The manufacturer should be able to provide sizing charts, which will show a relationship between flow rate and size or a similar curve that allows the user to determine the minimum or maximum valve size. The only difficult part generally is coming up with the data required to do the sizing. For water and steam services, the problem is generally easier since most equations and charts assume, in the case of liquid, water, and in the case of vapor, steam. For hydrocarbon services, especially gases, the problem can be more complicated because you, the user, may have to put in your own factors to compensate for the difference in density between, for example, hydrogen and steam. It will also often be necessary to assume a percentage open, for valves that could be operated partially open without harm. This sort of calculation is most commonly done on control valves, but it can occasionally be required on manually operated valves also.

The sizing of control valves, although generally a more technically challenging problem, is made somewhat easier by the fact that control valve sizing is done every day by thousands of engineers around the world, and many control valve manufacturers have provided sizing programs that run on microcomputers, just as there have been sizing programs that run on hand-held calculators, minicomputers, and mainframes, and before that, there were the charts and nomographs. All of these methods are still available to the user, of course, and the paper versions are still sometimes the easiest to use for the occasional user or those unequipped for computer calculations.

## Sizing Examples—Manual Valves

No attempt will be made here to provide sizing charts. For one thing, the manufacturers are really responsible for providing accurate data and charts, and books have a way of being static while engineering data are dynamic. Some examples are provided, mainly to show how the task is accomplished.

The first example will be the sizing of a boiler stop-check valve. Remember the principle of operation of a stop-check is that it provides the feature of automatic closure if the steam flow reverses, but also provides the ability to be manually closed to isolate the boiler from the steam header or vice versa. The stop-check uses a globe design to accomplish this dual purpose, but it is generally not used for throttling. It is sized to be fully open during normal operation.

Since the sizing chart relates velocity to pressure drop, the first step is to gather the information required to determine velocity. This includes design and operating conditions, pressure, temperature, and boiler steam rate in pounds per hour. From the design conditions comes the maximum flow rate, and sometimes the minimum flow rate, that the valve will be expected to handle. From the operating conditions comes the flow rate that the valve will be expected to see most of the time. It may be necessary to calculate maximum, minimum, and normal cases to be sure the same valve can handle all three. If one valve absolutely will not handle all three, it will be necessary to provide a bypass or some other design solution. If this situation appears, it would be a good idea to present the problem to the manufacturer and let the manufacturer determine whether or not it is safe to subject the valve to those conditions.

In addition to the charts, you will need access to a steam table to determine specific volume, which will be expressed in cubic feet per pound. Note that most steam tables show data in units of pounds per square inch *absolute,* and you will need to account for the 14.7 psi (1.03 bar) difference between absolute and gage pressures. The chart will allow you to determine the pressure drop across the valve by reading up from the value of flow rate times square root of specific volume and reading across from the curve to the pressure drop. There will be several curves on the chart, one for each of several valve sizes. The pressure drop produced by that flow rate is the value actually used to size the valve. You may find that a couple of different valve sizes are acceptable. You may also find that, if you calculate the same valve from different manufacturer's charts, one may come out to be a size larger or smaller than another. This does not happen often, and although it is possible that the charts are not terribly precise, it is more likely that the two valves are physically slightly different. If you get this result at the time you are calculating sizes prior to going out for bids on the valves, then you will want to recheck the calculations once you select a manufacturer. You probably will want the manufacturer to do the calculation anyway, as a cross-check on your calculation. The manufacturer will generally be willing to do this, since their valve will be subjected to the service and possible failure if it is mis-sized. A word to the wise: If you alone do the calculations, and the sizing is done wrong, then the manufacturer will not be sharing the blame for any resulting failure.

The *minimum* allowable pressure drop is generally about 1 psi or so. The reason for the minimum is that the $\Delta P$ is what holds the disc open. Pressure drops lower than this will cause the disc to try to drop shut, which will restrict the flow, which will raise the $\Delta P$, push the disc open wider, allowing the pressure drop to fall, which will cause the disc to try to close, and the result is at best a high-wear situation that causes the valve to fail in a few years and at worst a strangled flow that causes process failure somewhere in the system. The *maximum* pressure drop is, depending on the valve, about 10 psi or so. The maximum pressure drop is limited by wear on the seating surfaces by erosion, but it is possible that even 10 psi pressure drop is unacceptably large for the system hydraulics. The guiding principle of locating pressure drops is that the greatest pressure drop should be across the component that is controlling the flow, which is usually a control valve.

Besides the stop-check valve, the other component that needs this calculation is the check valve, especially one that might see variable flow rates. If variable flow rates are not present, then normally the only calculation that must be made is the one for minimum flow rate, which determines the maximum size of the valve. Maximum flow-rate calculations set a limit on the minimum size of the valve. The point of calculating the maximum size of the valve is to determine whether or not the line size is the maximum size. If the maximum valve size is equal to or greater than the line size, then use the line size for the valve size. This eliminates any problems with reducers or dimensional difficulties. But if the maximum size is smaller than the line size, then it will be necessary to abide by that maximum. In the literature, there will be mention of the fact that a smaller valve costs less than a line-size valve. This is true, but it is really beside the point unless your plant is on a very tight budget, since a smaller size valve plus two reducers and two extra welds may or may not be cheaper than one line-size valve.

The size calculation for check valves is especially important on valves in the discharge line from a reciprocating compressor. This is primarily true because the flow rate is constantly variable due to the pulsations. Sizing the valve is important, but it is also important to select a valve that has low movable mass, since regardless of your best efforts, it usually will not be possible to locate the valve beyond the effective reach of the pulsations. This is a fact of life regardless of the number of analog studies you run or the number of pulsation dampeners you install. The choice of valve comes first, and the preferred choices are either a leaf- or plate-type check valve with multiple flow passages or a tilting-disc check. Once you have selected the type, then you can calculate the size. Since the calculations are somewhat complex (they involve the resonance frequencies of the components, among other things) it would be better to provide your service data to the manufacturer or bidder and let them recommend the valve size for you.

For more conventional check valves, sizing of the valve is seldom necessary, but one factor should be considered. There is a certain minimum pressure required to initiate opening, known as cracking pressure. The cracking pressure is normally a fixed number for a given type and size of valve and is often very small, on the order of 0.1 to 0.5 psi or 2 to 15 in. of water (0.01 to 0.03 bar). In larger valves, the cracking pressure is generally smaller because of the increased force available relative to the mass of the components. Normally the only time cracking pressures need to be considered is in very low pressure or low differential pressure applications. The cracking pressures will be greater if the opening force is working against gravity, such as in vertical upflow lines.

In steam services and other very high pressure and very high cost valve locations, it may even be advantageous to size block valves for specific locations. The point of the calculation, after all, is to size the flow orifice or port diameter. But, as a general rule, in class 600 (PN 40) and lower carbon steel and stainless valving for hydrocarbon or water services, there is no point. One of the obvious reasons for this is that venturi valves are not normally made in lower ratings. Also, the damaging effects of high pressure drop are less severe when the system pressure is lower to begin with. Another reason is that steam, despite its not being terribly corrosive, is a fairly difficult service because of its nonlubricating nature and the combination of high pressure and high temperature involved. Another reason could be that power plant designers have historically been more used to calculating everything individually rather than relying on standardized approaches as practiced in some parts of the hydrocarbon processing industry.

The calculations required for continuous blowdown valves are somewhat different. The continuous blowdown valve is the one located near the boiler water level. It removes surface scum, oil, and dissolved impurities. The continuous blowdown valve is normally an angle needle type. Probably its closest relative is the choke used in oil and gas production on wellheads. The main difference in the sizing calculations is that for a continuous blowdown valve you are mostly sizing the orifice, or port size, and only incidentally the valve size. The valve size is usually determined by the blowdown line size, but once the blowdown valve body size is known, you can then select the correct chart to do the orifice sizing. You will need the design conditions, pressure, temperature, and pounds per hour of blowdown. There will be a special blowdown valve chart provided by the manufacturer. Note that only a small number of manufacturers make true boiler blowdown valves, and just a standard throttling valve is not acceptable for this service because of its severity. Since the blowdown valve is essentially a throttling valve, you will still need to consider rangeability. The ideal flow rate for a blowdown valve is about 50 to 65 percent of full capacity. So, knowing the design flow rate, you can calculate the full-capacity flow rate by dividing the design flow rate by (0.50 to 0.65). With the design flow rate, you can enter the sizing chart and find the point at which the design flow rate and the boiler pressure lines intersect. The orifice size curve that is next above this point is the correct choice.

The intermittent blowdown valve, which is the valve located near the bottom of the boiler to dispose of sediment and scale that collects on the bottom, also requires sizing, but sizing is not as critical as with the continuous blowdown valve. With the intermittent blowdown valve, the sizing refers to the body size. The code requires two valves, one for sealing and one for blowing, to protect the integrity of the seal. The blowing valve is opened last and closed first so that it takes the brunt of the pressure differential. The code requires other design provisions such as the requirement that the seat edge must be at least 25 percent of the diameter below the centerline of the valve. This effectively prohibits standard globe valves and puts some restrictions on the geometry of Y-pattern globes. Although a conventional valve may be suitable for intermittent blowdown after passing all these criteria, it is more common to use special valves, especially in high-pressure blowdown. Often a single valve body with two independent seats and discs is used. Another design used for the sealing valve is the seatless type, which is an angle pattern with the flow into the side port and out the bottom port, the opposite of conventional angle globes.

See ANSI B31.1 or the ASME Section I boiler code, whichever applies to your location, for more specific rules on boiler blowdown.

## Sizing Examples—Manual or Remote-Actuated Valves

All of the sizing calculations described above are ones that apply almost exclusively to manual valves. The next group of calculations are ones that can apply equally to manual or remote-actuated valves. Depending on circumstances and type of valve, almost all of these calculations might be done for actuated valves, but some, such as for butterfly valves, quite often are required for manual valves that perform some sort of throttling function either on a regular basis or a "set once and forget it" type of application.

Because control valves could be expected to be used in any fluid service, the place to start is with the fluid characteristics. The known information is generally the fluid flow in gallons per minute, the specific gravity and the allowable or maximum permissible pressure drop. The size of valve required can be determined from these data by the equation $C_v \doteq Q\sqrt{(G/\Delta P)}$ where $Q$ is the flow capacity in gpm, $G$ is the specific gravity relative to water, and $\Delta P$ is the differential available at the proposed location of the valve in psi. This is the basic liquid sizing equation. The specific gravity causes the required $C_v$ to increase proportionally with increasing specific gravity. Another fluid property, viscosity, also affects required $C_v$ but in a much less direct manner. In many cases viscosity can be ignored, especially in full turbulent flows. There is a published curve showing the correction factor required, based on Reynolds number. If the Reynolds number is not already known, it can be easily determined by a nomograph. At Reynolds numbers of approximately 4000 and higher, the viscosity correction factor essentially disappears. For low-flow or highly viscous services where the correction factor is significant, the calculated $C_v$ must be multiplied by the correction factor to yield a corrected $C_v$, which can then be used for selecting a valve size.

Sizing control valves for vapor flow is somewhat more complex because of the greater variation in compressibility and other physical factors. For a complete discussion of sizing, refer to the *Control Valve Handbook* or the *ISA Handbook.*

Remember that the first step, choosing the type of valve you need, has occurred outside of these equations by considering the type of fluid, the percentage curve and shutoff time and allowable leakage desired, the line size, and other more mechanical criteria. Once you have selected the type and style of valve, the sizing can proceed easily—as long as you are within the available body and trim sizing ranges. If not, step back to the choice of valve body and find something else more suitable. The sizing is done generally by reading the manufacturer's charts and selecting the body size, and trim size in the case of many common control valve types. The correct choice is the one that has the next larger $C_v$ than the calculated value, unless it is very much larger or there is one just a few percent smaller than the calculated requirement. To use a slightly smaller trim, go back and refigure the sizing to see how much pressure loss will arise and how much flow rate will result from the reduced flow capacity. It is in fact a good idea to calculate the predicted flow rate and pressure drop from each choice of valve and trim size. If this calculation is not standing alone, but is part of an entire line sizing calculation, these values are required for the complete flow calculation.

In general, the proper size for a control valve body is smaller than the proper size for the line. This is true for several reasons, one being that the object is to have as little pressure drop as possible in the piping and as much as possible, relatively speaking, in the valve. This is not true for actuated valves whose purpose is blocking in a system or uncontrolled venting of a system for emergency cases, or valves that are actuated because of their size or remoteness. Here, the valve itself is or should be the same as the other manual valves in the system, and the sizing concerns reduce to those of a manual valve, namely, that there be adequate flow capacity (line size). Valves whose purpose is to control flow should have as much pressure drop made available as possible, since the pressure drop or the differential pressure is one of the most common inputs for control signals to the valve. Making pressure drop available is achieved by minimizing pressure drop elsewhere in the system. Sometimes a rule of thumb can be used in which the control valve is one pipe size smaller than the line size. If the trim size permits, this selection gives a gradual reduction in flow area from the pipe cross section to the valve inlet port cross section to the trim size. There will, of course, always be cases where the trim size has to be small enough that the body needs to be two or more sizes smaller than the line, and cases where the trim size required is so large that it approaches the line size, but these are generally specialized applications.

# SUGGESTIONS FOR FURTHER READING

Many of the books and charts listed here are private publications. No attempt has been made here to list vendor catalogs. Any active specification engineer will quickly acquire shelves full of vendor catalogs, and you will find for yourself which ones are useful. All of the material here, even those publications compiled by valve companies, is published and available to anyone. Many of the sizing charts and equations have been computerized and are also available as published software.

## Books


*Check Valve Application Manual and User's Guide,* Rockwell International Inc. (now Edward Valves Inc.) Raleigh, N.C., 1988.

*Comprehensive Guide to Pneumatic Actuators for Valve Automation.* Wayne Ulanski. El-O-Matic USA Inc., Hackensack, N.J., 1987.

*Control Valve Handbook.* Fisher Controls International, Inc. Marshalltown, Iowa, 1977.

*Control Valve Sourcebook.* Fisher Controls International, Inc. Marshalltown, Iowa, 1988.

*Corrosion Data Survey, Metals Section.* National Institute of Corrosion Engineers. Houston, Tex., 1985.

*Corrosion Data Survey, Nonmetals Section.* National Institute of Corrosion Engineers. Houston, Tex., 1983.

*Equivalent Valves Reference Manual.* Oildom Publishing Co. of Texas. Houston, Tex., 1980.

*Flow of Fluids through Valves, Fittings and Pipe, Technical Paper No. 410.* Crane Company. King of Prussia, Pa., 1988.

*Glossary of Valve Terms.* Grove Valve & Regulator Company. Oakland, Calif., 1980.

*Hydraulic Valves and Controls, Selection and Application.* John J. Pippenger. Marcel Dekker, Inc., N.Y., 1986.

*ISA Handbook of Control Valves.* J. W. Hutchison, ed. Instrument Society of America, Research Triangle Park, N.C., 1976.

*Lyons' Encyclopedia of Valves.* Dr. Jerry Lyons. Van Nostrand Reinhold, N.Y., 1975.

*Lyons' Valve Design Handbook.* Dr. Jerry Lyons. Van Nostrand Reinhold, N.Y., 1982.

*MSS-SP-92 Valve Users' Guide.* Manufacturer's Standardization Society of the Valve and Fittings Industry. Arlington, Va., 1980.

*Piping Handbook.* Reno C. King, ed. McGraw-Hill, N.Y., 1973.

## Directories

*Thomas Register.* Thomas Publishing Company, N.Y., issued annually.

*USA Oilfield Service, Supply and Manufacturers Directory.* Pennwell Publishing. Tulsa, Okla., issued annually.

# APPENDIX
# Codes, Standards, and Specifications

This section lists some of the most useful codes, standards, and specifications that pertain to valves in some way. This is by no means all of the specifications you might need for a complete piping system. Only the standards that have some relevance to valve design, construction, or installation are listed.

For a more complete list, the best place to go is to your local library (public or in-house) to consult one of the private standards listing and publication services, such as VSMF. With these services you get, first, a paperback book more than an inch thick listing all the standards and specifications that they distribute, which also serves as an index, and second, a set of microfiche cards or microfilm rolls that contain the actual standards. This form is much cheaper than having paper copies of all these specifications. These standards publication services maintain up-to-date editions of U.S. codes and standards. From these organizations, in some cases, it is also possible to obtain "historic" or superseded/voided standards. There is a tendency for standards that are voided or deleted by their issuing organization to be referenced by users for 2 or 3 years afterward, in the same manner that revised standards sometimes do not get referenced in user specifications until 2 or 3 years after the fact.

No attempt has been made here to indicate multiple adoptions. Many of these standards have been adopted by the Department of Defense or referenced by the NRC and are so listed in the standard itself.

Some of the items listed here as standards are referred to by their own issuing organizations as specifications or something else, but they meet our definition of a standard.

## Codes

ASME

Boiler and Pressure Vessel Code

| | |
|---|---|
| Section I | Power Boilers |
| Section IIA | Material Specifications Part A |
| Section IIB | Material Specifications Part B |
| Section IIC | Material Specifications Part C |
| Section III | Nuclear Power Plant Components |
| Section V | Nondestructive Examination |
| Section VIII, Division 1 | Unfired Pressure Vessels |
| Section VIII, Division 2 | Unfired Pressure Vessels |
| Section IX | Welding and Brazing Qualifications |

| | | |
|---|---|---|
| ASME/ANSI | B31.1 | Power Piping |
| ANSI | B31.2 | Fuel Gas Piping |
| ASME/ANSI | B31.3 | Chemical Plant and Petroleum Refinery Piping |
| ASME/ANSI | B31.4 | Liquid Petroleum Transportation Piping Systems |
| ANSI | B31.5 | Refrigeration Piping |
| ASME/ANSI | B31.8 | Gas Transmission and Distribution Piping Systems |
| ASME/ANSI | B31.9 | Building Services Piping |
| ASME/ANSI | B31.11 | Slurry Transportation Piping Systems |

## Standards

| | | |
|---|---|---|
| ASME/ANSI | B1.1 | Unified Inch Screw Threads |
| ASME/ANSI | B1.5 | Acme Screw Threads |
| ASME/ANSI | B1.20.1 | Pipe Threads, General Purpose |
| ASME/ANSI | B1.20.7 | Hose Coupling Screw Threads |
| ASME/ANSI | B16.1 | Cast Iron Pipe Flanges and Flanged Fittings |
| ASME/ANSI | B16.5 | Steel Pipe Flanges and Flanged Fittings |

| | | |
|---|---|---|
| ASME/ANSI | B16.10 | Face-to-Face and End-to-End Dimensions for Ferrous Valves |
| ASME/ANSI | B16.11 | Forged Steel Fittings, Socket-Welding and Threaded |
| ASME/ANSI | B16.14 | Ferrous Pipe Plugs, Bushings and Locknuts with Pipe Threads |
| ASME/ANSI | B16.18 | Cast Bronze Solder Joint Pressure Fittings |
| ANSI | B16.20 | Ring Joint Gaskets and Grooves for Steel Pipe Flanges |
| ANSI | B16.21 | Nonmetallic Gaskets for Pipe Flanges |
| ASME/ANSI | B16.24 | Bronze Flanges and Flanged Fittings |
| ASME/ANSI | B16.25 | Buttwelding Ends |
| ASME/ANSI | B16.33 | Manually Operated Metallic Gas Valves for Use in Gas Piping Systems up to 125 psig |
| ASME/ANSI | B16.34 | Steel Valves, Flanged, Threaded and Welding End |
| ASME/ANSI | B16.37 | Hydrostatic Testing of Control Valves |
| ASME/ANSI | B16.38 | Large Metallic Valves for Gas Distribution (Manually Operated) |
| ASME/ANSI | B16.40 | Manually Operated Thermoplastic Gas Shutoffs and Valves in Gas Distribution Systems |
| ASME/ANSI | B16.41 | Functional Qualification Requirements for Power Operated Active Valve Assemblies for Nuclear Valve Plants |
| ANSI | B16.104 | Control Valve Leak Testing |
| ASME/ANSI | B18.2.1 | Square and Hex Bolts and Screws, Inch Series |
| ASME/ANSI | B18.2.2 | Square and Hex Nuts, Inch Series |
| ASME/ANSI | B18.3 | Socket Cap, Shoulder and Set Screws |
| ASME/ANSI | B46.1 | Surface Texture |
| ASME/ANSI | N278.1 | Self-Operated and Power-Operated Safety-Related Valves Functional Specification Standard, Reactor Plants and Their Maintenance |
| API | 6A | Specification for Wellhead Equipment and Christmas Trees |

| | | |
|---|---|---|
| API | 6D | Specification for Pipeline Valves (Steel Gate, Plug, Ball and Check Valves) |
| API | 6FA | Specification for Fire Test for Valves |
| API | 6FC | Specification for Fire Test for Valves with Selective Backseats |
| API | 594 | Wafer Type Check Valves |
| API | 595 | Cast Iron Gate Valves with Flanged Ends |
| API | 598 | Valve Inspection and Test |
| API | 599 | Steel and Ductile Iron Plug Valves, Flanged or Buttwelding Ends |
| API | 600 | Steel Gate Valves, Flanged or Buttwelding Ends |
| API | 601 | Metallic Gaskets for Piping—Double-Jacketed, Corrugated and Spiral-Wound |
| API | 602 | Compact Steel Gate Valves |
| API | 603 | Class 150 Corrosion-Resistant Gate Valves |
| API | 604 | Ductile Iron Gate Valves with Flanged Ends |
| API | 605 | Large Diameter Carbon Steel Flanges |
| API | 606 | Compact Carbon Steel Gate Valves with Extended Body |
| API | 607 | Fire Test for Soft-Seated Quarter Turn Valves |
| API | 608 | Metal Ball Valves, Flanged and Buttwelding Ends |
| API | 609 | Butterfly Valves, Lug Type and Wafer Type |
| API | 941 | Steels for Hydrogen Service at Elevated Temperatures and Pressures in Petroleum Refineries and Petrochemical Plants |
| API | 942 | Recommended Practice for Welded Plain Carbon Steel Refinery Equipment for Environmental Cracking Service |

Many of the ASTM standards listed below have been adopted in their entirety or with very slight modifications by ASME for use in Boiler Code work and are published by the ASME in part II of the B&PV under the same numeric designation, but with an SA or SB prefix replacing the ASTM's A or B prefix.

| | | |
|---|---|---|
| ASTM | A105 | Specification for Forgings, Carbon Steel, for Piping Components |
| ASTM | A126 | Specification for Gray Iron Castings for Valves, Flanges and Pipe Fittings |
| ASTM | A182 | Specification for Forged or Rolled Alloy Steel Pipe Flanges, Forged Fittings, and Valves and Parts for High-Temperature Service |
| ASTM | A193 | Specification for Alloy Steel and Stainless Steel Bolting for High-Temperature Service |
| ASTM | A194 | Specification for Carbon and Alloy Steel Nuts for Bolts for High-Pressure and High-Temperature Service |
| ASTM | A197 | Specification for Cupola Malleable Iron |
| ASTM | A216 | Specification for Steel Castings, Carbon, Suitable for Fusion Welding for High-Temperature Service |
| ASTM | A217 | Specification for Steel Castings, Martensitic Stainless and Alloy, for Pressure-Containing Parts Suitable for High-Temperature Service |
| ASTM | A307 | Specification for Carbon Steel Bolts and Studs |
| ASTM | A320 | Specification for Alloy Steel Bolting Materials for Low-Temperature Service |
| ASTM | A350 | Specification for Forgings, Carbon and Low-Alloy Steel, Requiring Notch Toughness Testing for Piping Components |
| ASTM | A351 | Specification for Steel Castings, Austenitic, for High-Temperature Service |
| ASTM | A352 | Specification for Steel Castings, Ferritic and Martensitic, for Pressure-Containing Parts Suitable for Low-Temperature Service |
| ASTM | A370 | Test Methods and Definitions for Mechanical Testing of Steel Products |
| ASTM | A395 | Specification for Ferritic Ductile Iron Pressure-Retaining Castings for Use at Elevated Temperatures |
| ASTM | A487 | Specification for Steel Castings Suitable for Pressure Service |

| | | |
|---|---|---|
| ASTM | A494 | Specification for Castings, Nickel and Nickel Alloy |
| ASTM | A522 | Specification for Forged or Rolled 8% and 9% Nickel Alloy Steel Flanges, Fittings, Valves and Parts for Low-Temperature Service |
| ASTM | A536 | Specification for Ductile Iron Castings |
| ASTM | A564 | Specification for Hot-Rolled and Cold-Finished Age-Hardened Stainless and Heat-Resisting Steel Bars, Wire and Shapes |
| ASTM | A723 | Specification for Alloy Steel Forgings for High-Strength Pressure Component Application |
| ASTM | A743 | Specification for Castings, Iron-Chromium, Iron-Chromium-Nickel, and Nickel-Base, Corrosion-Resistant for General Application |
| ASTM | A744 | Specification for Castings, Iron-Chromium-Nickel and Nickel-Base, Corrosion-Resistant for Severe Service |
| ASTM | B61 | Specification for Steam or Valve Bronze Castings |
| ASTM | B62 | Specification for Composition Bronze or Ounce Metal Castings |
| ASTM | B148 | Specification for Aluminum Bronze Castings |
| ASTM | B283 | Specification for Copper and Copper Alloy Die Forgings (Hot-Pressed) |
| ASTM | B367 | Specification for Titanium and Titanium Alloy Castings |
| ASTM | B381 | Specification for Titanium and Titanium Alloy Forgings |
| ASTM | B763 | Specification for Copper Alloy Sand Castings for Valve Applications |
| AWWA | C207 | Steel Pipe Flanges for Waterworks Service |
| AWWA | C500 | Gate Valves for Water and Sewerage Systems |
| AWWA | C501 | Cast-Iron Sluice Gates |
| AWWA | C502 | Dry-Barrel Fire Hydrants |

| | | |
|---|---|---|
| AWWA | C503 | Wet-Barrel Fire Hydrants |
| AWWA | C504 | Rubber-Seated Butterfly Valves |
| AWWA | C506 | Standard for Backflow Prevention Devices—Reduced Pressure Principle and Double Check Valve Types |
| AWWA | C507 | Ball Valves |
| AWWA | C508 | Standard for Swing Check Valves for Waterworks Service |
| AWWA | C509 | Resilient Seated Gate Valves for Water and Sewerage Systems |
| AWWA | C540 | Power-Actuating Devices for Valves and Sluice Gates |
| AWWA | C550 | Protective Interior Coatings for Valves and Hydrants |
| AWWA | C800 | Underground Service Line Valves and Fittings |
| CI | 6 | Recommended Practices for Valving in Chlorine Services |
| CGA | G4.4 | Industrial Practices for Gaseous Oxygen Transmission and Distribution Piping Systems |
| ISA | S7.3 | Quality Standard for Instrument Air |
| ISA | S7.4 | Air Pressures for Pneumatic Controllers, Transmitters and Transmission Systems |
| ISA | S20 | Specification Forms for Process Measurement and Control Instruments, Primary Elements and Control Valves |
| ISA | S75.01 | Flow Equations for Sizing Control Valves |
| ISA | S75.02 | Control Valve Capacity Test Procedure |
| ISA | S75.03 | Face-to-Face Dimensions for Flanged Globe-Style Control Valve Bodies (ANSI Classes 125, 150, 250, 300 and 600) |
| ISA | S75.04 | Face-to-Face Dimensions for Flangeless Control Valves (ANSI Classes 150, 300 and 600) |
| ISA | S75.05 | Control Valve Terminology |

| | | |
|---|---|---|
| ISA | S75.07 | Laboratory Measurement of Aerodynamic Noise Generated by Control Valves |
| ISA | S75.08 | Installed Face-to-Face Dimensions for Flanged Clamp or Pinch Valves |
| ISA | S75.11 | Inherent Flow Characteristics and Rangeability of Control Valves |
| ISA | S75.12 | Face-to-Face Dimensions for Socketweld-End and Screwed-End Globe-Style Control Valves (ANSI Classes 150, 300, 600, 900, 1500 and 2500) |
| ISA | S75.14 | Face-to-Face Dimensions for Buttweld-End Globe-Style Control Valves (ANSI Class 4500) |
| ISA | S75.15 | Face-to-Face Dimensions for Buttweld-End Globe-Style Control Valves (ANSI Classes 150, 300, 600, 900, 1500 and 2500) |
| ISA | S75.16 | Face-to-Face Dimensions for Flanged Globe-Style Control Valve Bodies (ANSI Classes 900, 1500 and 2500) |
| MIL | V-1189 | Valve, Gate, Bronze |
| MIL | V-5027 | Valve, Check, Oxygen, High Pressure |
| MIL | V-5529 | Valve, Hydraulic Directional Control, Rotary Selector |
| MIL | V-12003 | Valves, Plug: Cast Iron or Steel, Manually Operated |
| MIL | V-13803 | Valve, Gate: Double-Disk, Cast-Iron or Steel |
| MIL | V-16468 | Valves, Shutoff, Hydrocarbon Fuel Service, 100 psi and 150 psi Working Pressure |
| MIL | V-16556 | Valve, Solenoid, Three Way Bypass (Naval Shipboard Use) |
| MIL | V-16720 | Valve, Foot, Brass and Polyvinyl Chloride |
| MIL | V-16721 | Valve, Foot, Low Pressure, Cast-Iron |
| MIL | V-17501 | Valve, Globe, Hydraulic Operated, Piston and Diaphragm Type, Magazine and Hangar Deck Sprinkling |
| MIL | V-18110 | Valve, Gate: Cast or Forged Steel, and Alloy Steel, Outside Screw and Yoke (Sizes 2½ in. and Larger) |

| | | |
|---|---|---|
| MIL | V-18434 | Valves: Gate, Globe and Angle, Steel |
| MIL | V-18436 | Valves, Check: Bronze, Cast-Iron and Steel Body |
| MIL | V 22052 | Valves, Stop and Stop Check, Globe, Angle and Y Pattern, Cast or Forged Carbon or Alloy Steel, Outside Screw and Yoke (Sizes 2½ in. and Larger) |
| MIL | V-22687 | Valves, Ball, Naval Shipboard, for Air, Nitrogen, Helium, or Hydraulic Service (Sizes 2½ in. and Below) |
| MIL | V-24109 | Valves, Globe, Angle, Quick Change Cartridge Trim, High Pressure Hydraulic and Pneumatic |
| MIL | V-24232 | Valve, Pneumatic Four-Way, Two-Position Manually Operated |
| MIL | V-24509 | Valves, Flanged, Ball and Lubricated Plug for Combined Sewage and Seawater Service |
| MIL | V-24624 | Valves, Butterfly, Wafer and Lug Style, Shipboard Service |
| MIL | V-24630 | Valve, Check, In-Line, for Hydraulic Fluid and Lubricating Oil Fluid |
| MIL | V-46725 | Valve, Four-Way, Hydraulic, Manual |
| MIL | V-46727 | Valve, Three Position, Four Way, Hydraulic, Manual |
| MIL | V-58039 | Valve, Gate, Rising Stem, Double Acting, Aluminum |
| MSS | SP-6 | Standard Finishes for Contact Faces of Pipe Flanges and Connecting-End Flanges of Valves and Fittings |
| MSS | SP-9 | Spot-Facing for Bronze, Iron and Steel Flanges |
| MSS | SP-25 | Standard Marking System for Valves, Fittings, Flanges and Unions |
| MSS | SP-42 | Class 150 Corrosion Resistant Gate, Globe, Angle and Check Valves with Flanged and Butt Weld Ends |
| MSS | SP-44 | Steel Pipe Line Flanges |

| | | |
|---|---|---|
| MSS | SP-45 | Bypass and Drain Connection Standard |
| MSS | SP-53 | Quality Standard for Steel Castings for Valves, Flanges and Fittings and Other Piping Components, Magnetic Particle Examination Method |
| MSS | SP-54 | Quality Standard for Steel Castings for Valves, Flanges, Fittings and Other Piping Components, Radiographic Examination Method |
| MSS | SP-55 | Quality Standard for Steel Castings for Valves, Flanges, Fittings and Other Piping Components, Visual Method |
| MSS | SP-60 | Connecting Flange Joint Between Tapping Sleeves and Tapping Valves |
| MSS | SP-61 | Hydrostatic Testing of Steel Valves |
| MSS | SP-67 | Butterfly Valves |
| MSS | SP-68 | High Pressure–Offset Seat Butterfly Valves |
| MSS | SP-70 | Cast Iron Gate Valves, Flanged and Threaded Ends |
| MSS | SP-71 | Cast Iron Swing Check Valves, Flanged and Threaded Ends |
| MSS | SP-72 | Ball Valves with Flanged or Butt-Welding Ends for General Service |
| MSS | SP-78 | Cast Iron Plug Valves, Flanged and Threaded Ends |
| MSS | SP-80 | Bronze Gate, Globe, Angle and Check Valves |
| MSS | SP-81 | Stainless Steel, Bonnetless, Flanged, Wafer Knife Gate Valves |
| MSS | SP-82 | Valve Pressure Testing Methods |
| MSS | SP-84 | Steel Valves, Socket Welding and Threaded Ends |
| MSS | SP-85 | Cast Iron Globe and Angle Valves, Flanged and Threaded |
| MSS | SP-88 | Diaphragm Type Valves |
| MSS | SP-91 | Guidelines for Manual Operation of Valves |
| MSS | SP-92 | Valve Users Guide |

| | | |
|---|---|---|
| MSS | SP-93 | Quality Standard for Steel Castings and Forgings for Valves, Flanges and Fittings and Other Piping Components, Liquid Penetrant Examination Method |
| MSS | SP-94 | Quality Standard for Ferritic and Martensitic Steel Castings for Valves, Flanges, Fittings and Other Piping Components, Ultrasonic Examination Method |
| MSS | SP-96 | Guidelines on Terminology for Valves and Fittings |
| MSS | SP-98 | Protective Epoxy Coatings for the Interior of Valves and Hydrants |
| MSS | SP-100 | Qualification Requirements for Elastomer Diaphragms for Nuclear Service Diaphragm Type Valves |
| NACE | MR-01-75 | Sulfide Stress Cracking Resistant Metallic Materials for Oilfield Service |
| NFP | B93.7M | Hydraulic Fluid Power—Valves—Mounting Interfaces |
| NFP | B93.9M | Symbols for Marking Electrical Leads and Ports on Fluid Power Valves |
| NFP | B93.33M | Interfaces for 4-Way General Purpose Industrial Pneumatic Directional Control Valves |
| NFP | B93.40M | Series of Mounting Interfaces for 4567 Maximum psi Four Port Hydraulic Fluid Power Directional Valves |
| NFP | B93.49M | Hydraulic Fluid Power—Valves—Pressure Differential—Flow Characteristic—Method of Measuring and Reporting |
| NFP | B93.55M | Hydraulic Fluid Power—Solenoid Piloted Industrial Valves—Interface Dimensions for Electrical Connectors |
| UL | 125 | Valves for Anhydrous Ammonia and LP Gas (Other than Safety Relief) |
| UL | 144 | Pressure Regulating Valves for LP Gas |

| | | |
|---|---|---|
| UL | 260 | Dry Pipe and Deluge Valves for Fire-Protection Service |
| UL | 262 | Gate Valves for Fire-Protection Service |
| UL | 312 | Check Valves for Fire-Protection Service |
| UL | 429 | Electrically Operated Valves |
| UL | 789 | Indicator Posts for Fire-Protection Service |
| UL | 1091 | Butterfly Valves for Fire-Protection Service |
| UL | 1106 | Marine Manually Operated Shutoff Valves for Flammable Liquids |
| UL | 1121 | Marine Through-Hull Fittings and Sea-Valves |
| UL | 1468 | Direct-Acting Pressure-Reducing and Pressure-Control Valves for Fire-Protection Service |
| UL | 1486 | Quick Opening Devices for Dry-Pipe Valves for Fire-Protection Service |
| UL | 1726 | Automatic Drain Valves for Standpipe Systems |

## Recommended Practices

| | | |
|---|---|---|
| API | RP550-I-6 | Control Valves and Accessories |
| ISA | RP75.06 | Recommended Practice for Control Valve Manifold Designs |

# GLOSSARY OF VALVE TERMS

The following terms are ones that are used in a specific sense in the context of this book, especially as they relate to valves, piping, or materials. The common meaning, or the standard dictionary definition, of these terms may be a little different. Common scientific terms are not included.

When you are working within the rules of a given design code, that code may define some of these terms in a much more specific manner than the definition used here. The definitions in this book are not intended to be in conflict with definitions as used in common codes and standards, but rather are intended to acquaint the reader with the nomenclature as commonly used in conversation and general written correspondence about valves.

Other good glossaries of valve terms are contained in the *Valve Handbook,* by Dr. Jerry Lyons, and several books published by manufacturers, including the *Glossary of Valve Terms*, published by Grove Valve and Regulator Company.

**AARH** Arithmetic average roughness height—a measure of the roughness of a surface expressed in microinches. The higher the number, the rougher the surface. Numerically very close to RMS (root mean square) roughness height, often used interchangeably. The designation Ra will be replacing these terms.

**acetal** A fairly rigid slippery plastic, used as a seat material for valves in high-pressure, noncorrosive applications. One trade name is Delrin.

**ACI** Alloy Castings Institute—the organization, now defunct, that developed the numbering system for cast grades of alloy steel, such as CF8M or CN7M. Was a division of the Steel Founders' Institute.

**Acme thread** A thread pattern in which the height of the thread is approximately the same as its width, but trapezoidal with a flat top and 29-degree included angle on the two sides. Mostly used for transmitting force. It is the most common thread form for stems.

**AFNOR** Association Française de Norme—French national standards organization.

**AIChE** American Institute of Chemical Engineers—organization concerned with the useful application of chemical reactions, such as would be used in fluid processing.

**AISI** American Iron and Steel Institute—organization that developed the numbering system for steels, commonly applied to rolled or wrought products, such as 1018 or 4340.

**alloy steel** A steel consisting primarily of iron, with some percentage of one or more other elements such as chromium, nickel, manganese, or vanadium deliberately added to enhance its properties.

**AMS** Aerospace Materials System—set of standards for materials of interest to aircraft and space vehicle design, certain of which are also useful for fluid power and general industrial work.

**angle valve** A valve in which the inlet and outlet end connections are at right angles to each other. Not to be confused with a Y-pattern valve.

**annealing** Heating to a given temperature, then slow cooling, in order to produce some desired microstructure or mechanical properties. Generally annealing is used to soften a steel.

**ANSI** American National Standards Institute—standards organization responsible for coordinating the work of U.S. standards writing groups with each other and with other national standards organizations. Known as ASA until 1967.

**API** American Petroleum Institute—organization that develops standards for materials and articles used in the petroleum- and gas-gathering (production) industry, and also the hydrocarbon-processing industry.

**aramid** A high-strength fiber, used in place of asbestos fibers in some applications. Some other fibers have better chemical resistance.

**arcuate** A curve that parallels the arc of a circle and joins to it at each end.

**arc welding** Any one of a group of processes that uses melting created by the heat of an electric arc between an electrode and the base metal, with or without the addition of filler metal.

**ASA** American Standards Association—former name of what is now ANSI. Renamed United States of America Standards Institute in 1967, then American National Standards Institute in 1969.

**ASM** American Society for Metals—organization that publishes a great deal of basic data on metals, metal properties, and metal fabrication.

**ASME** American Society of Mechanical Engineers—organization responsible for maintaining several codes and large numbers of standards, covering numerous different industries.

**ASTM** American Society for Testing and Materials—organization responsible for maintaining standards covering materials, testing methods, and in some cases such as plastics, the dimensional and manufacturing standards for finished products.

**austenite** A microstructure found in iron and nickel alloys, in which the crystal structure is face-centered cubic. Austenite does not undergo phase changes with temperature.

**austenitic stainless steel** The common stainless steels, where the primary microstructure is austenite and the composition is primarily iron but also includes both chrome and nickel. These steels are designated as 300 series, such as 304, 304L, 316, or 347.

**average wall** The standard way of referring to pipe wall thicknesses, in which the mill tolerances are expressed as a percent plus and minus, above and below the specified wall thickness.

**AWS** American Welding Society—organization responsible for standards covering welding materials and codes for structural and sheet metal welding.

**backflow** Flow opposite to the normal direction of flow.

**backseat** In many rising-stem valves, the seating surface at the top of the body cavity, which a shoulder on the stem seats against to prevent leakage up through the packing area.

**ball valve** A valve that has a spherical, or section of a spherical, closure element that opens and closes by rotating one-quarter turn.

**bar** Unit of pressure equal to 14.5 psi or 100 kPa.

**BB** Bolted bonnet—a valve closure construction in which the primary closure (bonnet or cap or cover) is retained by bolts that resist the internal pressure forces.

**Belleville spring** or **Belleville washer** A disc-shaped conical spring, resembling a washer, which provides spring force along its own centerline. Also sometimes known as a cup spring.

**bending moment** The resultant of a force and its distance from a given point, whose action would tend to cause the bending of the item in question.

**bevel** A surface prepared for welding by machining, grinding, or cutting at an angle to the adjacent material surfaces. May have any one of several standard or special contours.

**BHN** Brinell Hardness Number—a dimensionless indicator of hardness. The Brinell hardness test is defined in ASTM-E10. The higher the number, the harder the material. Also known as HB.

**blanket order** A purchase order that allows the release of various quantities and types of material over a period of time, all at an agreed price per item as contained on published price sheets, rather than a lump sum individual purchase of items.

**block and bleed** A valve configuration in which the flow through the valve, from the inlet port to the outlet port, is blocked, while another small port is provided for the purpose of bleeding down (draining or depressuring) the cavity in between. Sometimes a single valve so equipped, sometimes an assembly of two or three valves.

**blowdown** In boiler terminology, the practice of discharging from the boiler small quantities of water from regions rich in dissolved or suspended solids, in order to keep the solids from concentrating in the boiler water and fouling the inside surfaces of the boiler. Also, in safety relief valves, refers to the discharge of fluid through the valve when it opens.

**body** The part of a valve that encloses the internals and to which the topworks are attached. Normally the end connections are an integral part of the body.

**bolt** Fastener consisting of a threaded bar section, with or without one integral nut, with or without a short unthreaded section.

**bolt circle** The diameter of the circle that the bolt hole centerlines occur on, in flanges and other circular bolted connections.

**bonnet** The cover or removable top component of a valve, containing the packing gland and stem opening. Generally gate and globe valves are considered to have a bonnet. The same part on other valves is called the cover or cap.

**bore** The same as inside diameter, but generally said of the ends of butt-weld valves and of flanges where the inside diameter is actually bored or finish machined. Tolerances can be much closer than the inside diameter of a cast body.

**bright anneal** An anneal in a controlled (low-oxygen) atmosphere to prevent discoloration of the surfaces. Used on stainless steels.

**BSI** British Standards Institution—organization responsible for standards in Great Britain and Northern Ireland.

**bubble-tight** A seat leakage condition in which, during the allotted time of the test, no perceptible leakage comes past the seat being tested. Applies to air-under-water testing—the same test using water-under-air is referred to as **drop tight**. Defined in API 598, MSS-SP-61, and ANSI B16.104.

**Buna-N** Common term for nitrile rubber. One trade name is Hycar (trademark of Goodrich Chemical Co.).

**bush** Same as bushing. Preferred term in British usage.

**bushing** Hollow, short cylindrical component surrounding a shaft or stem to guide or support it.

**butterfly valve** A valve that has a circular disc-shaped closure element that pivots one-quarter turn about its vertical centerline to open and close.

**butt weld** Weld where the adjoining edges to be welded are parallel and facing each other. The weld preparation for a butt weld can be a single- or double-angle bevel, a J-bevel or U-bevel, or parallel without bevel for very thin welds.

**bypass** A smaller line containing a valve that comes off of a larger line just upstream of a major valve and rejoins the same line just downstream of the valve. The line provides a way to bypass the main valve.

**cap screw** A threaded fastener that has an integral head and threads all the way up to the head.

**carbide** A compound of carbon and a metal, such as iron carbide or chromium carbide.

**carbide precipitation** Migration of carbides to the grain boundaries in a metal, coming out of solid solution. These carbide-rich areas are more susceptible to brittle fracture and the carbide-depleted areas are more ductile.

**carburizing** A method of impregnating the surface of a steel part with carbon, forming carbides and increasing the surface hardness.

**cast** or **casting** A product or the act of producing a product made by pouring molten metal into a mold and allowing it to solidify, thus taking the shape of the mold. The mold can be produced in any of a number of ways.

**cavitation** The phenomenon in which the local pressure at a point in a flowing fluid becomes lower than the vapor pressure of the fluid, thus causing small bubbles to form. The bubbles then implode when the local pressure rises again, causing shock waves that are very destructive to the walls of the passageway or the valve trim in the area of the cavitation.

**cementite** Common name for iron carbide, found in steels.

**CGA** Compressed Gas Association—organization responsible for standards covering fittings on gas bottle connections and other aspects of specialty gas handling and safety.

**chain wheel** Device that is shaped like a handwheel but has a chain guide around its periphery. Either attached to or in place of a handwheel for operating valves higher than a person's reach.

**characterized** A special shape cut into a seat or disc such that the flow varies in a specific manner as the valve opens and closes.

**Charpy test** Test to measure notch toughness, or ability to resist fracture from loads on surface discontinuities, by imposing an impact load on a specimen opposite from a V-shaped notch. Used primarily for low-temperature work on materials that exhibit a pronounced drop in toughness below a certain temperature. Charpy test is defined in ASTM-E23.

**check analysis** The analysis taken from a heat of steel after it has cooled and has been semifinished.

**check valve** Valve whose purpose is to prevent flow from returning in the direction from which it came. Most check valves have no means of outside actuation, depending on flow forces, gravity, or both.

**chevron packing** Packing that consists of endless rings of molded plastic or fiber, with a chevron-shaped cross section such that pressure from under the chevron tends to force the edges of the packing ring more tightly against the walls of the packing chamber, thus increasing the seal.

**chloride stress cracking** A phenomenon in which the presence of chlorides causes cracking in metals in regions where the material experiences high stress forces. Most often seen in stainless steels, since it most severely affects chromium-bearing materials.

**CI** Cast iron—the common term for cast gray iron, or iron containing flake carbon in the range of ½ to 2½ percent. Cast iron is brittle, exhibiting very little ductility before fracturing.

**CIF** Cost, insurance, and freight—commercial term referring to the total cost associated with an item at the point where it is received by the purchaser.

**clapper** Another name for the disc in a swing check valve.

**class** A pressure rating expressed as a dimensionless number. The class rating charts give actual pounds per square inch (psig) or barg maximum allowable pressure at a given temperature.

**closure** The device or object that is placed into or across an opening in a pressure-retaining body for the purpose of closing it off.

**code** A set of rules compiled by an engineering standards–making organization, addressing basic questions of safety and reliability, that applies to all components of a physical installation, such as a plant or processing unit, and that may also have the force of law.

**cold working** Mechanical processing, such as rolling or drawing, that takes place well below any metallurgical phase transformation temperature. Cold working of many metals increases their hardness and decreases their ductility.

**control valve** Valve that has an automatic actuator that responds to signals sent by pneumatic, electrical, or other means for the purpose of controlling or varying the fluid flow in the pipe line. In common usage, any valve equipped with an actuator.

**corrosion** Deterioration of a metal caused by chemical action. There are a number of distinct types of corrosion with different effects on the metal.

**corrosion resistant** A material that is not affected by corrosive action. The term "corrosion-resistant steel" usually means a stainless steel.

**creep** Slow yielding of steel, occurring at elevated temperatures under stresses close to the yield strength.

**crevice corrosion** Corrosion that occurs only in crevices and stagnant areas, such as the inside of socket welds.

**critical temperature** The temperature above which the crystal structure in steels transforms into austenite.

**cryogenic** Referring to temperatures significantly below normal atmospheric temperatures. Definitions vary, but often any temperature below −50°F (−45°C) is considered cryogenic.

**CS** Carbon steel or cast steel—iron containing carbon in the form of carbides, about 0.1 to 0.3 percent carbon with no other alloying elements other than the sulfur, phosphorus, and other elements present in almost all steels. Steels are fairly ductile, exhibiting about 20 to 30 percent ductility or stretching before fracturing.

**CSA** Canadian Standards Association—Canadian national standards organization.

**$C_v$** Flow coefficient expressed as the number of gallons of water that would flow through an opening, such as a valve port, in 1 minute under a differential pressure of 1 psi.

**CWP** Cold working pressure—a pressure rating that indicates the maximum allowable pressure under nonshock (steady-state), ambient-temperature conditions.

**cylinder operator** A valve actuator in which the force used to operate the valve is provided by a hydraulic or pneumatic cylinder.

**decarburization** The phenomenon in which carbon is driven out of solution into the free spaces at the grain boundaries in steels. Occurs at elevated temperatures and leads to a loss of strength.

**Delrin** Trademark of E.I. du Pont de Nemours Co, Inc. for acetal plastic.

**delta $P$ or $\Delta P$** The differential pressure between two points, numerically equal to $P_1 - P_2$.

**design pressure** The pressure used in calculating required wall thicknesses, operating torques, flange ratings, and other variables. Generally the design pressure is set at a value higher than the operating pressure, to include all the reasonable allowances for surge pressures and variation in operating conditions. The MAWP (maximum allowable working pressure) may be higher than the design pressure since it is based on the actual or as-built thicknesses, while the design pressures are used to determine the specified thicknesses.

**design temperature** The temperature that is used to determine allowable stresses for the purposes of design calculations. Generally, the design temperature is set at a value higher (or farther from ambient) than the operating temperature, and includes allowances for upsets and variation in operating conditions.

**destructive test** A mechanical or chemical test that results in the consumption or destruction of the part being tested.

**diaphragm** A thin, flexible elastomer or metal sheet used to apply forces or pressures through.

**die cast** A casting process in which a metal mold is used. The metal mold is produced by machining or engraving and is reusable many times. It is made in two or more pieces that separate to release the cast part.

**differential pressure** Same as delta *P* or $\Delta P$.

**DIN** Deutsche Industrie Norme—West German national standards organization.

**disc** (also spelled **disk**) The circular or roughly circular closure element in gate, globe, swing check, and butterfly valves.

**DN** Nominal diameter—standard abbreviation for pipe size used in ISO standards.

**DOT** Department of Transportation—government agency regulating pressure vessels and valves that are part of transportation vehicles, including tank trucks and tank cars.

**double block and bleed** Either a pair of block valves with a section of piping between them that contains a bleed point, or a single valve that seals in two locations with a bleed point between the two. The bleed may or may not be valved itself, and may be normally open or normally closed.

**drain** An outlet at the bottom of an enclosed fluid space such as a pipe line, which when open allows liquids to flow out.

**drive pin** or **drift pin** A slightly tapered pin, forced or hammered into a hole drilled into two mating pieces for the purpose of pinning the two together and preventing their relative motion.

**dry lube** A lubricant in solid or powdered form, often used as a coating. Common lubricants are moly disulfide or Teflon coatings.

**ductile iron** A cast iron in which the carbon is in spheroidal enclosures, formerly called nodular iron, developed in the late 1940s. It has about 18 percent ductility, less than cast steels, but more than cast (gray) iron or malleable iron.

**ductility** The ability of a material to deform plastically. Percent ductility is measured by the reduction in area of the cross section of a specimen stretched until it breaks.

**duplex steel** A high-alloy steel that contains significant quantities of ferrite in an austenitic matrix. Duplex steels possess as much or more corrosion resistance as austenitic stainless steels, plus the stress corrosion cracking resistance of ferritic steels.

**Durometer** A device for determining hardnesses of resilient materials, or the hardness measured using the device, trademark of Shore Instrument Co. Hardnesses are expressed in Shore units, defined in ASTM-D2240.

**dye penetrant** *See* PT.

**dynamic seal** A seal in which one of the members moves with respect to the other.

**elastomer** A polymer or long molecular chain material having significant elastic properties.

**elbow** or **ell** Pipe fitting having two end connections at some angle to each other, either 90 or 45 degrees. Used for changing the direction of a pipe line.

**elongation** The amount of deformation or permanent stretch in a steel. Percent elongation is measured in relation to an original gage length marked on a specimen before it is tensile tested.

**end to end** The dimension from one end to the opposite end of a valve. Usually used to refer to valves that do not have flange faces.

**end piece** A removable component that allows access to the internals of a valve through the end. May be internal or external to the body.

**EPDM** Ethylene propylene dimonomer—the generic name for an elastomer made from ethylene and propylene. One trade name is Nordel.

**erosion** Deterioration of a material by wear, from fluid action with or without entrained solids.

**ERW** Electric resistance welded—pipe made from steel strip, which in a continuous process is rolled into a circular cross section, then the edges of strip are welded together using the resistance heating caused by a high current passing through the steel at the line where the two edges meet.

**ETA** Estimated time of arrival—a commercial term indicating the date that an item is scheduled to arrive at the location where it is needed.

**examination** The act of examining a part or a finished product or system to determine if it is in compliance with requirements. Any examination method may be used, including the review of any paperwork associated with the component or system. Examination may be performed by the party that performed the work in question.

**explosion-proof** A degree of protection of electrical components in which the component can be installed in an explosive atmosphere, one where the stoichiometric ratio of flammable vapor to air is such that an explosion could occur, with the area where arcing takes place being effectively isolated from the atmosphere.

**export packing** A method of packaging an item for shipment suitable for the extra rigors of long-distance ocean voyages, multiple transshipments, and longer transit times than are required for domestic shipments.

**fabricated** Built by some combination of metalworking processes that involves joining components together.

**face-to-face** The dimension of a flanged valve, from the mating surface (raised face or flat face) of one end flange to the same surface on the opposite end flange. Valves with ring-joint flanges are dimensioned from the outermost surfaces, although that is not the mating surface.

**fail-closed** A condition in which a valve or other component that is normally in some position, open, closed, or in between, will close if power or signal is lost.

**fail-last** Also **fail-locked**—a condition in which a valve or other component that is normally in some position, open, closed, or in between, will freeze in its last position if power or signal is lost. Generally less desirable than either fail-closed or fail-open.

**fail-open** A condition in which a valve or other component that is normally in some position, open, closed, or in between, will open if power or signal is lost.

**fail-safe** A design for equipment or valves such that, if a failure or loss of power occurs, the item will position itself in a configuration that limits further damage to itself, other equipment, or nearby personnel.

**FAS** Free alongside ship—commercial term meaning that the seller takes responsibility for the item and its transportation and transportation costs up to the point of loading aboard ship, referring to orders destined for export or water carriage.

**FDA** Food and Drug Administration—U.S. government organization that regulates, among other things, the manufacturing of products for human consumption, including foods, cosmetics, health aids, and prescription drugs. Plants manufacturing such items are inspected and validated by the FDA prior to startup and periodically thereafter.

**female thread** A recessed threaded connection with internal threads.

**FEP** Fluorinated ethylene propylene copolymer—fluoropolymer used for linings, seats, and seals, similar to PTFE but with somewhat different properties.

**ferrite** A microstructure found in steels at room temperature, having a body-centered cubic structure. Ferrite undergoes several phase changes with increasing temperature.

**ferritic stainless steel** Steel containing iron and chromium, with no nickel, and having a ferritic crystal structure. These steels are designated as 400 series, such as 405 or 416.

**FF** Flat face—a flange surface in which the gasket sealing area is the entire surface from the inside diameter to the outside edge of the flange. The surface is flat, normally smooth without serrations.

**fire resistant** Term referring to an item that has been designed to survive, at least for a while, under normal fire conditions in a plant. May be either intrinsically resistant to fire damage, designed with fire-resistant backup systems such as fire-safe, soft-seated valves, or externally protected with insulation or intumescent (charring) materials.

**fire-safe** A valve design that is capable of passing a fire test, with specified limits on leakage to the atmosphere and downstream after being closed subsequent to the fire exposure. Several fire test specifications are in common use. Also said of enclosures built around valve actuators and supply lines or cable trays leading to such valves, more properly referred to as fire resistant.

**fire tested** Same as fire-safe.

**fitting** Any component, other than valves, used with pipe as part of the pressure system, but normally referring to items that are covered by national standards.

**flange** A pipe fitting or a portion of a flanged piping component, designed to be mated to a companion flange or flanged surface by means of bolts inserted through holes in the flange face.

**floating ball** A ball valve design in which the ball is not rigidly held on its rotational axis.

**flow coefficient** The number of gallons of water that will flow through an orifice, such as a valve, in 1 minute when the differential pressure from one side of the valve to the other is 1.0 psi. Abbreviated $C_v$.

**fluid** Any substance that can be made to flow. Gases, liquids, slurries, and most powdery and granular substances qualify.

**fluoroplastic** or **fluoropolymer** Any of a class of polymers made from molecules with structures similar to the common hydrocarbon molecules but with fluorine atoms replacing part or all of the carbon atoms. One of the most common is PTFE.

**FM** Factory Mutual Association—an organization that certifies products used in fire- or other safety-related industrial installations.

**FOB** Free on board—commercial term that means that the seller has responsibility up to the point of delivery to a common carrier.

**forging** A metalworking process that involves hammering or squeezing, with or without a die, at hot working temperatures.

**fracture toughness** A measure of the toughness, or resistance to fracture, of a material. Commonly of concern at atmospheric or lower temperatures, with steels that lose fracture toughness rapidly below a certain temperature and become brittle. Measured by Charpy test.

**free machining steel** Any steel that has had lead or sulfur added to make machinability easier. Free machining steels can be somewhat less corrosion resistant and are sometimes prohibited for wetted parts.

**full bore** or **full port** A valve or other component in which the seat area has substantially the same cross section and cross-sectional area as the end connections have. Depending on what the valve will be used for, sometimes having the same area is sufficient, but sometimes the same internal diameter all the way around is required.

**full penetration** A weld through the complete thickness of the parts being joined, from either one or both sides.

**galling** A physical phenomenon wherein two mating or closely pressed surfaces, such as the threaded surfaces of a nut and a bolt, momentarily bond or weld together in microscopically small areas, then are torn apart. The resultant surfaces are very rough and damaged.

**galvanic corrosion** Corrosion that occurs where dissimilar metals are in close proximity in an electrolytic solution, such as water. The anode side is the one that loses metal (is less noble).

**gasket** A component whose purpose is to seal a joint between two larger components, softer than the surfaces of the joint being sealed, and usually squeezed by means of bolting or clamping to effect a seal.

**gate valve** A valve in which the closing member moves up, out of the way of the flow, and basically slides along the seat(s), like the gate in a castle wall.

**gauge** A device that performs some measuring function and displays the results, either through a dial display or, in the case of make-up gauges such as thread gauges or tubing nut gauges, by either going or not going a specified distance. Dial gauges can be either analog (with a pointer that moves across a dial proportionately to the change in the gauge input) or digital (with the value of the gauge input displayed as a numeric quantity).

**gland follower** The component that is used to hold down or retain the gland on a packing chamber.

**gland** or **gland bushing** The component that serves as the top or outermost surface or closure on the packing chamber or stuffing box.

**globe valve** A valve, originally somewhat globe shaped, in which the closing member is circular in cross section and moves along a line concentric with the seat axis to open and close.

**Grafoil** Trademark of Union Carbide Corp. for a flexible graphite foil used for making gaskets and packing.

**grain size** A measure of the average grain size in a metal, defined by ASTM-E112.

**grease fitting** A device that permits injection of grease into a bearing surface. The fitting is threaded into a hole drilled into the bearing support member and contains a small spring-loaded ball check valve. Commonly known as a Zerk fitting.

**handle** An operating device for quarter-turn valves, intended to be grasped by hand to turn the stem, usually retained with a nut, sometimes including a latching device.

**handwheel** A wheel-shaped valve-operating device, intended to be grasped with one or both hands, which allows turning the valve stem or operator shaft that it is attached to.

**hardfaced** or **hardfacing** Treatment that results in a region on the surface of a part that is harder than the interior or substrate. Normally done by weld deposition of a harder alloy such as Stellite (trademark of Stoody Deloro Stellite Inc.)

**hardness** The ability of a material to resist indentation.

**HAZ** Heat-affected zone—the zone of base metal immediately adjacent to a deposited weld that, although it has not melted, has reached a substantially elevated temperature such that it may have undergone some phase changes and it may have some constituents of the melted weld metal diffused into it.

**header** A large line from which smaller lines branch out.

**heat treat charts** Charts made as a permanent record of the temperature cycle experienced by an item during heat treatment, made by temperature recorders.

**heat treatment** A set of heating and cooling operations intended to create or improve some desired properties in a material.

**holiday** A void or unattached area in a region under a coating such as a paint or a wrapping where the coating is not adhered to the substrate.

**hot tap** A connection made to a pipe line while the line is under pressure or in service. Special procedures are required to make an opening in the pipe without leaking any of the line contents.

**HRB** or **HRC** Rockwell B or C hardness—hardnesses measured on scales comparing the sizes of indentations made in the tested material. The larger the number, the harder the material. Defined by ASTM-E18.

**Huey test** A corrosion-resistance test for stainless steels, most useful for predicting resistance to intergranular corrosion.

**HV** Vickers hardness—hardness measured on a scale comparing the size of diamond-shaped indentations made in the tested material. Defined by ASTM-E92.

**HVAC** Heating, ventilating, and air-conditioning—discipline concerned with fluid handling in buildings, with services including those needed for comfort and habitability but not specialized process services.

**hydraulic** Relating to any liquid or liquid flow properties, not necessarily water, as opposed to pneumatic (vapor).

**hydraulic motor** A device whose driving force comes from fluid and whose output is a continuous torque on a shaft.

**hydrostatic test** A pressure test using water, in which the item or system to be tested is filled with water and pressurized, using a pump, to some pressure, to detect leaks. Often the pressure used is 1 ½ times the design pressure, plus a factor to correct for lower allowable stresses at operating temperature, plus water head.

**Hypalon** Trademark of E.I. du Pont de Nemours Co. Inc. for a polyethylene-based rubber.

**IBBM** Iron body, bronze mounted—common term for valves with iron body and bonnet (pressure-retaining parts) and bronze trim (seats, stem, bushings).

**IBBT** Iron body, bronze trim—same as IBBM.

**ID** Inside diameter—the distance, either nominal or actual, from the inside wall of an annular surface to the opposite inside wall.

**impact test** A test to measure toughness, such as Charpy test. The force required to break a specimen is measured. The higher the force, the more tough the material.

**inclusion** A particle or solid impurity trapped in a weld, casting, or ingot during solidification.

**Incoloy** Trademark of International Nickel Co. for any of several nickel-iron-chrome alloys.

**Inconel** Trademark of International Nickel Co. for any of numerous nickel-chrome-iron alloys.

**inspection** The act of examining a part, product, or system by a person representing the purchaser, either a direct employee, an agent, or a third-party inspector, but not by the entity that performed the work being examined, except in the presence of the purchaser's inspector and under the inspector's direction.

**intergranular corrosion** Corrosion that occurs in the boundaries between grains. The surface will have an "etched" or "crazed" appearance.

**investment cast** Also known as lost wax cast, in which a pattern is made from an easily melted material (usually duplicated from a master pattern), then encased in a hard plasterlike mold material. The pattern is then melted out, either before or during the introduction of molten metal, and the mold is then broken to release the cast part. Finishes are very clean and tolerances are very close in this method of casting.

**IPS** Iron pipe size—same as nominal size. Applies to steel and, formerly, wrought-iron pipe with nominal outside diameter, or other pipe made to the same outside diameter as steel pipe. This term is becoming obsolete.

**ISO** International Standards Organization—worldwide standards coordinating organization.

**ISRS** Inside screw, rising stem—common term for any valve design in which the stem threads are exposed to the fluid, below the packing, and the stem rises up through the packing when the valve is opened.

**Izod test** Another test for toughness or fracture resistance, similar to a Charpy test but using a keyhole-shaped notch in the specimen.

**JIS** Japan Industrial Standard—designation for standards published by the national standards organization of Japan.

**Kel-F** Trademark of 3M Company for a fluorocarbon rubber.

**keyway** Slot machined in a shaft or hub for the insertion of a key to hold two components together.

**killed steel** Steel that is killed, or deoxidized, by the addition of aluminum or silicon to chemically combine with the gases present during solidification.

**knife edge corrosion** Corrosion characterized by sharp, narrow cavities that resemble cracks. Often found at the edges of welds or in heat-affected zones.

**Kynar** Trademark of Pennwalt Co. for polyvinylidene fluoride.

**KSI** Thousand pounds per square inch—unit of measure commonly used in work concerning stresses.

**ladle analysis** Chemical analysis of a heat of steel taken from the ladle just prior to pouring.

**laminar flow** A flow of fluid, usually liquid, inside a confined passageway with velocity so low that very little turbulence occurs. Flow lines, or the path taken by any given molecule in the flow stream, are essentially straight.

**lamination** A defect in metal that takes the shape of a smooth crack, often running parallel to the surface of plate steel or forgings.

**lateral expansion** A measure of ductility of steel undergoing fracture toughness testing. Measured as a percent of the transverse dimension of the specimen after fracturing to its original transverse dimension.

**lever** An operating device for quarter-turn valves, more correctly known as a handle. Another type of lever is the device used on some valves to pull the stem straight out to open. This type meets the classical definition of a lever in which there is a fulcrum, a point to apply force and a point where the lever exerts force.

**lifting lugs** Devices attached to a larger unwieldy object, either welded on or cast integrally, with eyes or openings to allow easy attachment of hooks or slings for lifting.

**limit switch** Electromechanical device that detects the arrival of an object such as a valve stem indicator at a predetermined point, by means of the detected object touching a movable part of the limit switch and making or breaking a circuit. The presence/absence of this current constitutes a signal that is in turn detected by some other device and causes some other action to happen, such as turning off a drive motor. A **proximity switch** performs the same function but is an electromagnetic device.

**line** A term referring to a conduit of fluid or pressure that begins at one point and ends at some other point.

**liquid penetrant** *See* PT.

**LNG** Liquified natural gas—natural gas (mostly methane, with some small quantities of heavier components) cooled below its boiling point.

**lockhopper** A valve that controls the flow of fluids consisting of solid granular or powdery materials, usually in a downward direction. It is an on-off device by virtue of its actions, but can be considered a modulating device because of its ability to regulate flows by the frequency with which it cycles.

**locking device** Any valve attachment whose purpose is to prevent the operation of the valve by unauthorized persons.

**LPG** Liquid petroleum gas—a mixture of hydrocarbon components under some pressure in a liquid state at ambient temperature.

**mA** Milliampere—measure of electrical current used in control signaling.

**Magnaflux** Trademark of Magnaflux Corp. for the first magnetic particle inspection process. *See* MT.

**male thread** An external or projecting thread, for instance on the end of a pipe.

**malleable iron** A cast iron possessing some ductility, on the order of 5 to 10 percent. Used for pipe fittings, but rarely for valves.

**manifold** A short header or section of pipe that serves as a junction point for several lines coming together or diverging.

**MAOP** Maximum allowable operating pressure—same as MAWP.

**martensite** A microstructure found in iron alloys and steels. Formed by rapid cooling from temperatures where austenite is the primary crystal structure. Greater proportions of martensite make steels more hard and more brittle.

**MAWP** Maximum allowable working pressure—the maximum pressure (must be stated in conjunction with a temperature) that a component can be subjected to on a continuous basis. Could be greater than the design presure, since the MAWP is determined by the actual installed thickness or as-built condition of the component.

**mechanical tubing** Hollow steel stock, round or other shapes, used for structural or mechanical purposes but not intended for pressure-retaining services as is pressure tubing or pipe.

**metal to metal** A seating design characterized by the lack of any soft deformable seating material. Metal-to-metal seats can withstand much higher pressures and temperatures than soft seats, but leakage rates are usually greater except in special valve designs.

**meter prover** A device that circulates a known quantity and type of fluid through a flow meter to calibrate it.

**meter run** A section of special pipe, selected for minimum roughness, out of roundness and constant diameter and installed on both sides of a flow meter.

**MFD** Mechanical flow diagram—same as P&ID.

**MIL** Designation for United States military standards.

**mill certificate** *See* MTR.

**minimum wall** Method of specifying wall thicknesses so that all the tolerances are on the plus side, as opposed to average wall.

**MMCF** Thousand thousand cubic feet—term used to describe volume of gas in pipeline and distribution industry.

**modulus of elasticity** The ratio of stress applied to a material over the resulting strain or deformation, in the elastic range where no permanent deformation occurs.

**mold** The cavity that a casting is poured into. Molds can be sand with resin binder, or hard plaster for investment casting, or machined steel for die casting.

**Monel** Trademark of International Nickel for several 67 percent nickel, 33 percent copper alloys.

**MRO** Maintenance and repair order—*see* blanket order.

**MSS** Manufacturers Standardization Society of the Valve and Fitting Industry—trade organization responsible for issuing voluntary standards for valves and piping material. MSS withdraws its standards when another standards-issuing body such as ASME issues one covering the same subject.

**MT** Magnetic particle test or magnetic particle inspection - an examination technique using magnetic fields or an electromagnetic technique to pass a magnetic field through a magnetizable steel to detect flaws. Location of flaws is made visible by sprinkling iron filings over the surface of the part, which tend to move to the location of defects that interrupt the magnetic force lines, including defects that lie up to $\frac{1}{16}$ in. (2mm) or so under the surface. Defined in ASTM-E709.

**MTR** Mill test report—a document showing the results of chemical and physical testing done on a product, usually on the raw material. The information required is the properties specified in the ASTM or other specification for the material being manufactured. Normally supplied by the mill or the original producer of the raw material.

**NACE** National Association of Corrosion Engineers—organization responsible for publishing recommended practices and data on corrosion and corrosion resistance.

**NDE** Nondestructive examination—general term for any examination or inspection peformed on materials that does not involve destroying the item being tested.

**NDT** Nondestructive testing—same as NDE.

**needle valve** Small valve in which the sealing element is small and sharply tapered or needle shaped.

**NEMA** National Electrical Manufacturers Association—trade organization responsible for standards for certain electrical equipment such as motors.

**Neoprene** Formerly a trademark of E.I. du Pont de Nemours Co. Inc. for chloroprene rubber.

**NFPA** National Fire Prevention Association—standards group responsible for standards on fire prevention and fire-related safety.

**NHT** National Hose Thread—standard straight thread for fire hoses and threaded-end fire equipment. Not interchangeable with NPT threads.

**nipple** Short piece of pipe, threaded or unthreaded, less than a foot (300 mm) long.

**nitriding** Surface hardening process involving impregnation of the surface with nitric acid to form iron nitrides.

**nitrile** A type of synthetic rubber. One trade name is Hycar.

**nominal size** Anything, such as pipe, that is designated by a dimensionless number and not a size in units of measure such as inches or millimeters.

**nondestructive test** Any test method that can be used directly on the item or part in question without destroying it or rendering it unfit for its intended service.

**nonrising stem** A valve stem designed so that as it turns, its threads engage the closure element to move it up and down, but does not itself move along its axis.

**normalized** A condition of steels, induced by heat treating to about 100°F above the upper critical temperature and slow cooling, in which the impact toughness is improved and grain size is improved.

**normally closed** A state in which a valve or other component stays closed in the absence of a signal or manual intervention. Any such outside action will open the valve.

**normally open** A state in which a valve or other component stays open in the absence of a signal or manual intervention. Any such outside action will close the valve.

**NPS** Nominal pipe size—dimensionless number used as designator for sizes of pressure pipe.

**NPT** National Pipe Thread—standard tapered thread for pressure pipe.

**NST** National Straight Thread—standard straight thread for pressure pipe. Normally requires a gasket or other backup for tight seal.

**OCMA** Oil Companies Materials Association—British standards organization that did pioneering work on fire-safe valves, among other things. Now dissolved.

**OD** Outside diameter—actual diameter of a part or a pipe, measure to the outside surface. In pipe, the outside diameter is its specified diameter in actual units. For instance, NPS 2 (or "2-in.") pipe is 2.375-in. (60.3 mm) nominal outside diameter.

**oil field** or **oilpatch** Term for the onshore oil and gas drilling and gathering industry.

**operating nut** The square, tapered nut that fits on top of a valve shaft and allows it to be turned by a tee-handle from above. Usually installed on buried valves.

**operating pressure** The pressure that a component normally sees during the course of day-to-day operation. This pressure, plus any other factors such as upsets that may occur, is used to determine the design pressure.

**operating temperature** The temperature that a component normally sees during day-to-day operation. This temperature, plus any other factors such as excursions that may occur, is used to determine the design temperature. The exception is that insulation thicknesses are usually calculated from operating temperature.

**operator** A person who operates valves and other equipment in an industrial setting, or a device attached to a valve that provides force or torque to the valve to make the closure element move. Can be either automatic or manual.

**O-ring** A sealing device consisting of a continuous ring, usually round in cross section and made of an elastomeric material.

**OS&Y** Outside screw and yoke—valve design in which the stem threads are above the packing gland, or outside the valve body, and there is a yoke or bracket to support the top or outer end of the stem.

**oxidation** Chemical reaction of any material with oxygen. Scale, rust, and discoloration are physical evidence of oxidation of iron alloys.

**packing** Any soft substance used to seal the area of a shaft where it protrudes from inside a pressure boundary, such as a valve or pump.

**P&ID** Piping and instrument diagram—generally, the drawing that shows schematically all the equipment, piping, instrumentation, and control devices in a plant, but without showing geographic locations or orientation.

**packing gland** A device installed on a packing chamber to retain the packing and to maintain pressure on it.

**passivation** A treatment that changes a chemically active metal surface to a chemically inactive state.

**pattern** A shape used to make molds from. Patterns are the same configuration as the raw casting will be, and molds are the negative of that configuration. The molds are formed around the pattern, and the pattern is either removed or destroyed, depending on the casting process.

**PEEK** Polyetheretherketone—high-temperature semirigid elastomer, developed by ICI and trademarked Victrex.

**PFA** Perfluoroalkoxy—fluoropolymer used for linings, seats, and seals, similar to PTFE but with somewhat different properties.

**PFD** Process flow diagram—generally, the drawing that shows the basic process flows in a plant, with major equipment and controls shown, but without showing specifics of line sizes, minor connections, instrumentation, or controls.

**PFI** Pipe Fabrication Institute—trade group of pipe fabricators, or companies engaged in the production of spool pieces (semifinished piping assemblies suitable for job site installation) responsible for issuing standards on piping fabrication.

**phenolic** A thermosetting material used for electrical insulation.

**phosphatizing** A surface treatment of steel parts consisting of a dip into an acid phosphate bath that intimately coats the surface with a phosphate-rich layer that retards rusting and provides better coating adhesion.

**photomicrograph** A photograph of a microscopic image, one magnified more than ten times, to show microstructure characteristics.

**pickling** A treatment that removes surface oxides and scale from steels, usually stainless steels, by immersion in an acid solution.

**pig** Also known as a **scraper**—the solid object placed in a pipe line, to be pushed along by line pressure, for the purpose of separating one fluid from a different one being shipped through the same line or to clean (scrape) the walls of the pipe.

**pinion** A small-diameter gear, meshing either with a much larger gear or with a flat toothed rack.

**pipe** A hollow cylindrical product intended for containing internal pressure, made to a standard set of dimensions, normally standard OD and wall thickness.

**pitch** The distance between corresponding points on adjacent threads. A single-threaded shaft will drive a mating piece a distance equal to the thread pitch for each complete revolution.

**pitting corrosion** Corrosion characterized by small cavities or pits in the metal surface, which may combine to form large pits.

**plastic** Generic term for one of two types of material, thermosetting and thermoplastic. Thermoplastics, which lose strength with increasing temperature and regain it as temperature decreases over a normal range, are the ones most commonly used for plastic piping materials and valve components.

**PLCA** Pipe Line Contractors Association—trade group of companies engaged in the laying or installation of cross-country pipelines.

**plug valve** A type of valve in which a cylindrical or tapered cylindrical (truncated cone) section turns one-quarter turn to close and open the flow passageway.

**PMI** Positive material identification—any scheme for cross-checking the identity of a piece of material, often but not necessarily using a nuclear analyzer, going beyond the minimum step of comparing the heat numbers marked on the material with the material test reports.

**PN** Nominal pressure—standard abbreviation for pressure rating used in ISO standards.

**pneumatic** Relating to vapor or gas or vapor or gas flow properties in general, but specifically to air.

**polyethylene** Flexible thermoplastic used for pipe and for valve seats, among other things.

**polypropylene** Semiflexible thermoplastic used for pipe and for pipe linings.

**porosity** A defect found in castings or welds, consisting of gas bubbles or voids in the solidified metal.

**port** An opening in a valve body or closure element, usually used in control valve terminology.

**position indicator** A device that displays, by its angular orientation or displacement, the approximate position of the valve closure element.

**positioner** A feedback device that aids in the accurate positioning of a valve closure or throttling element with respect to the magnitude of the signal being received.

**potentiometer** An electrical device that produces a variable potential, or voltage, in an output wire depending on the position of a contact along a coil.

**precipitation hardening** A method of heat treatment used on certain supersaturated solid solution alloys. Produces a material both corrosion resistant and possessing very high strength.

**pressure** The force exerted by a fluid on the surfaces containing it.

**product analysis** Chemical and physical analysis performed on a finished material, such as rolled plate or rough forging.

**proof test** Any test used to demonstrate the integrity of a finished component, such as a pressure test.

**PSI** Pounds per square inch—measure of force, either pressure force or tensile or compressive stress in a material.

**PSIA** Pounds per square inch, absolute—pressure force expressed without reference to ambient pressures.

**PSIG** Pounds per square inch, gage—pressure force expressed with reference to standard atmospheric pressure. Standard atmospheric pressure is defined as 14.7 psia.

**PT** Penetrant test or dye penetrant test or dye penetrant inspection—an examination technique that uses a penetrating dye sprayed on the surface of a part, followed by a developer that washes off the dye from all areas except pits, cracks, and other surface defects. The dye is either bright red or fluorescent, and examination is done either under room light or ultraviolet light. This test works on all nonporous surfaces but can detect surface artifacts only. Defined in ASTM-E165.

**PTFE** Polytetrafluoroethylene—*See* Teflon.

**pup** A short piece of pipe used to extend the end of a component out farther. Like a nipple, only larger in diameter.

**PVDC** Polyvinylidene chloride—*See* Saran.

**PVDF** Polyvinylidene fluoride—*See* Kynar.

**quenched** A heat-treating method in which an item, having been heated to some temperature, is cooled suddenly by being immersed in a bath of ambient-temperature water or oil, or sometimes subjected to a high-velocity stream of ambient air, as opposed to being allowed to cool slowly. This "locks in" some metallurgical phase that would only occur ordinarily at that elevated temperature by not allowing it time to transform back to its ambient-temperature phase.

**rangeability** The spread of minimum to maximum signal that can be recognized and acted on by a control device.

**RAS** Required at site—a commercial term meaning the date at which an item is needed at the point where it will be used.

**reduced port** A valve port opening that is smaller than the line size or the valve end connection sizes.

**reducer** A pipe fitting designed to be the transition from one pipe size to another size. The term **increaser** is used only for fittings with end connections that are one-way, such as male and female or bell and spigot, and the fitting is designed to be installed in the direction that the line size increases.

**regular port** A port opening that is the size regularly furnished on that type of valve, normally a reduced port.

**regulator** A valve that regulates the pressure of the fluid passing through it such that its output remains fairly constant although the input fluid pressure is variable, as long as the input is higher than the output pressure that the valve is set for.

**relief valve** A valve designed to open when the pressure under the seat reaches a preset value, by means of a spring or a poppet or any one of several other devices. All relief valves, by definition, are self-actuating.

**resilient seat** Any seating configuration that includes a soft, resilient seating element, either on the fixed seat or on the closure element but not usually both.

**RF** Raised face—a flange sealing surface in which the gasket seating area is a portion of the diameter covering the region from the inside diameter to some radius lying just inside the bolt holes, with that portion raised slightly above the remainder of the flange surface. This increases the effective load on the gasket and increases the sealing effectiveness. The gasket seating surface is usually serrated or grooved, either with concentric or phonographic (spiraling) serrations.

**RFP** Request for proposal—commercial term for the document that the buyer, owner, or contractor sends out to solicit bids on a given piece of work. Some organizations refer to all such requests as RFPs, others confine the term to requests for services only. *See* also RFQ.

**RFQ** Request for quotation—commercial term for the document that the buyer, owner, or contractor sends out to solicit bids on a given item or piece of work. Sometimes requests for design services or other labor are referred to as RFPs, while the RFQ refers to manufactured goods to be purchased.

**rising stem** A valve stem with threads arranged so that as the stem turns, the threads engage a stationary threaded area and lift the stem along with the closure element attached to it.

**RJ** or **RTJ** Ring joint or ring-type joint—a flange sealing surface in which the gasket seating area is two narrow lines of metal-to-metal contact along a metal ring, softer than the flange, that is set into a groove in each flange face.

**RMS** Root mean square—measure of surface roughness in units of microinches. Numerically close to AARH (arithmetic average roughness height) and often used interchangeably.

**road box** or **buffalo box** A cylindrical protective cover, usually with lid, buried over an underground valve and surrounding the valve topworks, allowing access to the valve although most of the valve body is actually buried.

**Rockwell hardness** *See* HRB or HRC.

**rodding** The practice of inserting a rod or bar into the open end of a valve or pipe to clean out or dislodge material accumulated inside.

**RT** Radiography or radiographic inspection—an examination procedure using X rays or, more usually, gamma rays from a radioactive source to produce an image on film of internal artifacts in an item. These images depict cracks, inclusions, porosity, and other defects that materially affect the apparent density in the direction that the view was taken.

**rupture disc** A device consisting of a thin, usually metal, sheet clamped between two flanges and scored or otherwise prepared to rupture cleanly when the pressure under the disc reaches a predetermined amount.

**SAA** Standards Association of Australia—Australian national standards organization.

**SAE** Society of Automotive Engineers—organization responsible for various specialized tubing and valve standards used for hydraulic work, among other things.

**safety valve** A valve design to relieve overpressure. Mechanically the same as a relief valve, but the reasons for locating one versus the other are not always the same.

**Saran** Trademark of Dow Chemical Co. for polyvinylidene chloride.

**SB** Screwed bonnet—a bonnet or cover that is attached to the body by threads, where the bonnet must be turned to engage the threads.

**SCC** Stress corrosion cracking—damage or failure caused by cracking from the combined effects of localized stress and chemical action. Stresses can be live, from pressure or loads, or residual such as from welding. Chlorides on stainless steels and sulfides on many materials are the most common corrodents.

**schedule pipe** Pressure pipe with wall thicknesses described as a schedule, such as schedule 40.

**scotch yoke** A device for translating linear motion into rotary motion, such as in a valve actuator, by means of a rotating yoke driven back and forth by means of a shaft that the yoke is pinned to.

**screwed** Any connection that is made by engaging two threaded-end components.

**SDR** Standard dimension ratio—a method of designating wall thickness of plastic pipe, similar to schedule in steel pipe except the SDR is an inverse relationship, where larger SDR numbers are lighter wall thickness. A given SDR number has still thicker wall as the pipe size increases.

**sealant** A high-viscosity semisolid, such as a grease, used to fill a cavity and seal off the passage of more fluid substances.

**seal weld** A weld that does not contribute anything to the mechanical integrity of an assembly, but is made purely to seal or prevent leakage from, for instance, a threaded joint.

**seat** The fixed component, mounted in the valve body, that the closure element contacts in order to close off flow.

**self-relieving** Any design that provides for automatic depressuring of an enclosed space without outside intervention.

**sensitization** A condition of stainless steels in which the material becomes susceptible to intergranular corrosion. Other conditions can also cause intergranular corrosion.

**short pattern** When two different end-to-end lengths for a given product, such as a ball or plug valve, become established by custom or by standards publication, the shorter of the two is known as short pattern. The longer of the two is usually known as regular pattern and is usually more common.

**shutoff valve** A valve whose primary purpose is to act as a main block valve, usually as an emergency shutdown but not necessarily integrated into the control system as a true emergency shutdown valve would be.

**sintered** The process of producing a material by mixing finely ground or powdered ingredients, then pressing and sometimes heating without approaching the melting point of any of the constituents. Usually employed to produce materials that could not be alloyed together by melting, including mixtures of metals and nonmetals.

**slab gate** A type of through-conduit valve in which the gate is cut from a single piece of steel plate.

**slurry** A mixture of solids in a liquid, with a high percentage of solids, not dissolved but being transported by the liquid as it flows.

**SNG** Synthetic natural gas—gas made by a synthesis process from other raw materials such as coal but usable in lieu of natural gas. Also called syngas.

**snubber** A device used to absorb shocks, sometimes similar to an automotive shock absorber.

**socket end** An end connection configured for socket-welding.

**socket weld** A strength or attachment weld in which the two components are assembled in a male-female configuration and a fillet weld is applied at the juncture of the two components.

**solenoid valve** A small valve in which the motive power to open and close the valve is obtained from the motion of a solenoid.

**solution anneal** A form of heat treatment used on stainless steels, in which the material is heated to approximately 2000°F so that carbides can go into solid solution, followed by a rapid cooling so that these components are trapped in solution.

**solvent-welded** Essentially a welding process that takes place by chemical action in a thermoplastic material. A solvent is applied to two surfaces to be mated or allowed to flow between two mating surfaces, which softens the two and allows them to fuse together into one common structure. Cementing is a common term for a lesser action in which the two surfaces do not necessarily fuse.

**sour** A crude oil or gas stream containing significant quantities of sulfur compounds.

**specification** A document that defines the requirements that a finished product must conform to, normally including chemical and mechanical properties, tolerances, marking, and shipping.

**spindle** Same as stem or shaft. Preferred term in British usage.

**spline** A set of grooves somewhat similar to gear teeth, for the purpose of interlocking male and female members such as a shaft and an internal splined coupling.

**spool piece** A fabricated section of piping, with or without flanges, that is assembled on site into a larger piping system. Named for the first flanged piping components, which were cast iron or steel with integral flanges, and resembled a thread spool.

**spur gear** A gear or set of gears in which the gear axes are parallel and the gear teeth on each are oriented with the same pitch.

**SS** Stainless steel—any of a number of types of iron alloy with chrome, nickel, or other elements that does not oxidize in free air.

**stabilize anneal** A form of heat treatment used on stainless steels, by first heating the material to approximately 1625°F, depending on the alloy, which causes carbides to precipitate, then holding at that temperature until chromium or added stabilizing elements diffuse into the areas rich in carbides, to reduce the risk of intergranular corrosion.

**stabilized stainless steel** An austenitic stainless steel that has been stabilized, or prevented from suffering carbide precipitation, by the addition of elements such as columbium (niobium) in type 347 stainless or titanium in type 321 stainless, in very small amounts that preferentially combine with the carbon in the structure.

**standard** A document that defines requirements.

**static seal** A seal that takes place between two components that do not move in relation to one another. Static sealing is generally easier to accomplish than dynamic sealing.

**Stellite** One of a number of hardfacing alloys, usually applied by weld deposition but sometimes investment cast as a solid component. Hardfacing alloys are covered in AWS A5.13, the common Stellite #6 being alloy CoCrA.

**stem** The shaft, rod, or bar that is rotated or translated from outside a valve and causes movement in the closure element attached to it.

**stem nut** The threaded nut that surrounds a reciprocating valve stem and causes the stem to move when the nut is rotated.

**stop collar** A clip or bracket that surrounds and is attached to a shaft, preventing further motion of the shaft when it contacts a fixed surface on the valve.

**strain** The deformation that results from applying a stress to a material. Strain can be elastic or plastic or a combination.

**street ell** An elbow that has one male and one female end, as opposed to the normal elbow that has two female ends. Refers to elbows with threaded ends.

**stress** The tendency to cause deformation resulting from a force or combination of forces applied to a material.

**stud** or **stud bolt** A bolt that has no integral head, but requires a nut on each end or is designed to be threaded into a tapped hole on one end with the other end projecting out and requiring a nut. The center section may or may not be threaded.

**stuffing box** Also known as packing box, the volume surrounding a shaft at the area on the shaft where it emerges from a pressurized or isolatable space, used to contain the packing.

**subheader** A smaller header that branches off from a larger header, but still has smaller branches coming off of it.

**swage** In piping terminology, a component that is smaller on one end than the other, to join pipe of two different sizes together. Generally, a swage has male ends and is longer than a reducer of equivalent size.

**SWB** Seal-welded bonnet—a bonnet or cover design that is mounted to the body in some manner, usually by threading, and in addition has a seal weld around its periphery to prevent leakage.

**sweet** Gas or oil with no or very low levels of sulfur or sulfur compounds.

**swing check** A check valve in which the closure element is suspended from the top and swings out of the way.

**SWP** Steam working pressure—an obsolete rating system that incorporated the effects of temperature by rating components in terms of the pressure of saturated steam they were capable of.

**tack weld** A small weld in the root of a larger weld gap, generally removed prior to the completion of the weld, whose purpose is to hold or tack the two mating components together while the larger weld is being deposited.

**tailpiece** In end-entry valves, the component that functions as a bonnet would in allowing access to the internals of the valve.

**TAPPI** The (Technical) Association of the Pulp and Paper Industry—trade association of pulp and paper manufacturers responsible for standards on materials used in the pulp and paper industry.

**tee** Pipe fitting having three end connections. Used for joining or splitting two flows.

**Teflon** Trademark of E. I. du Pont de Nemours Co. Inc. for polytetrafluoroethylene and related fluoropolymers.

**TEMA** Thermal Exchanger Manufacturers Association—trade association responsible for standards pertaining to heat exchangers of various types.

**tempering** A form of heat treatment in which a part, previously heat treated in some manner, is reheated to a temperature lower than sufficient to cause any change of state, followed by a slow cooling.

**tensile strength** The maximum load, expressed in force per unit area such as pounds per square inch, that a material can sustain before failing.

**tensile test** A test of materials performed by stretching or pulling a specimen until it breaks. This test provides information about tensile, yield, and ultimate strengths, deformation, and elongation. Defined in ASTM-E6 and ASTM-E8.

**test certificate** A document containing the results of chemical, mechanical, or other tests performed on a material.

**3A** International Association of Milk, Food and Environmental Sanitarians—technical organization responsible for standards on sanitary services for food processing, specifically dairy.

**throttling** The act of reducing the pressure or flow rate of a fluid passing through a valve.

**through conduit** A gate valve that in the open position presents a smooth continuous circular cross section to the flow, suitable for passing pigs, scrapers, or other solid objects.

**thrust** The net force on the end of a rotating or reciprocating shaft.

**topworks** Term that refers to anything on or above the bonnet of a valve, including operator, yoke, position indicator, and so forth.

**torque** The rotational force imposed on or through a shaft.

**torque switch** A switch that is activated by the exertion of a certain amount of torque against it.

**torsion** A twisting action in which the stresses produced in the part under stress are shear stresses.

**toughness** The ability of a material to absorb impact energy by deforming without fracturing.

**T-pattern** A valve body shaped like a T, where the inlet and outlet are on the same centerline and the valve stem and topworks are perpendicular to the piping axis.

**transflow** A multiport (more than two-port) valve design in which the flow passing between the common port and one of the other ports is not completely shut off before the third port begins to open. In other words, some flow is always present as the flow is being switched.

**transition piece** A special piping component whose purpose is to make a special transition between different piping materials, for instance between a heavier, lower strength pipe and a lighter, higher strength one, or between two materials that cannot easily be welded directly to one another.

**transmission line** A cross-country pipeline.

**trim** Any of the assorted small wetted parts in a valve, typically made of a higher alloy material for increased wear or corrosion resistance, analogous to the trim painting on a house.

**trunnion** A very short heavy shaft that serves to support or to act as a bearing for a larger part.

**tube** Hollow cylindrical shape not dimensioned to pipe standard sizes.

**turbulent flow** Flow in which the flow lines are not uniform and smooth, but rather has constant mixing and higher velocity than a typical laminar flow. Reynolds numbers greater than about 3500 are turbulent flow.

**UB** Union bonnet—type of valve construction in which the bonnet or cap is held on by a union nut with threads on the body.

**UHMWPE** Ultra-high molecular weight polyethylene—form of high-density polyethylene used for valve seats and for pipe. It is most commonly used in butterfly and ball valve seats for applications where Teflon is not suitable.

**UL** Underwriters Laboratories—organization that tests materials intended for use in fire and safey applications.

**ultimate strength** Same as tensile strength.

**UNI** Unificazione Italiana or Ente Nazionale Italiano de Unificazione—Italian national standards organization.

**unibody** A type of valve body in which the body itself consists of one piece, such that it is impossible to disassemble the valve while it is in line. Normally this refers to ball valves, in contrast to two-piece and three-piece threaded and bolted bodies.

**union** A pipe connection that consists of two end pieces with an external nut, retained on one end piece and threading onto the other, to make the connection tight.

**UNS** Unified Numbering System—system for uniquely identifying metals and metal alloys of engineering interest.

**USASI** United States of America Standards Institute—name taken by the American Standards Association in 1967, changed to American National Standards Institute in 1969.

**USDA** United States Department of Agriculture—government agency responsible for regulating, among other things, the storage and processing of food products related to agriculture, such as meat, grains, and dairy products.

**UT** Ultrasonic test or ultrasonic inspection—an examination method that uses high-frequency sound waves passed into a material and reflected back to a detector. The reflected signals can be interpreted to determine the depth of any artifacts that produce a difference in sonic reflectivity, including the opposite side of the piece, laminations, some inclusions, or voids.

**valve** A device used to control the flow of fluid contained in a pipe line. "Control" can mean regulating, slowing, stopping, or diverting.

**vent** A connection on the top or high point of a pipe line to allow air to escape. May or may not be equipped with a valve.

**venturi** A section of pipe line or valve that has a smaller cross section than that of the pipe on either side, with a smooth transition from one cross section to the other. Venturi valves have a smaller cross-sectional area in the seat area than at the end connections.

**viscosity** A measure of the inherent resistance to flow of a fluid. More viscous fluids require more differential pressure and more energy to be propelled than less viscous fluids do.

**Viton** Trademark of E. I. du Pont de Nemours Co. Inc. for a fluoroelastomer rubber.

**VT** Visual test or visual inspection—an examination method that consists of an experienced person carefully looking at the object in question for visual evidence of defects.

**wall thickness** The thickness of the pressure-retaining shell of pipe or valve body/bonnet. See also nominal wall thickness and average wall thickness.

**water hammer** A phenomenon in which sudden closing of a valve or shutoff of a pump produces severe shock waves in the liquid inside the piping. Water hammer can be heard by bystanders, but is primarily undesirable because of the mechanical damage it can do, not the noise.

**WE** Weld end—having ends prepared for butt-welding.

**weatherproof** A degree of protection of electrical components in which the component can be installed outdoors, exposed to rain or atmospheric moisture, without damage, but not necessarily resistant to dust, chemical vapors, or submersion.

**wedge gate** A type of gate valve in which the gate or disc is wedge shaped, thinner at the bottom, to wedge itself tightly between the two seats when closed.

**welch plug** A stamped metal plug, rounded or domed, that is forced or hammered into an opening just slightly smaller than the outside diameter of the plug, designed to protect or cover some part such as a bearing inside the opening.

**weld neck** A type of flange having a welding neck, where the hub is fairly thick at the point where it attaches to the flange and tapers down on the outside to the same diameter as the mating pipe. It is attached to the pipe by a circumferential butt weld (groove weld).

**wier** A special flow cross section, normally open topped.

**WOG** Water-oil-gas—one of the early rating designations, still in use today for small valves, chiefly in low ratings. Also called nonshock rating. Normally this rating is meant to be the maximum working pressure at ambient temperature (32 to 100°F).

**worm gear** A gear set in which the input shaft is offset from and perpendicular to the output shaft, and the driving gear is very small and perpendicular to the driven gear.

**WP** Working pressure—synonym for operating pressure.

**wrench** A removable actuating device or handle for quarter-turn valves.

**X ray** Common term meaning radiography. In some cases, X rays are used, but gamma radiation is much more commonly used as the penetrating rays.

**yield strength** The amount of force required to initiate yielding of a material. For many metals, yielding is defined as the point where a force produces a 0.2 percent offset. Above the yield point, the material suffers plastic deformation.

**yoke** The portion of a valve's topworks that acts as a bracket to support the top or outer end of the stem and stem bearing.

**Y-pattern** A design of globe or check valve in which the stem is inclined at 45 to 60 degrees from the axis of the end connections, as opposed to 90 degrees as in most valves. The two end connections are still on the same centerline. Not to be confused with an angle valve.

# INDEX